Handbook of Palladium-Catalyzed Organic Reactions

T
b
n

Handbook of Palladium-Catalyzed Organic Reactions

Synthetic Aspects and Catalytic Cycles

J.-L. Malleron
Rhône-Poulenc Rorer, Vitry, France

J.-C. Fiaud
Université Paris-Sud Orsay, France

and

J.-Y. Legros
Université Paris-Sud Orsay, France

ACADEMIC PRESS
Harcourt Brace & Company, Publishers
San Diego London New York Boston Sydney Tokyo Toronto

ACADEMIC PRESS INC
525B Street
Suite 1900
San Diego, CA 92101, USA

ACADEMIC PRESS LIMITED
24-28 Oval Road
LONDON
NW1 7DX, UK

This book is printed on acid-free paper

A catalogue record for this book is available from the British Library

ISBN 0-12-466615-9

Typeset by Fraser Williams (Scientific Systems) Ltd.
Printed and bound in Great Britain by Halstan & Co. Ltd., Amersham, Bucks

96 97 98 99 00 01 EB 9 8 7 6 5 4 3 2 1

Table of Contents

Foreword

Palladium chemistry and its application to the synthesis of organic compounds already has a long history. It was involved in the shift of raw C_2 materials from coal chemistry via acetylene to oil chemistry via ethylene. For years the Wacker process (which is palladium catalyzed) provided several million tons per year of acetaldehyde from ethylene. From the 1960s, homogeneous catalysis by palladium complexes started to be applied to more complicated organic molecules (Tsuji-Trost allylic substitution, Heck, Stille, Neghishi or Suzuki coupling reactions, Bäckvall functionalization of dienes, etc.). In the last 10 years, organic transformations catalyzed by palladium complexes have become one of the most active fields in homogeneous organometallic catalysis as judged by the number of publications per year. They are used routinely in multi-step total syntheses, as well as in fundamental research (e.g. the recent developments in asymmetric catalysis). Mechanistic details of the various catalytic cycles are reasonably well understood.

It is therefore opportune that this handbook is offered to those synthetic organic chemists wishing to take advantage of the rich chemistry of palladium for solving their problems. This handbook has been prepared for practitioners by Dr. J.-L. Malleron, an industrial chemist (Rhône-Poulenc-Rorer Co., Vitry, France), and Prof. J.-C. Fiaud and Dr. J.-Y. Legros from Université Paris-Sud (Orsay, France). For each class of reaction, a mechanism taken from the most recent literature is included which aids understanding and optimization of a given process. This handbook originates from a computerized database of more than 3,000 references which has been created and organized by Dr. Malleron.

The aim of the book is to provide the user with the main organic transformations catalyzed by palladium complexes, with a choice of references, in order to find the closest example for solving a given problem. To that end, structural features of the reactants as well as some experimental parameters are indicated in the tables. The reactions are numbered from RXN1 to RXN84. This large number of transformations shows the power of palladium homogeneous catalysis. This handbook offers the synthetic organic chemist in either industry or academia an impressive source of inspiration and information. The manipulation of this handbook is not difficult and is helped by the instructions given by the authors. It is also very useful for the reader to have access to the corresponding database on diskette. This greatly extends the scope of the book since it allows searching by keywords.

In conclusion it is very fortunate that this important work of selection and classification of data on palladium-catalyzed reactions originally from an industrial chemist, is now available to the scientific community. I congratulate the authors and wish them every success in their enterprise.

H. KAGAN

Abbreviations and Symbols

[Red.]	reducing agent
[Ox.]	oxidizing agent
L	ligand
BQ	benzoquinone
H_2BQ	dihydrobenzoquinone
dba	dibenzylideneacetone
dppb	bis(diphenylphosphino)butane
Tf	trifluoromethylsulfonyl (triflyl)
M, M'	metal
Nu	nucleophile
E	electrophile
$( )_n$	$(CH_2)_n$
RF	polyfluorinated alkyl chain
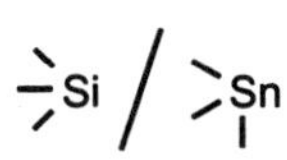	RR'R"-Si / RR'R"-Sn
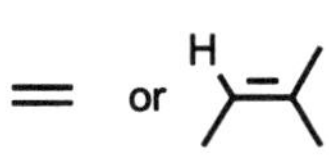	alkene
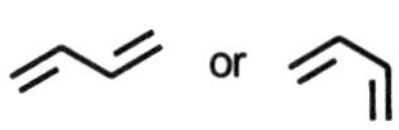	1,3-diene
	allene
	alkyne
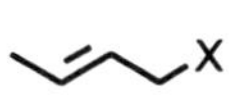	allylic derivative
	allylic epoxide
	diphosphine

NOTES

The references are given with abbreviations from the CAS source index. They are indicated as follows:

name of the main author, abbreviation of the journal, year, first page

or, when necessary,

name of the main author, abbreviation of the journal, year, volume, first page

I- INTRODUCTION

Organometallic chemistry is widely used by organic chemists and palladium chemistry is probably one of their most versatile tools. For synthetic and economic reasons, catalyzed reactions are preferred to stoichiometric reactions. It is very important to have a good knowledge of the synthetic aspects of organic chemistry, promoted by palladium complexes and of the catalytic cycles involved. In spite of excellent reviews and series having been published on this topic, it appears difficult to quickly and easily find complete information in this area.

This review provides a compilation of the most well-known catalytic reactions related to organo-palladium chemistry. The references have been covered up to November 1995. The mechanisms are generally described with respect to the data in the literature and by the use of information found mainly in publications. For each catalytic cycle, the chemical parameters are listed and one or more related references are given. As an example, for the cross-coupling reactions between RX and R'M, the identity of R, R', X, and M parameters are defined. The references which are quoted contain at least one example of these parameters . A few palladium-catalyzed reactions are not reported in this review because of insufficient data concerning the mechanism or because the reaction affords a complex mixture of products.

For classification, a name has been assigned to each reaction related to a reaction number (RXN) and these have been listed. With this classification, for any general mechanism presented here, it is possible to identify very quickly to which class the reaction belongs from the table of contents.

The first chapter presents the graphical abstracts for all the reactions (reaction numbers) which are described in this handbook. Thus it is easy to search for a reaction type and then look at the literature references in the tables.

In addition, the catalytic cycles and the corresponding chemical parameters have been entered into a database of palladium-catalyzed reactions. As well as the various examples quoted in the publications, it is possible to view either the general catalytic cycle (RXN) related to each reference or all the chemical parameters related to a reaction number. For each reference, one or more examples have been selected with respect to the reaction. ISIS software (version 1.2) from MDL, EXCEL software (version 5.0c) and WORD software (version 6.0c) from MICROSOFT have been used. At the present time, this database is made available to many research centres by the Rhône-Poulenc company. An example of the database form is shown over the page.

Any remaining errors and inaccuracies are solely the responsibility of the authors.

23/03/1995	*PALLADIUM DATABASE*	Chem. Parameters	Zoom	Mechanism

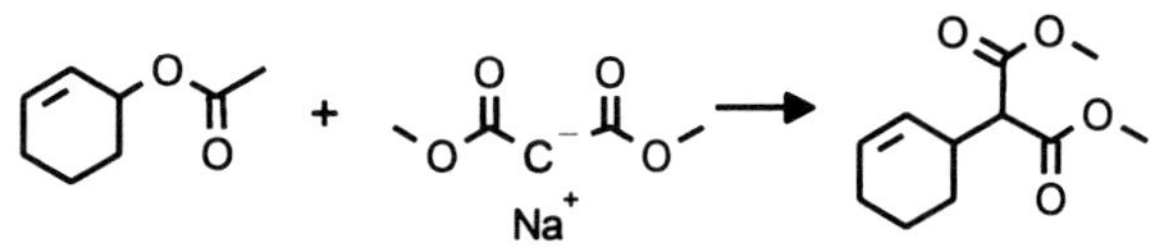

Reaction # :	2467
Variation :	1
Yield (%) :	75.0
Exp. Conditions :	N_2/ 20°C/ 48h

Solvent #	Solvent Name	Catalyst #	Catalyst Name
1	THF	1	Pd(dba)2 dppe

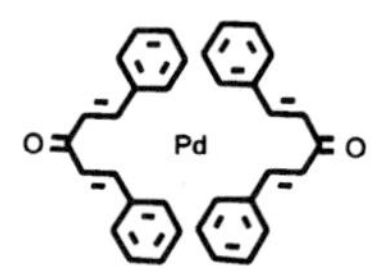

Step:	1
RXN Code:	RXN32

Comments:

Several examples.

References:

FIAUD J.-C., MALLERON J.-L., TETRAHEDRON LETT., 1980, 21, 4437

RXN type:

Substitution, Addition and Elimination on Pro-π-Allyl Substrates

Substitution, Addition and Elimination on Pro-π-Allyl Substrates

RXN32

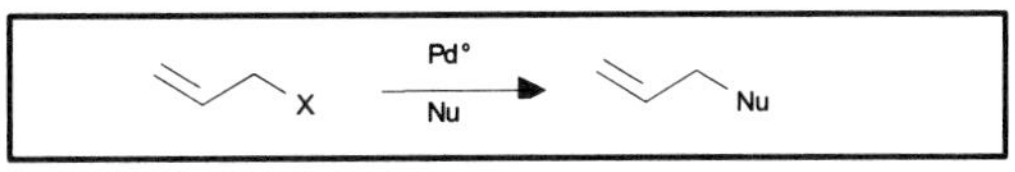

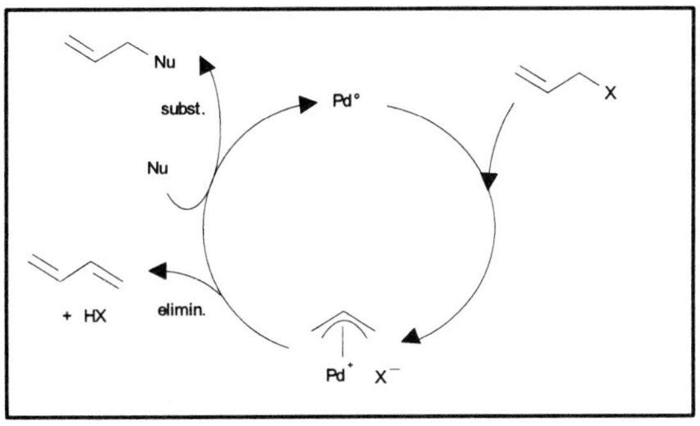

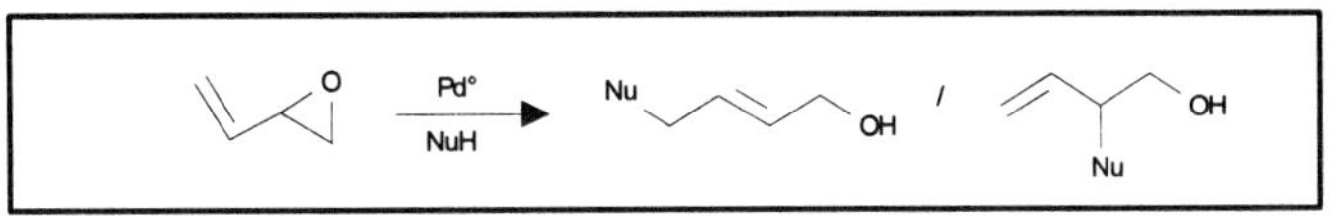

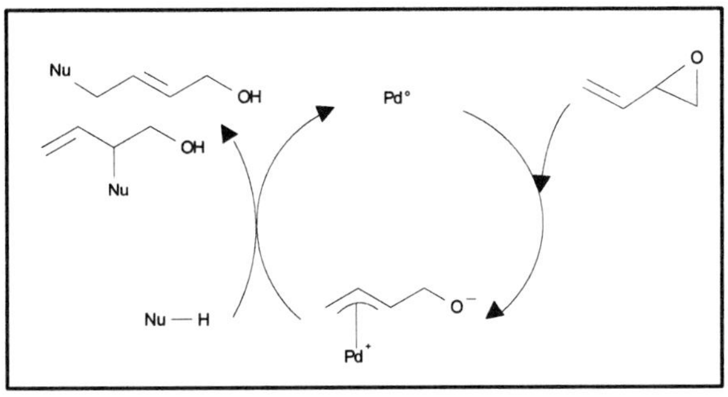

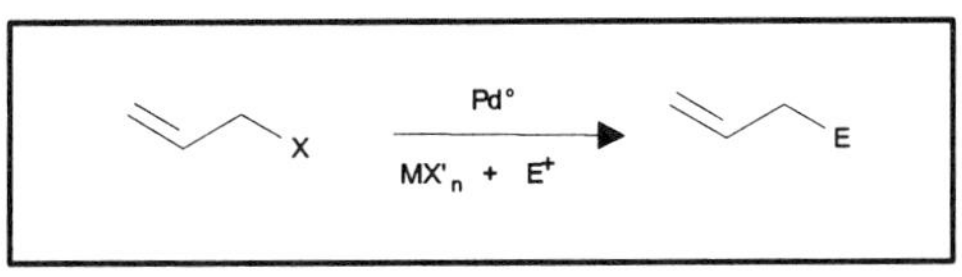

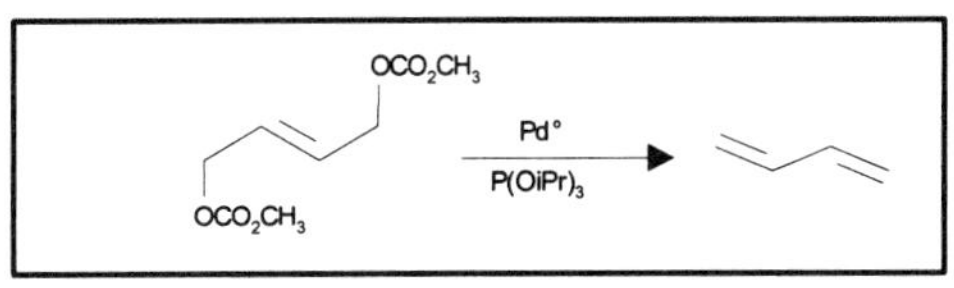

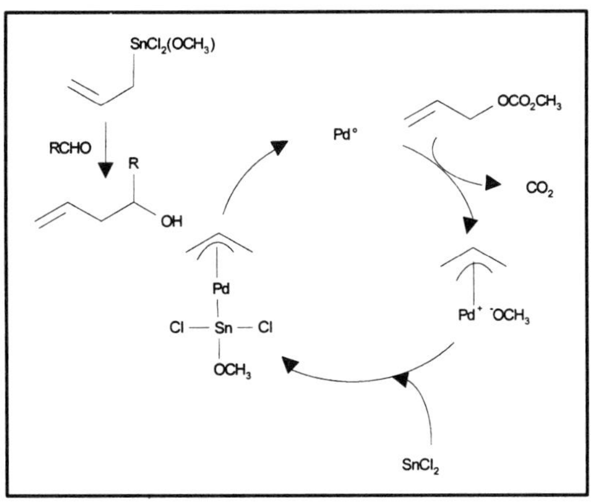

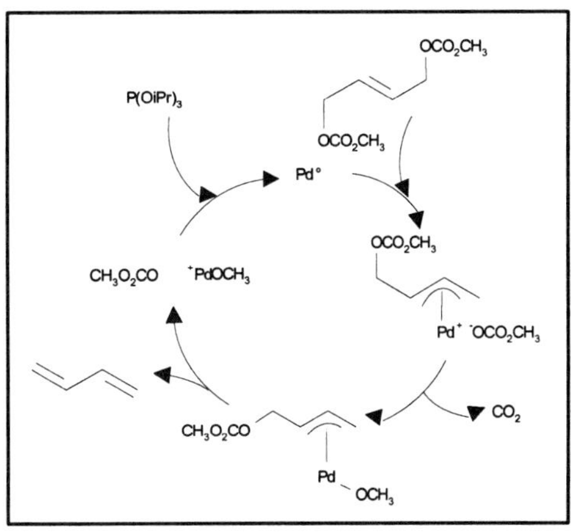

Leaving Group	References
OH	Atkins K.E., *Tetrahedron Lett.* , 1970; 3821 Bergbreiter D.E., *J. Chem. Soc., Chem. Commun.* , 1989; 883 Miura M., *J. Chem. Soc., Perkin Trans. 1* , 1992; 2833 Sakakibara M., *Tetrahedron Lett.* , 1994; 8013
OR	Klumpp G.W., *Tetrahedron Lett.* , 1988; 3579 Takahashi K., *Bull. Chem. Soc. Jpn.* , 1972; 230 Kusama T., *Chem. Pharm. Bull.* , 1992; 1718
OC_6H_5	Fiaud J.C., *J. Organomet. Chem.* , 1978; 154, 175 Takahashi K., *Bull. Chem. Soc. Jpn.* , 1972; 230 Tsuji J., *Tetrahedron Lett.* , 1978; 2075 (elimination)
OAc	Trost B.M., *J. Am. Chem. Soc.* , 1976; 630 Sinou D., *Tetrahedron Lett.* , 1991; 2025 Trost B.M., *J. Am. Chem. Soc.* , 1994; 4089 (asymmetric catalysis) Andersson P. G., *Organometallics* , 1995; 14, 1 (elimination) Trost B.M., *Tetrahedron Lett.* , 1979; 2301 (elimination)
OCOtBu	Fiaud J.C., *J. Org. Chem.* , 1990; 4840

Acknowledgements: We are grateful to Mrs. GIRARDET, Mrs. M'HOUMADI, Mrs. T-P-H. N GUYEN, Mr. L. BAUDRY, Mr. P. DESMAZEAU for their contributions to the preparation of this manuscript.

II- GRAPHICAL ABSTRACTS OF REACTION NUMBERS

Reaction number	Graphical abstracts
RXN1	
RXN2	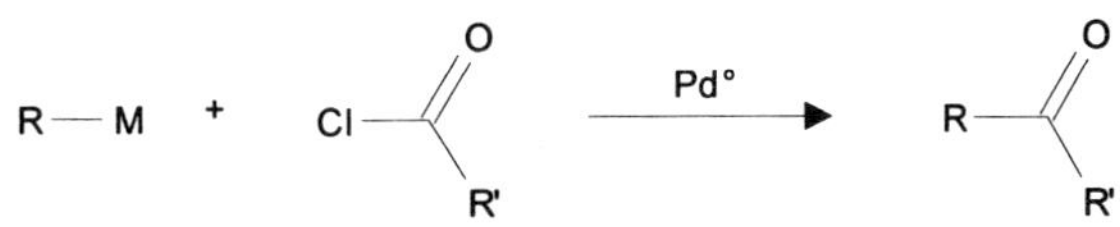
RXN3	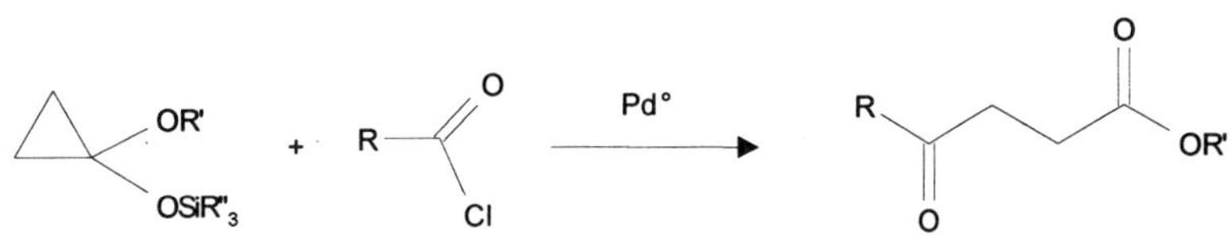
RXN4	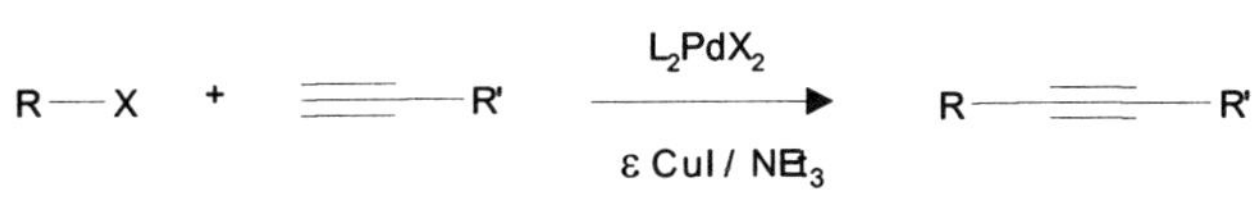
RXN5	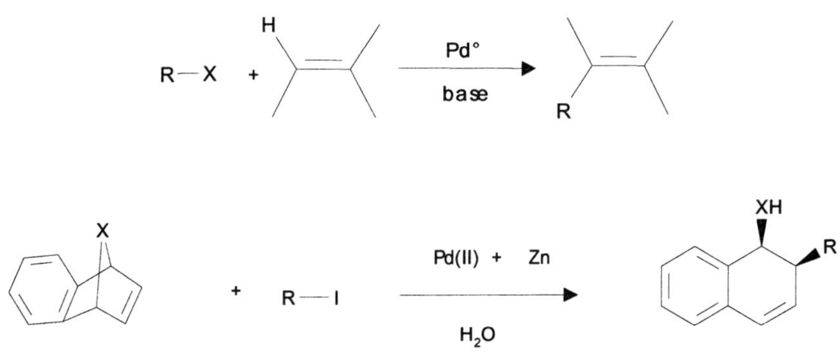

RXN6	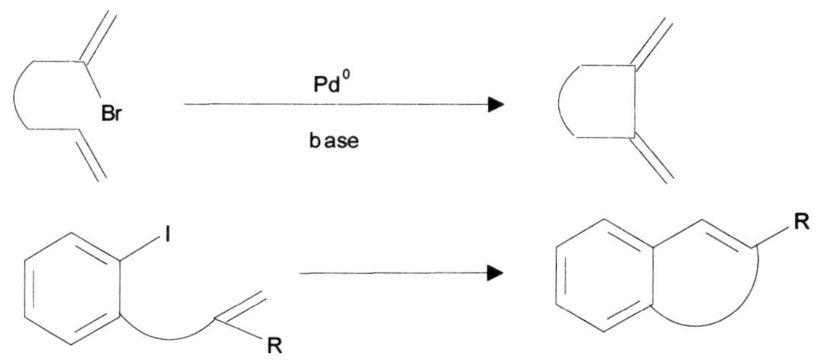
RXN7	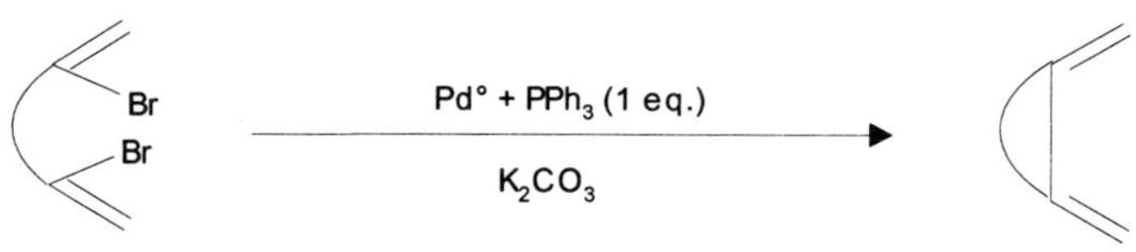
RXN8	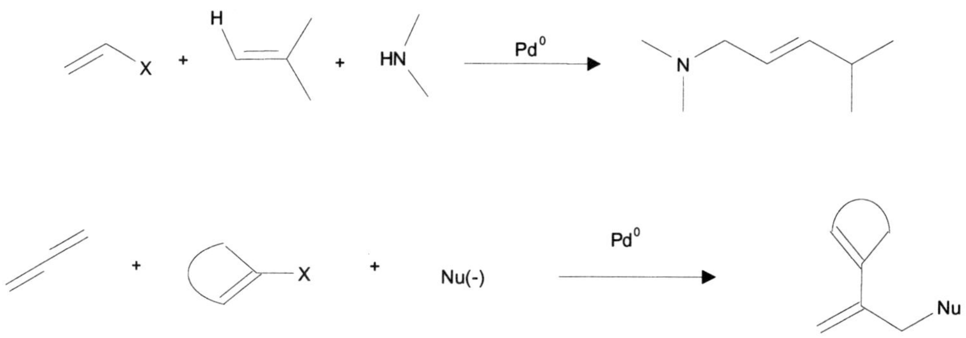
RXN9	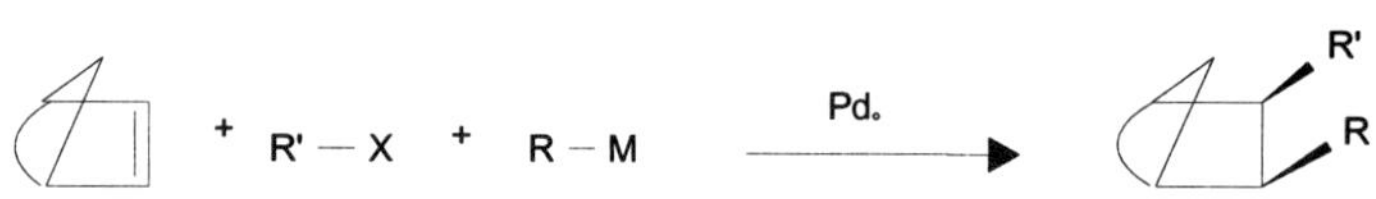
RXN10	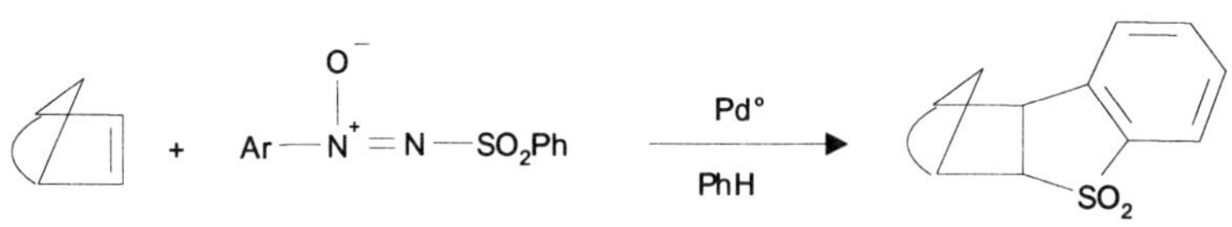

RXN11	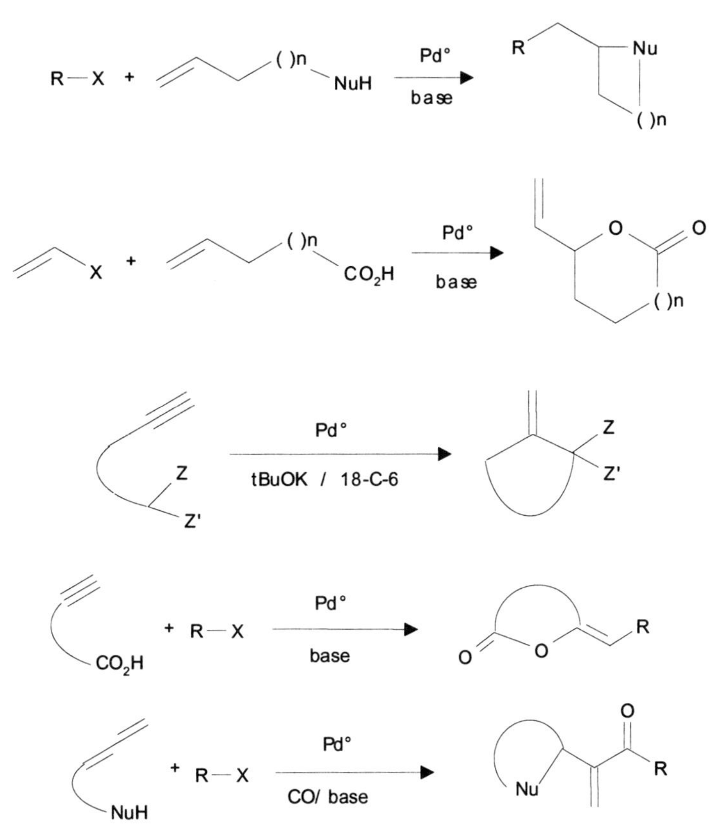
RXN12	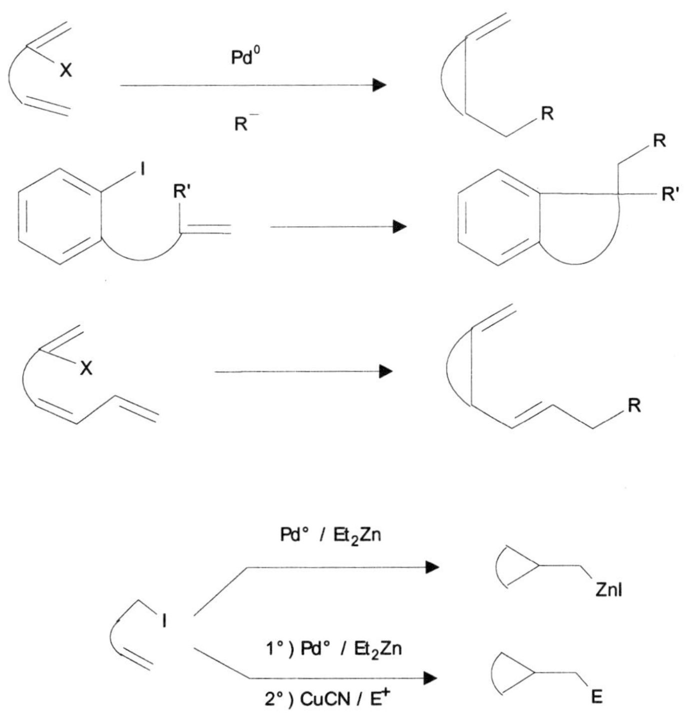

RXN11:
R—X
NuH
( )n
Pd°
base
R
Nu
X
CO2H
O
tBuOK / 18-C-6
Z
Z'
R—X
CO/ base
RXN12:
X
Pd0
R−
I
R'
R
Pd° / Et2Zn
ZnI
1°) Pd° / Et2Zn
2°) CuCN / E+
E

RXN13	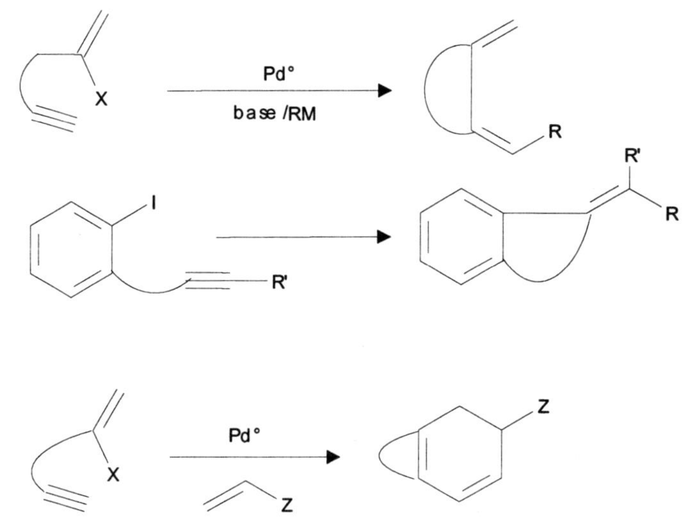
RXN14	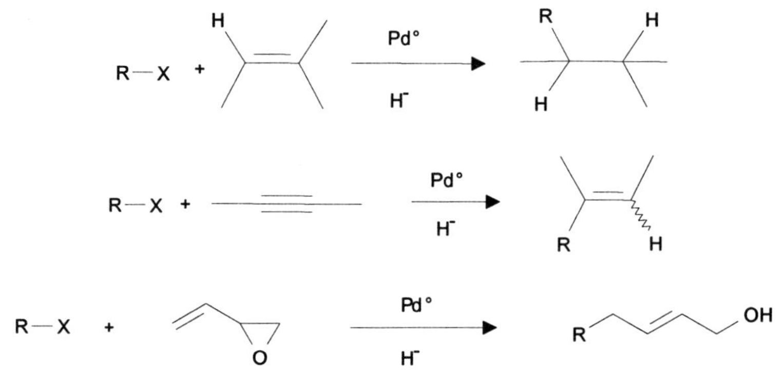
RXN15	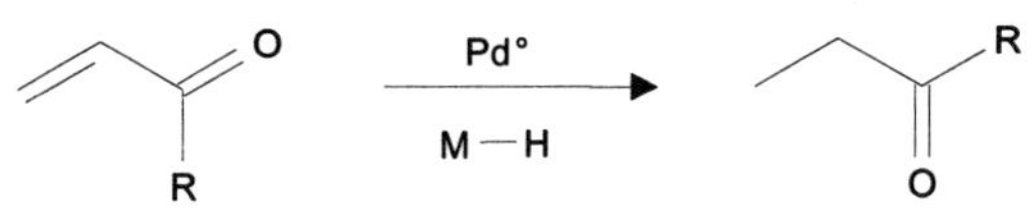
RXN16	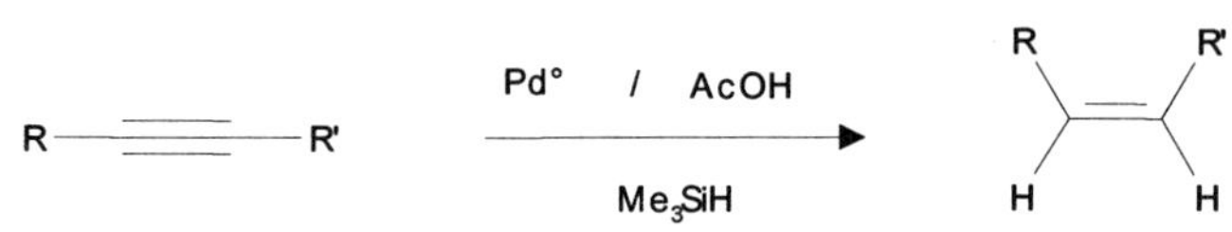
RXN17	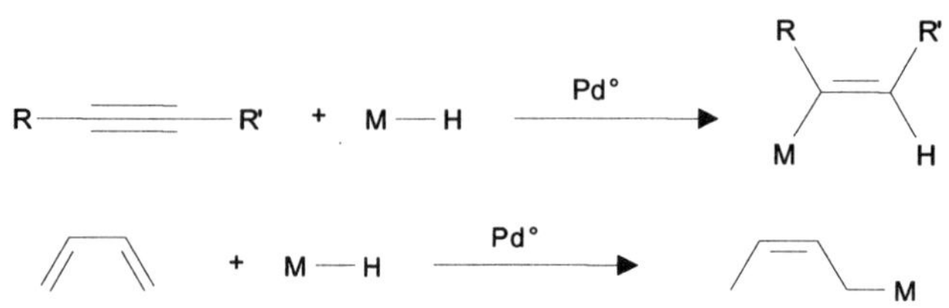
RXN18	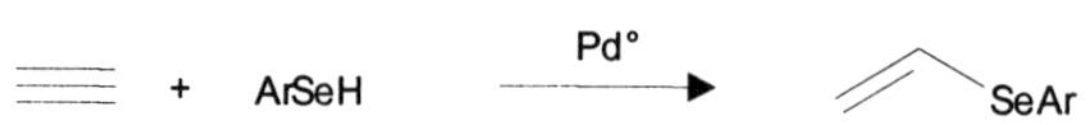

RXN19	
RXN20	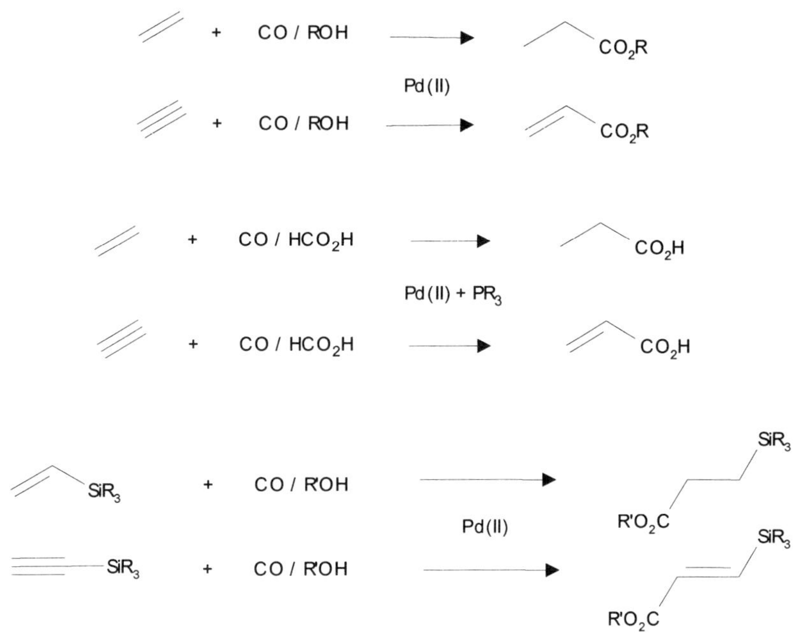
RXN21	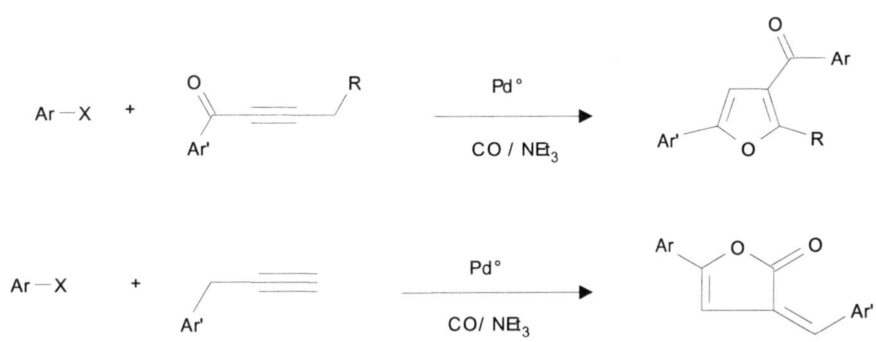
RXN22	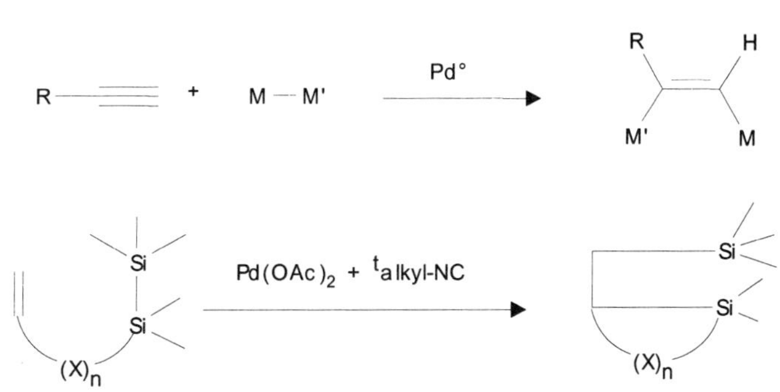

RXN23

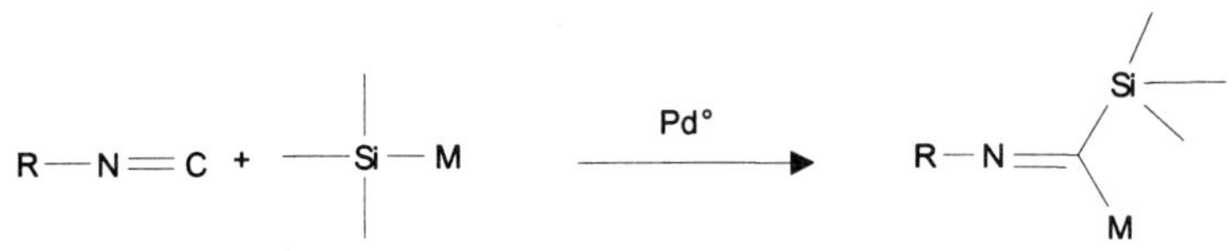

RXN24

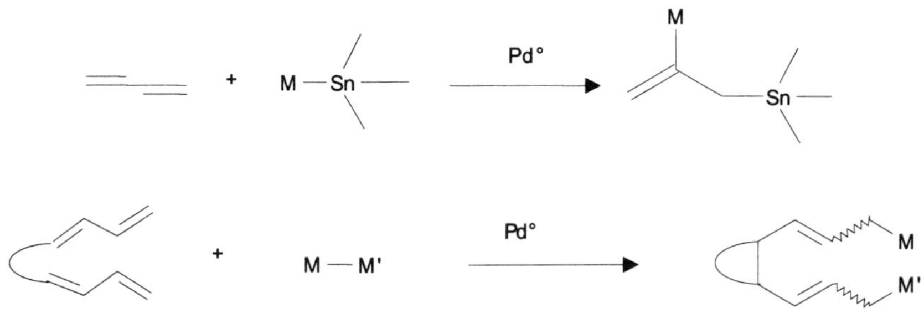

RXN25

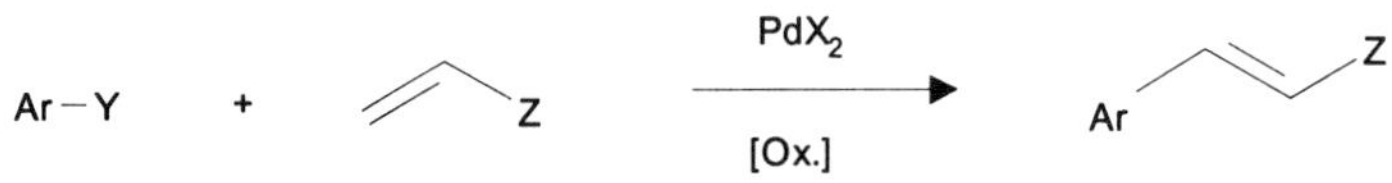

RXN26

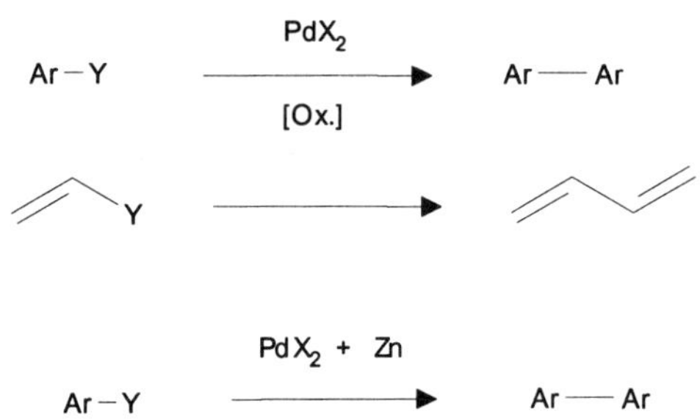

RXN27

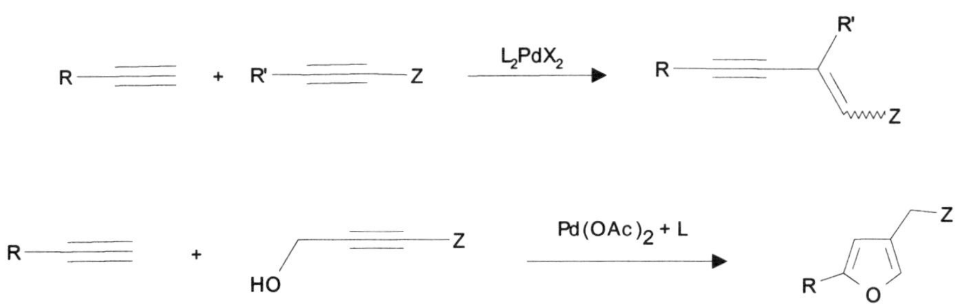

RXN28

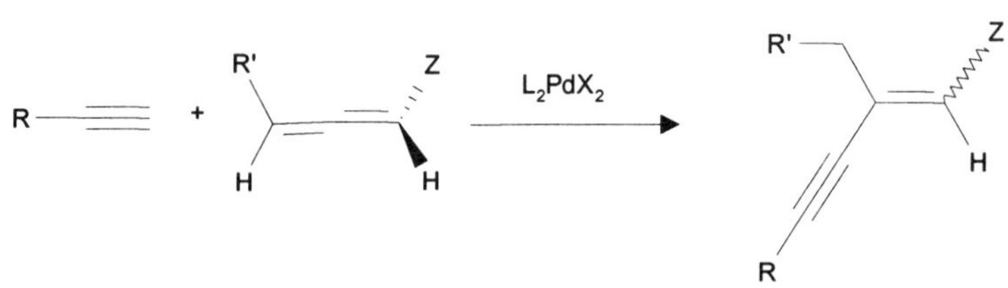

RXN29	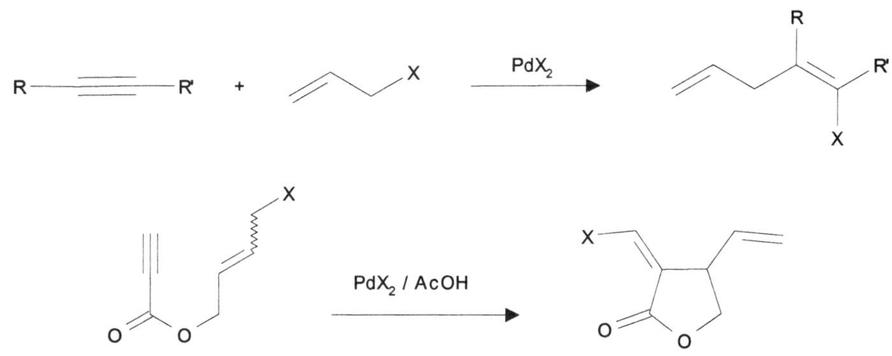
RXN30	
RXN31	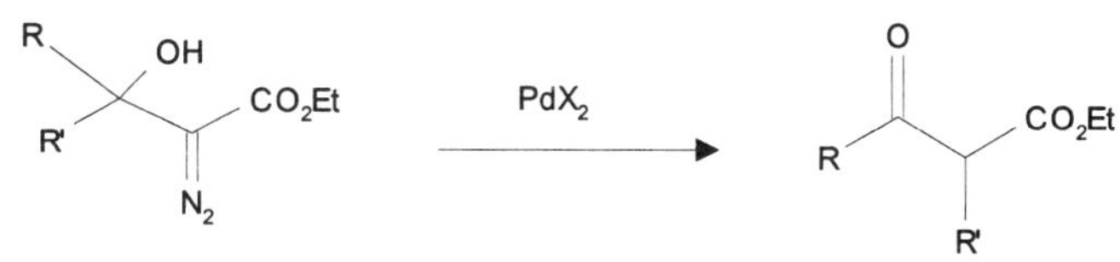
RXN32	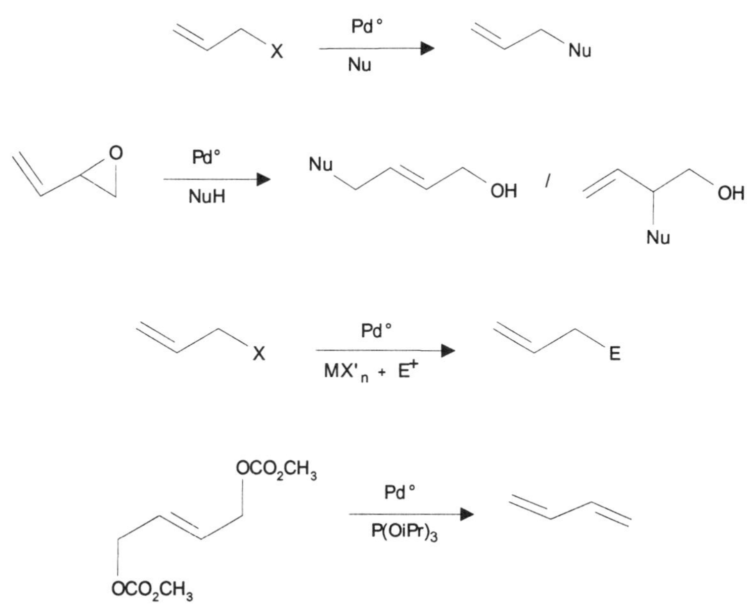
RXN33	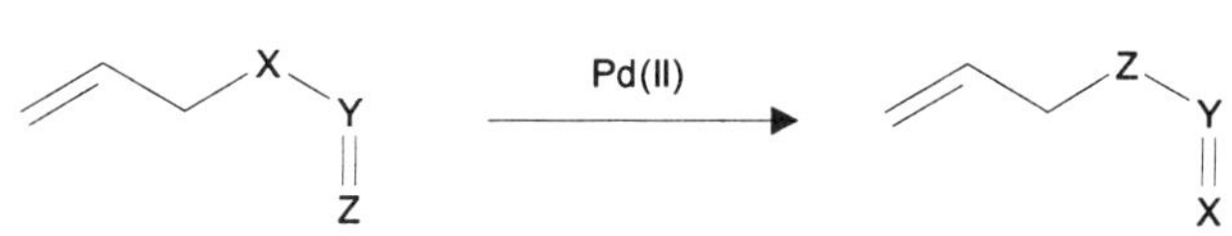
RXN34	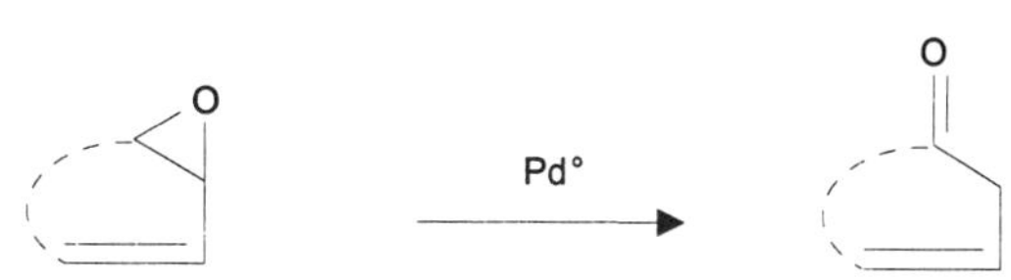

RXN35

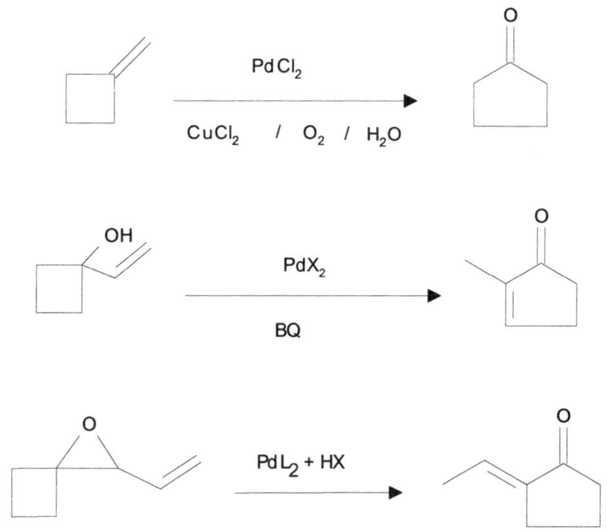
PdCl$_2$
CuCl$_2$ / O$_2$ / H$_2$O
O
OH
PdX$_2$
BQ
O
O
PdL$_2$ + HX
O

RXN36

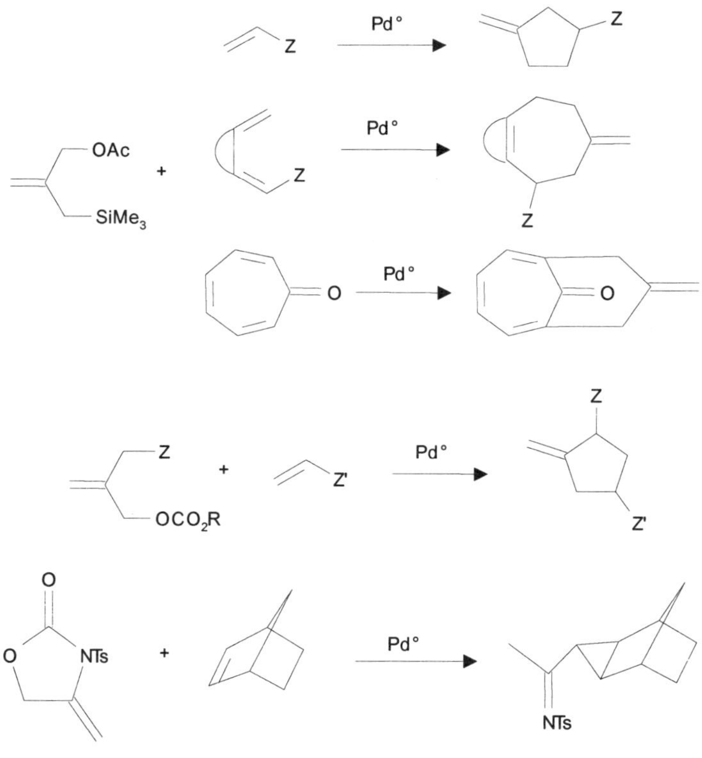
Z
Pd°
Z
OAc
SiMe$_3$
+
Z
Pd°
Z
O
Pd°
O
Z
+
Z'
Pd°
OCO$_2$R
Z
Z'
O
O
NTs
+
Pd°
NTs

RXN37

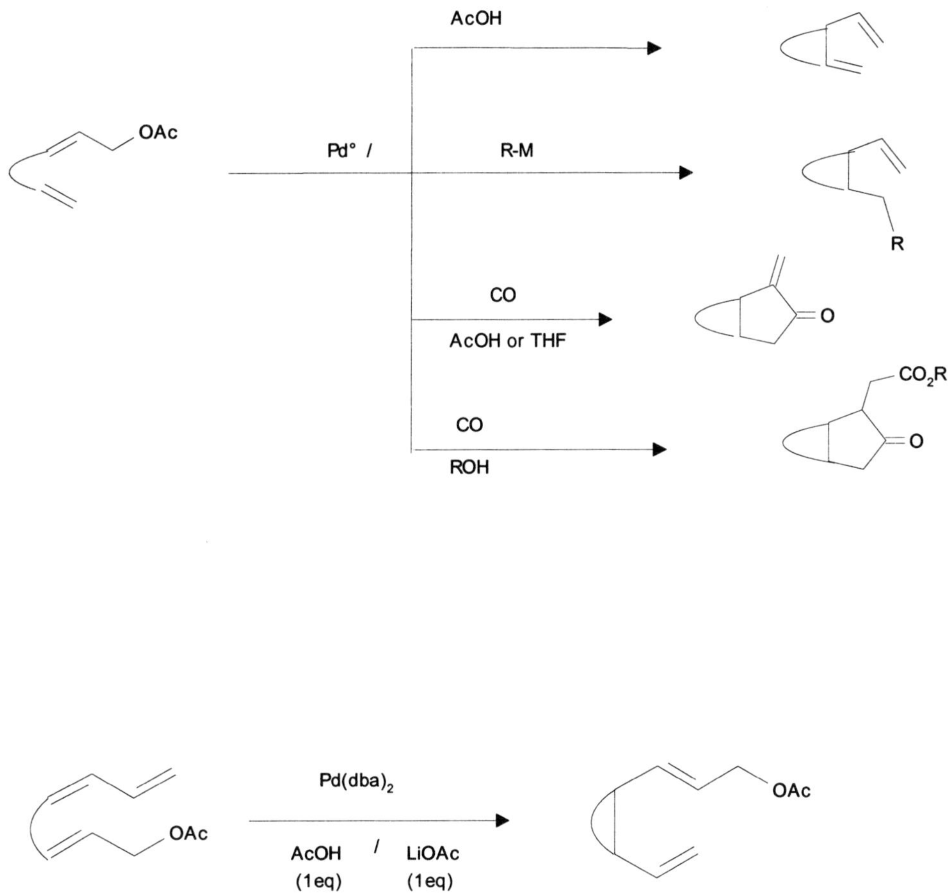

RXN38

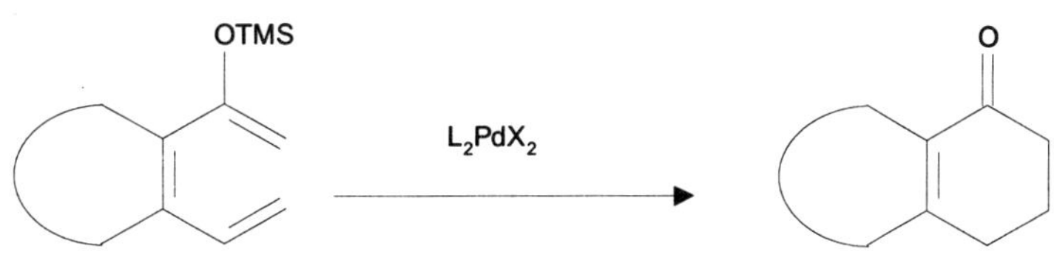

RXN39

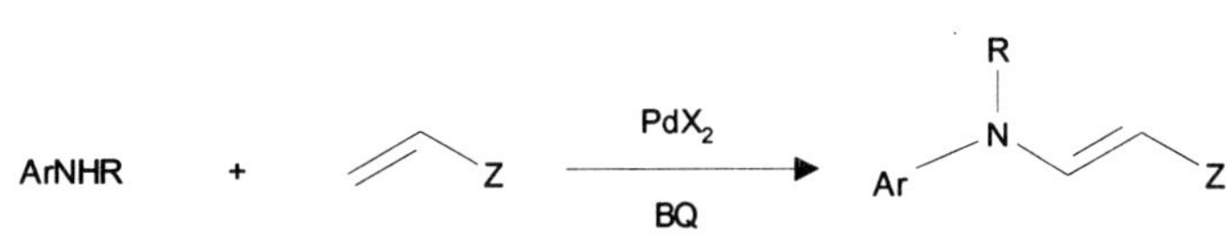

RXN40

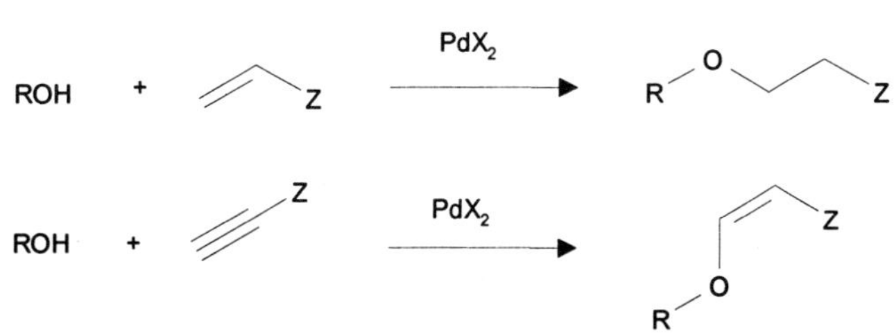

RXN41

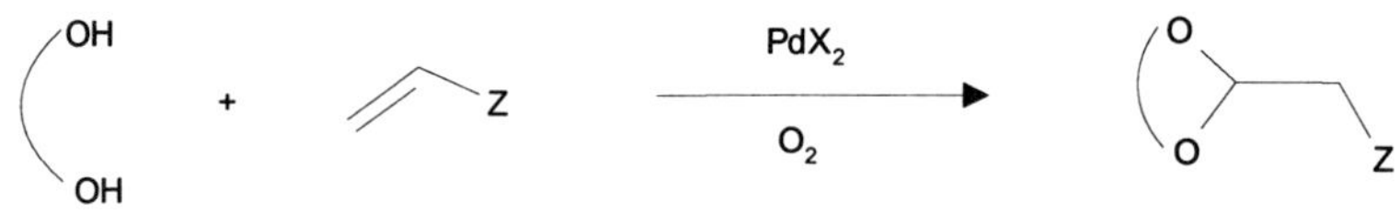

RXN42

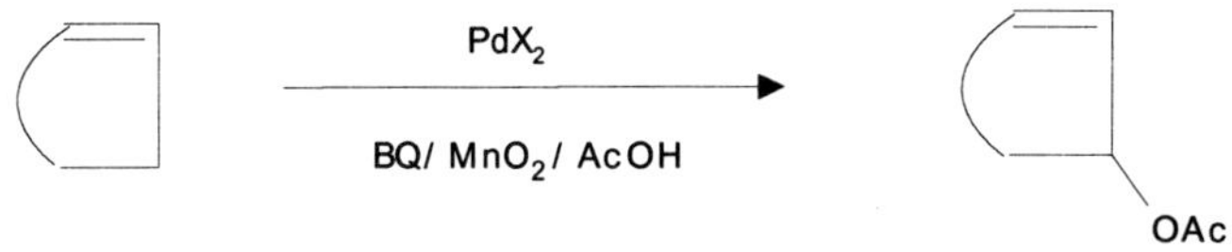

RXN43

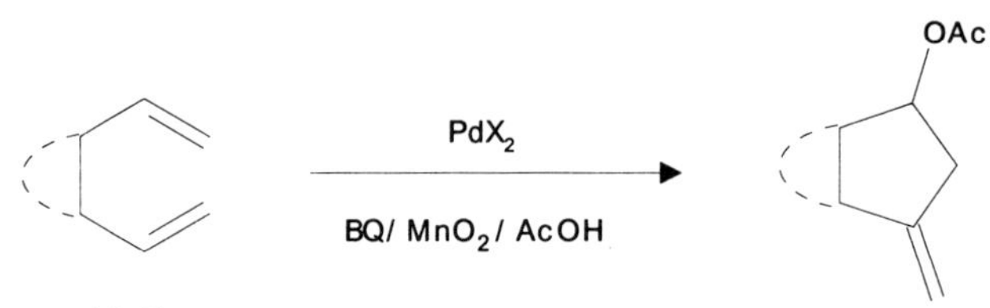

1,5-diene

RXN44

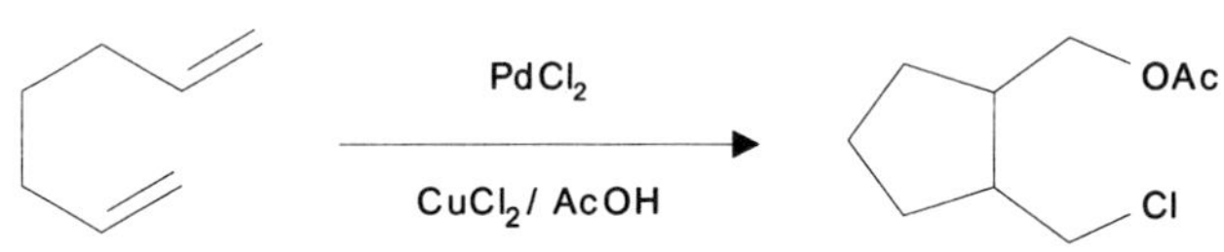

1,6-diene

RXN45

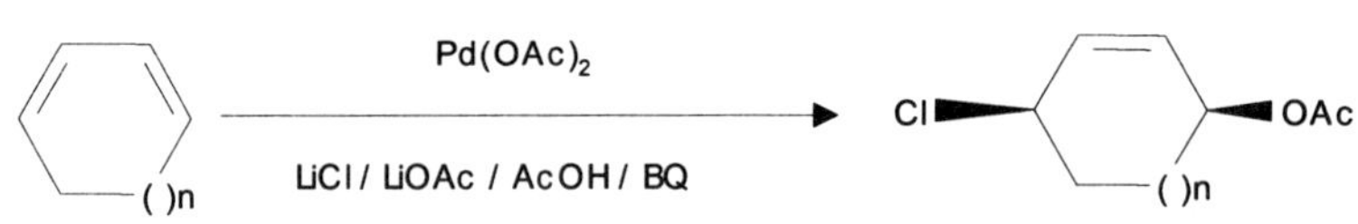

RXN46

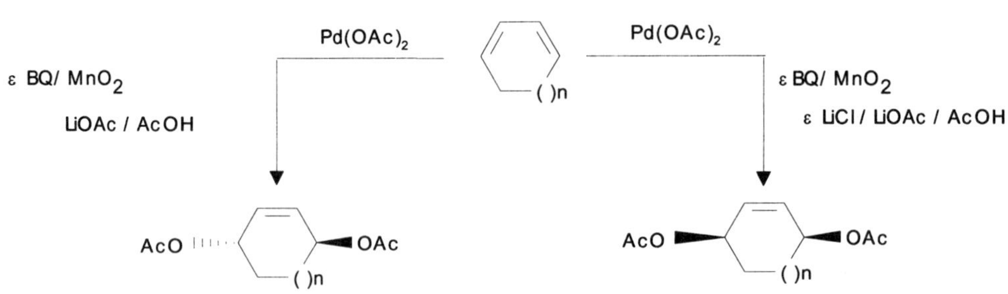

RXN47

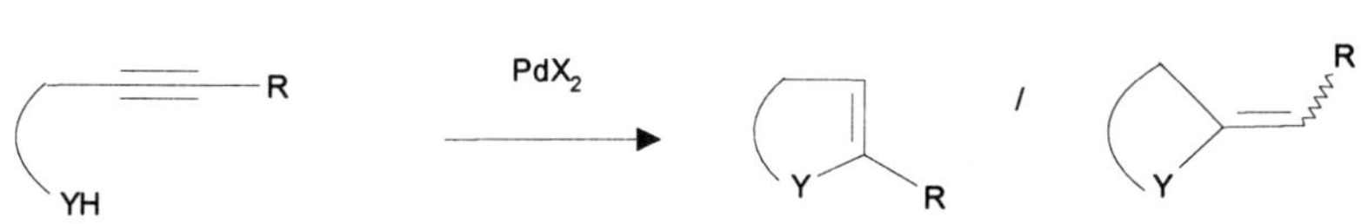

RXN48

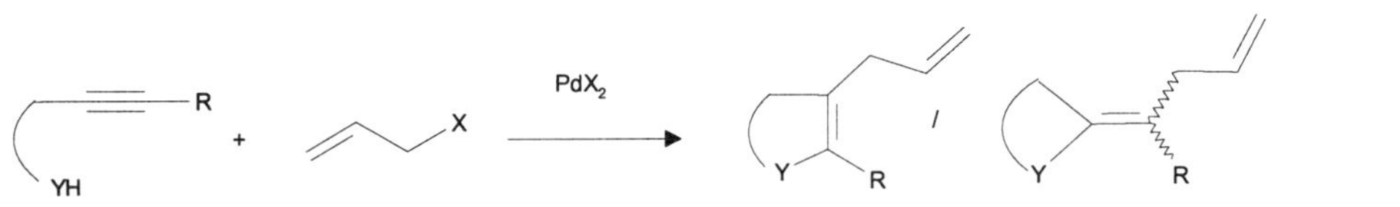

RXN49

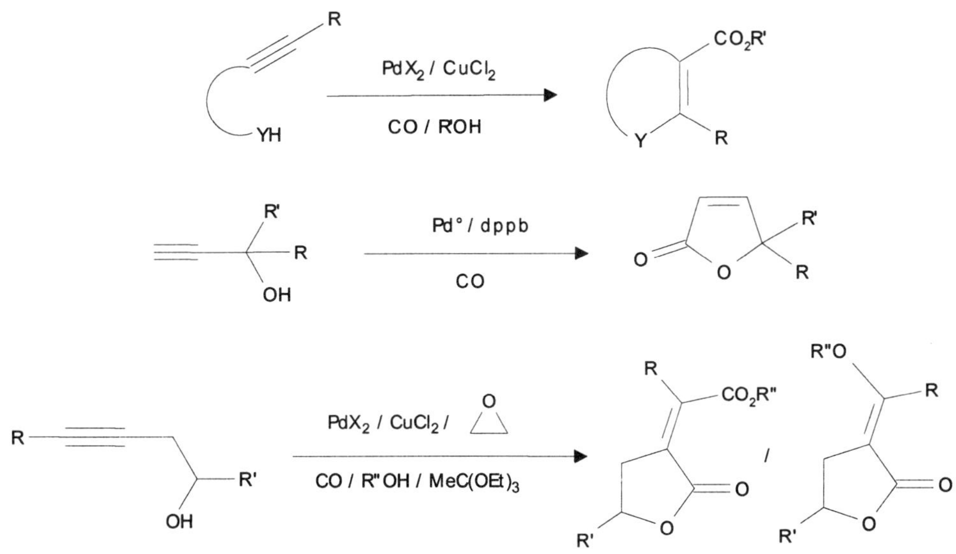

RXN50

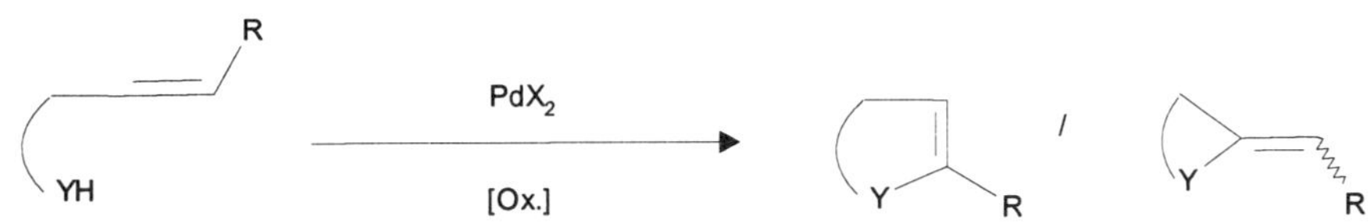

RXN51

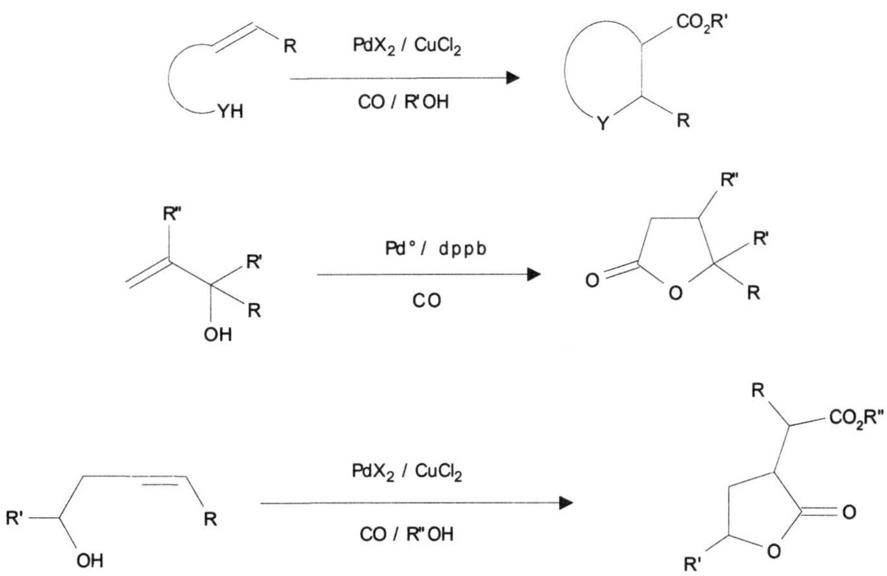

RXN52	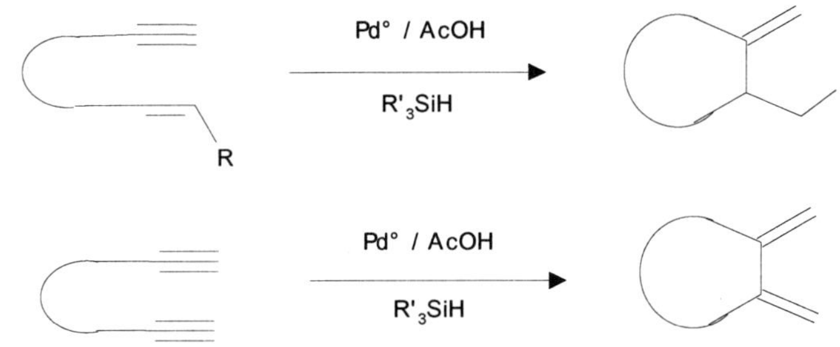
RXN53	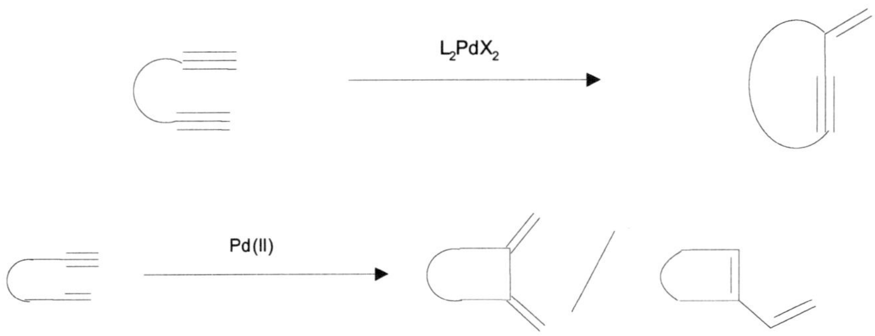
RXN54	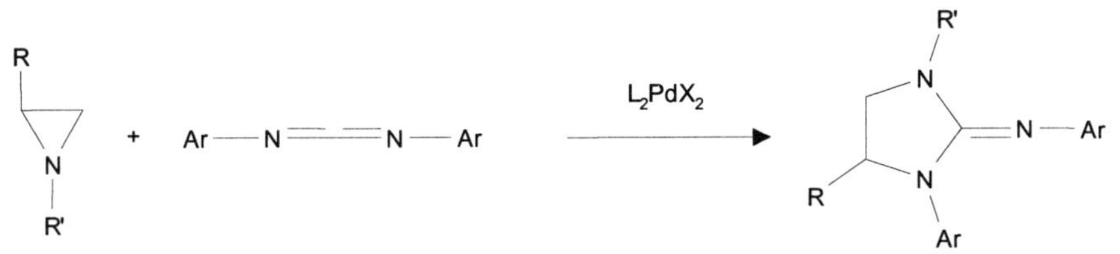
RXN55	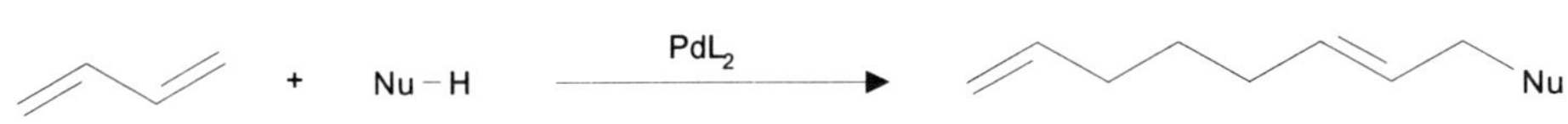
RXN56	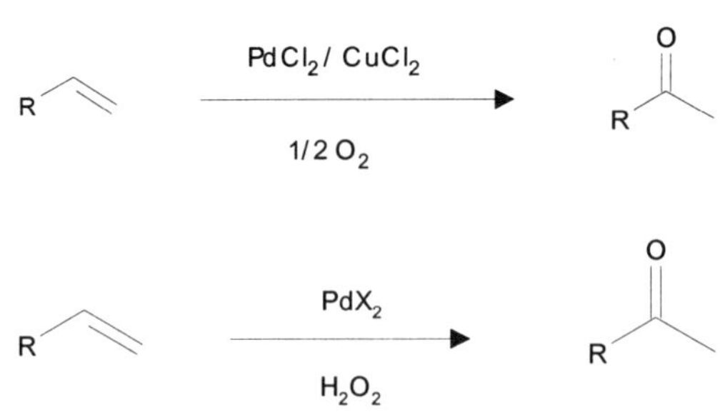

RXN57	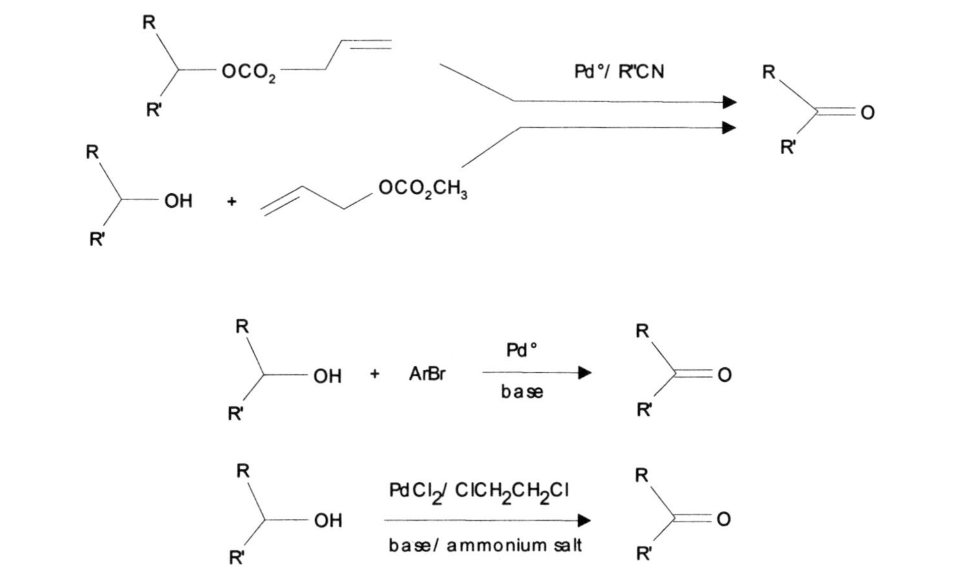
RXN58	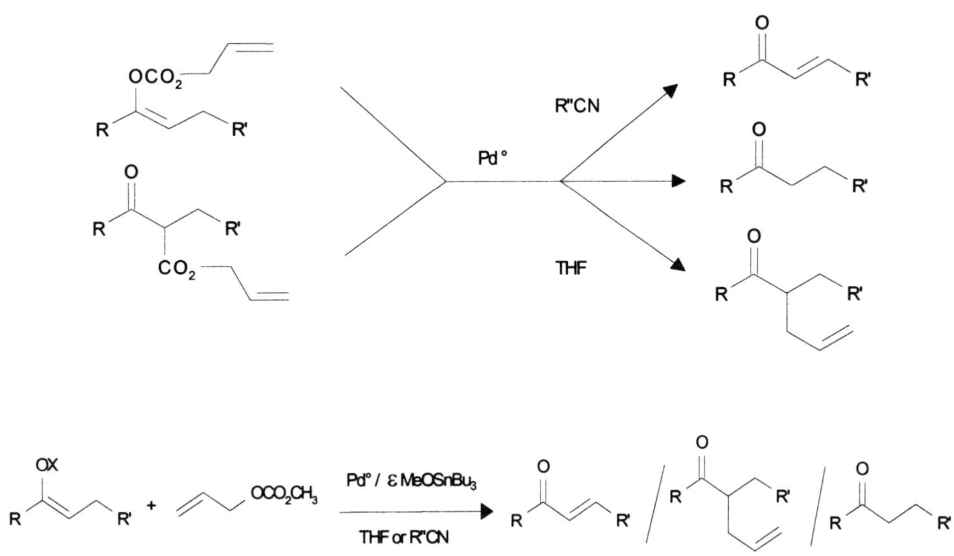
RXN59	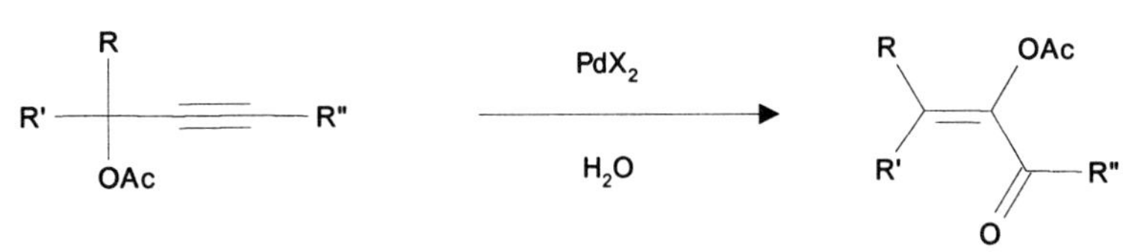
RXN60	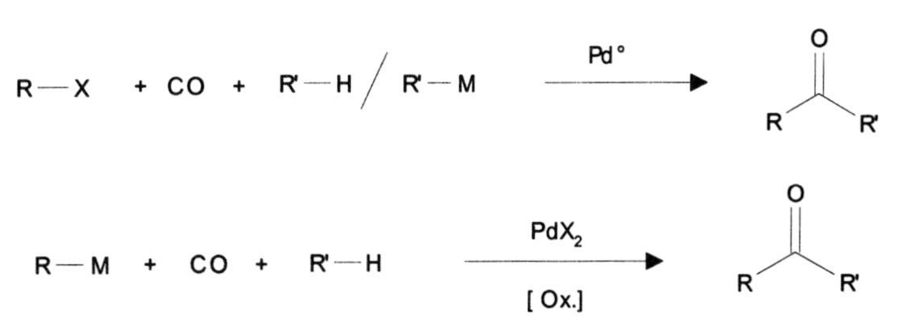

RXN61

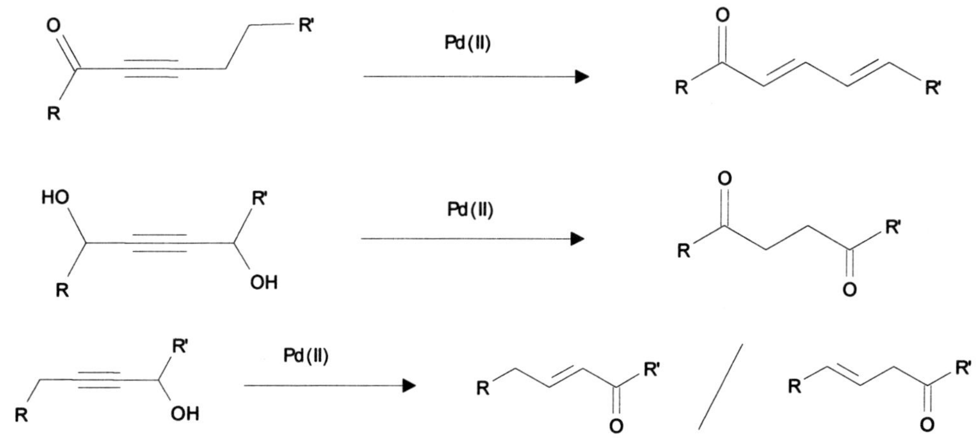

RXN62

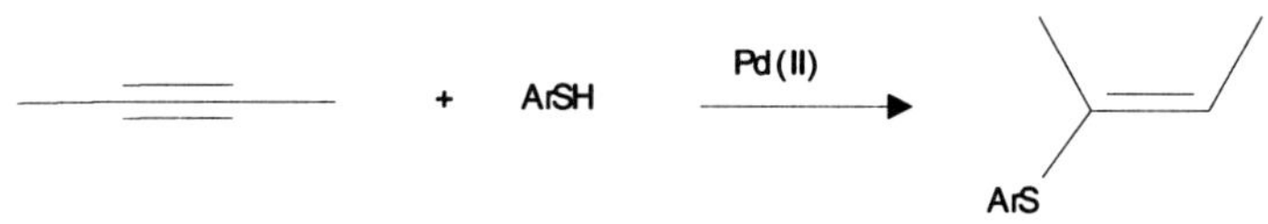

RXN63

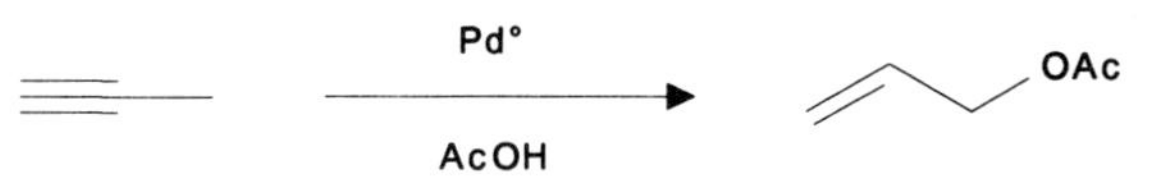

RXN64

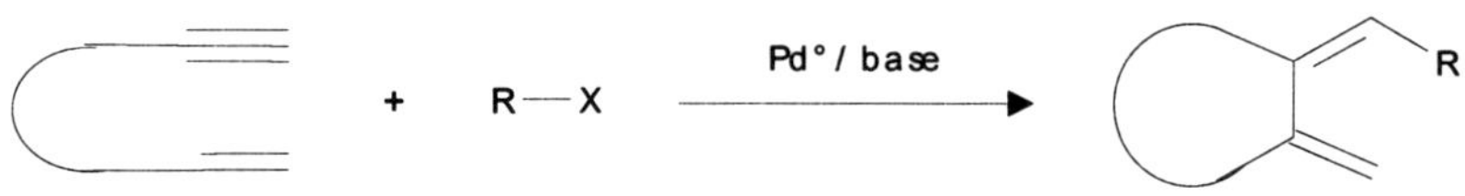

RXN65

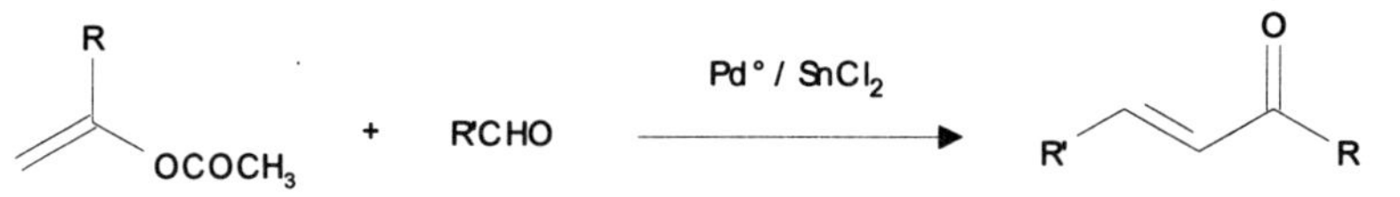

RXN66

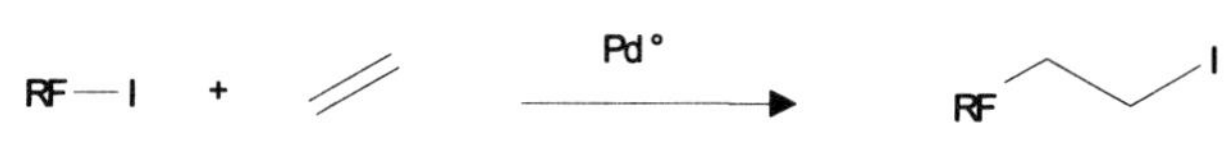

RXN67

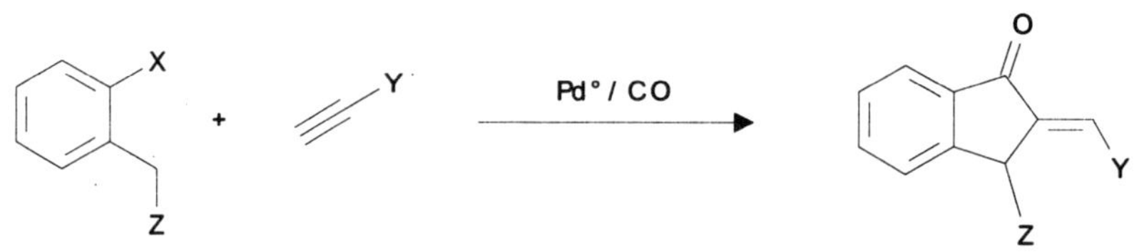

RXN68

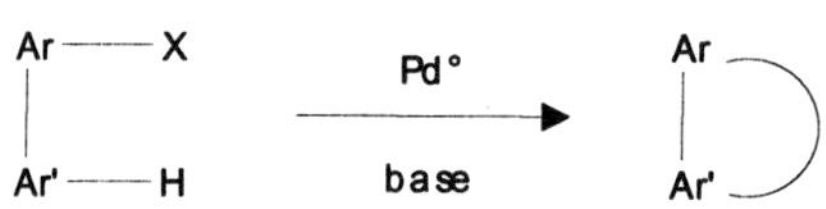

RXN69

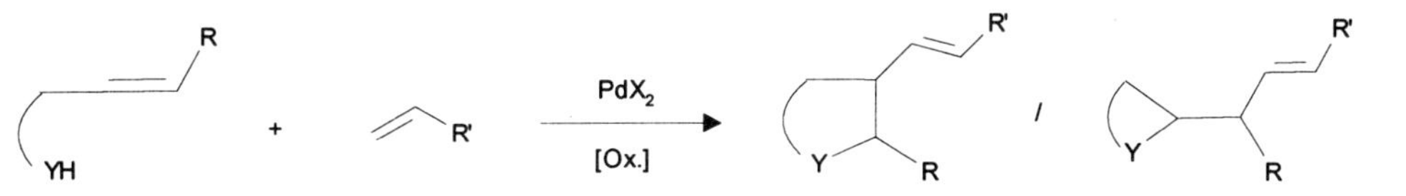

RXN70

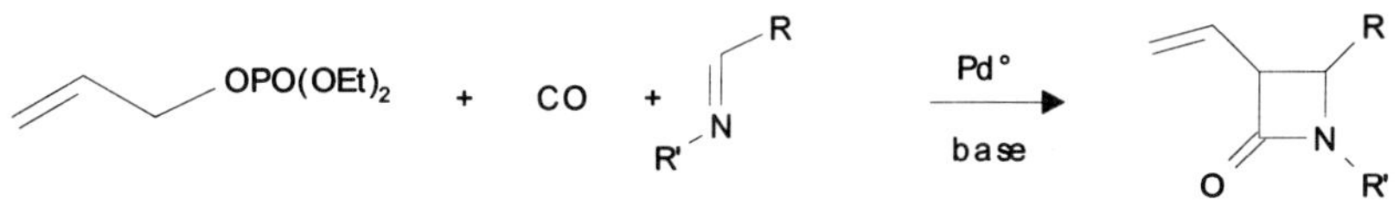

RXN71

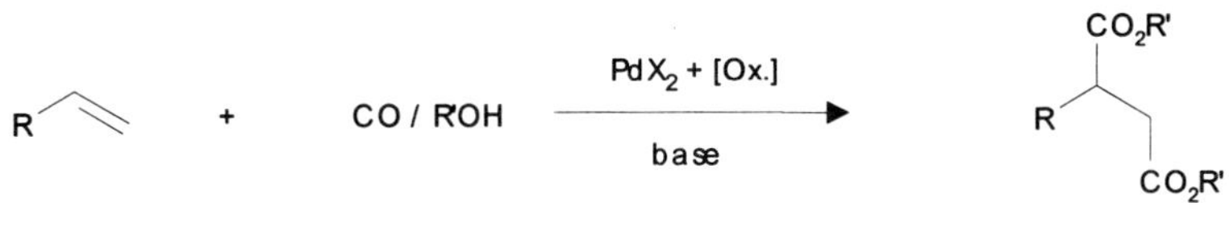

RXN72

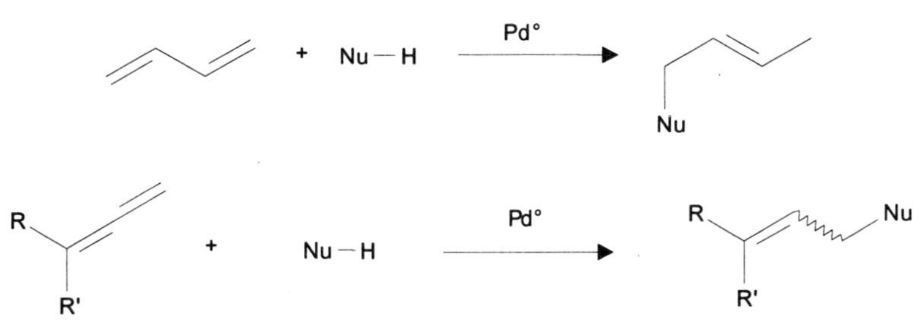

RXN73

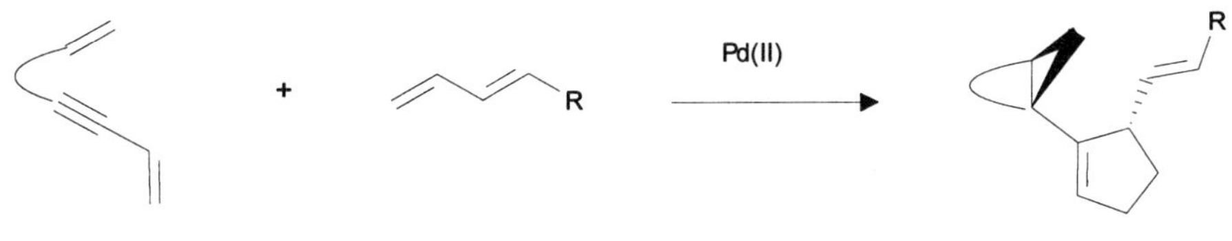

RXN74

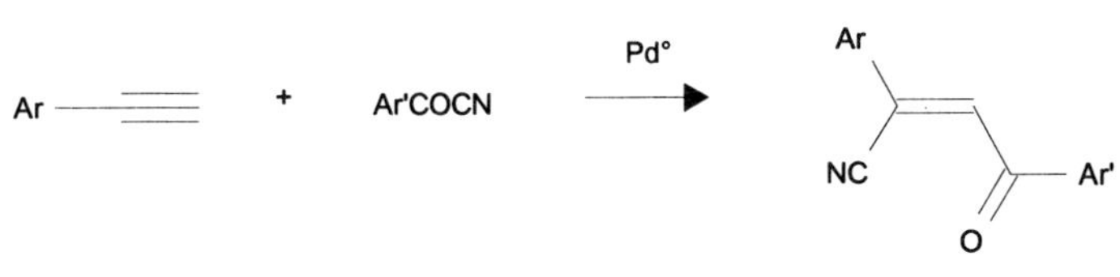

RXN75

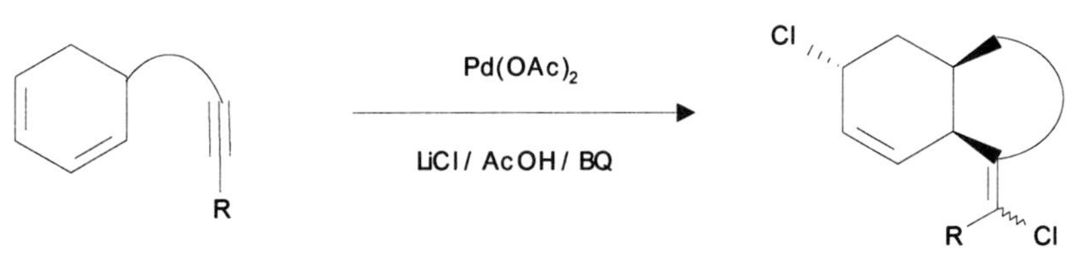

RXN76

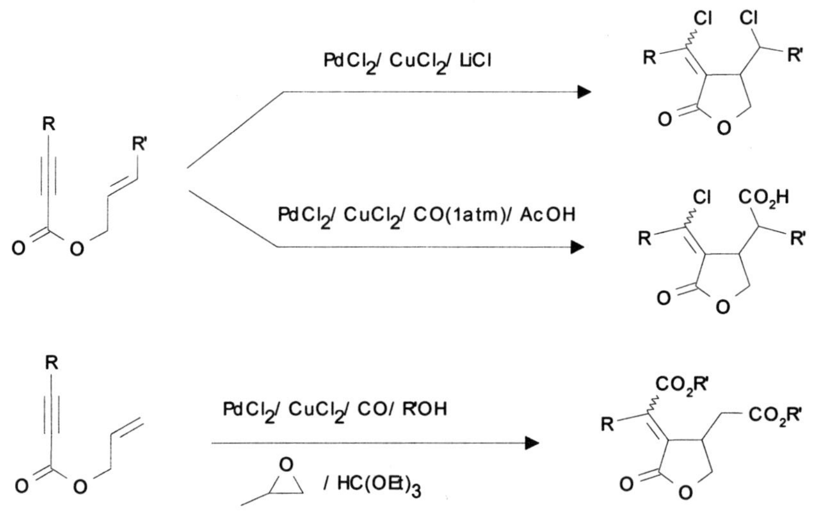

RXN77

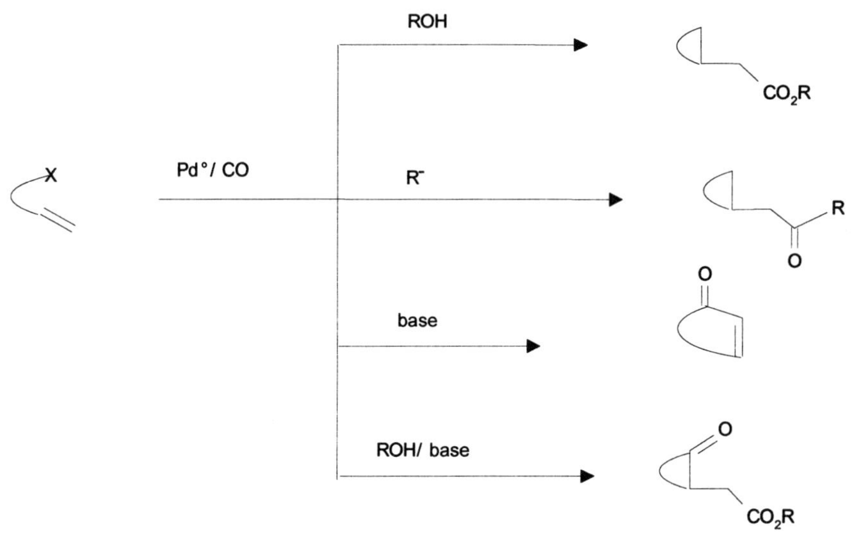

RXN78

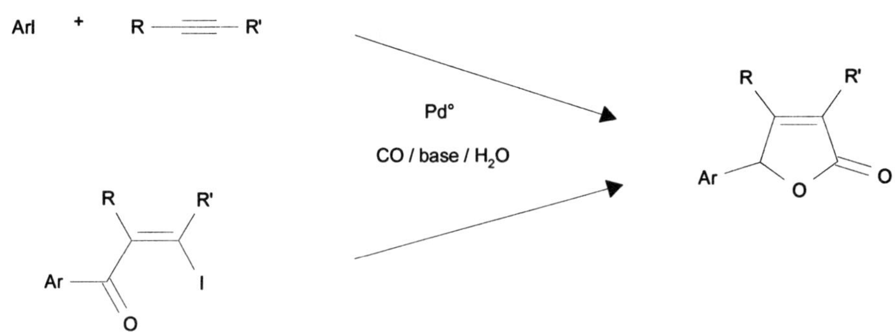

RXN79

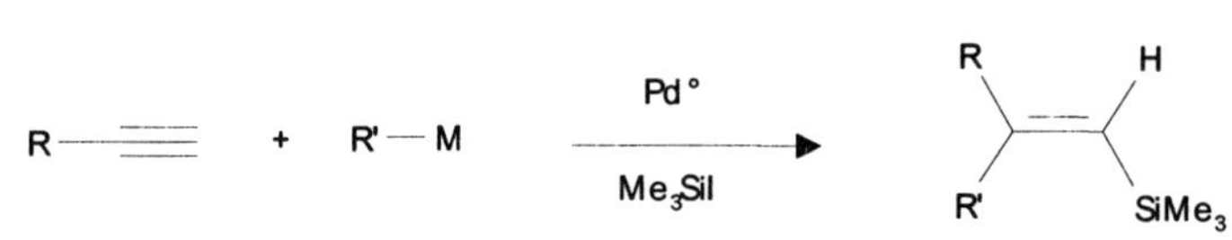

<table>
<tr><td>RXN80</td><td>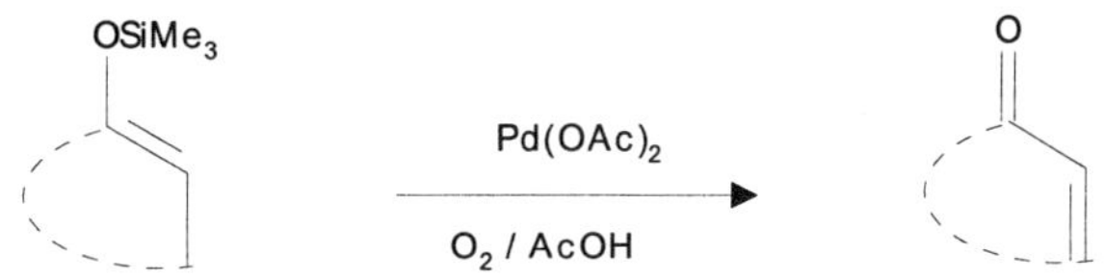
</td></tr>
<tr><td>RXN81</td><td>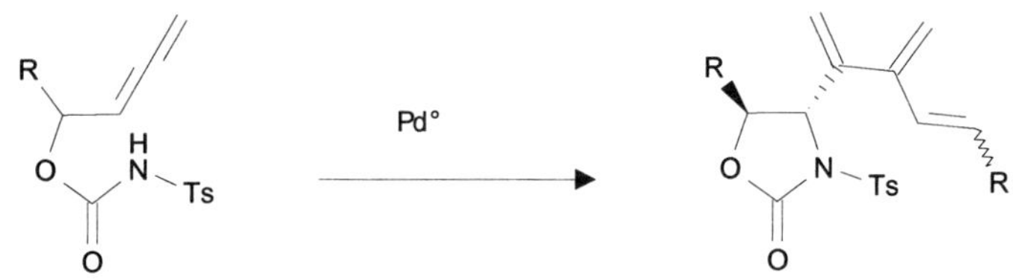
</td></tr>
<tr><td>RXN82</td><td>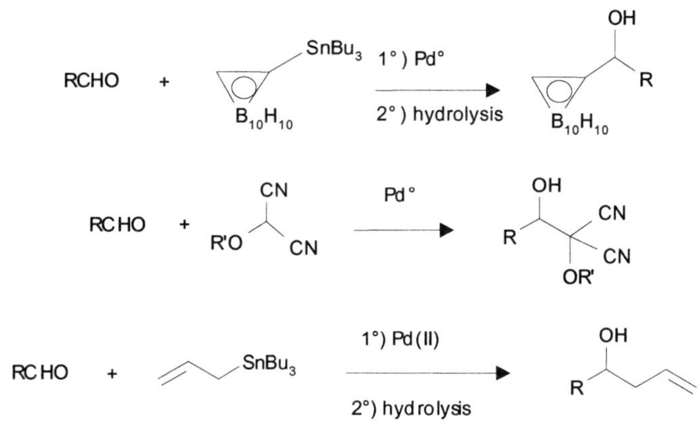
</td></tr>
<tr><td>RXN83</td><td>
</td></tr>
<tr><td>RXN84</td><td>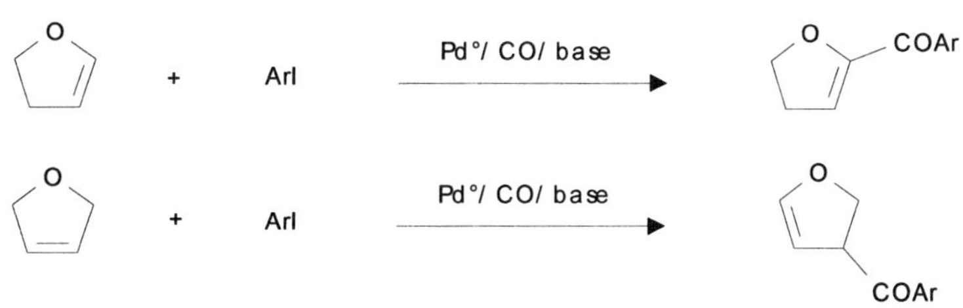
</td></tr>
</table>

III- REACTIONS CATALYZED BY PALLADIUM COMPLEXES

RXN1 Cross-Coupling of Organometallics with RX Derivatives

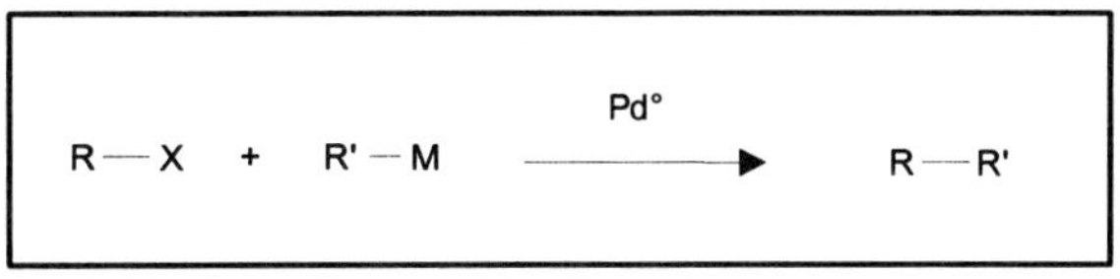

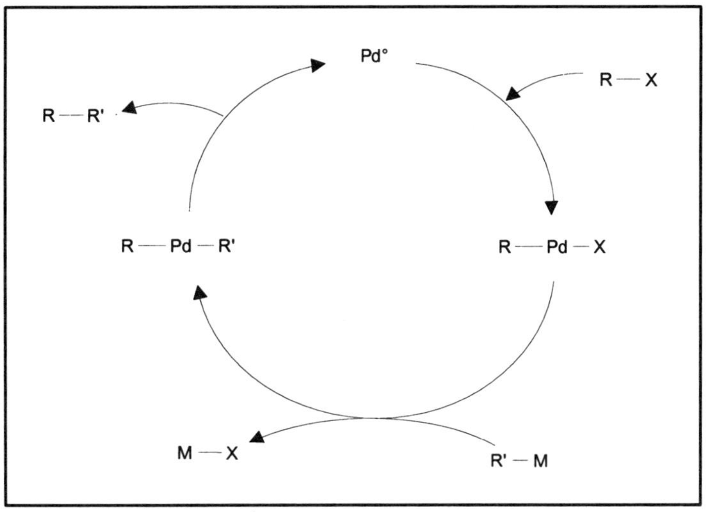

R	X	R'	M	References
C=C	X	R'	M	Yamamura K., *Tetrahedron Lett.*, 1977; 4429 Murahashi S.I., *J.Org.Chem.*, 1979; 2408
C=C	Br	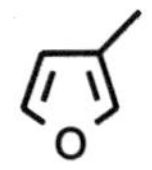	B	Keay B.A., *Synlett.*, 1994; 625
C=C	I	Phenyl	Zn	Luo F.T., *Tetrahedron Lett.*, 1992; 6839
$(CH=CH)_n$-C=C-	I	$(CH=CH)_n$-C=C-	B	De Lera A.R., *Tetrahedron*, 1995; 2435
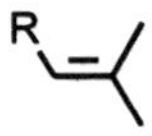	Br	R'	B	Soderquist J-A., *Tetrahedron Lett.*, 1995; 3119 (intramolecular)
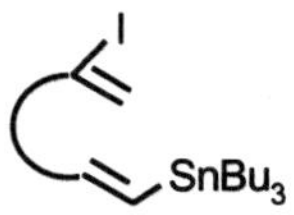	I	-	-	Hodgson D. M., *Synlett.*, 1995; 267 (intramolecular)
C=C-CHO	Br	C=C	B	Urdaneta N., *J. Organomet. Chem.*, 1994; 464, C33
C=C-CO	I	C=C	Sn	Johnson C.R., *Tetrahedron Lett.*, 1992; 919
C=C-CO	I	Ar	Sn	Johnson C.R., *Tetrahedron Lett.*, 1992; 919
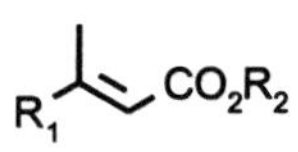	OSO_2CF_3	R	B	Gibbs R.A., *Tetrahedron Lett.*, 1995; 5669
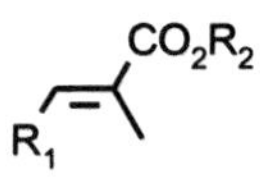	Br	R'	Sn	Rossi R., *Tetrahedron*, 1994; 12029
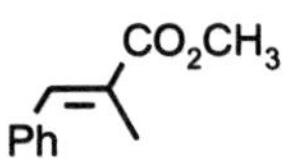	Br	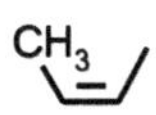	Sn	Zapata A.J., *J. Organomet. Chem.*, 1994; 479, C6
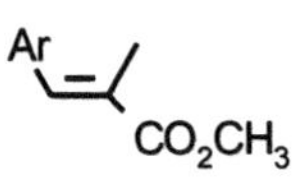	Br	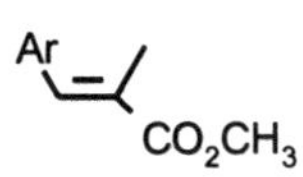	Sn	Horikawa H., *Synthesis*, 1995; 582
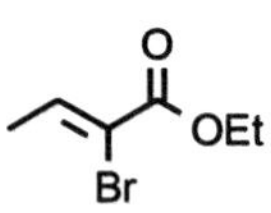	Br	Ar	Zn	Rossi R., *Tetrahedron Lett.*, 1994; 6913 Rossi R., *Tetrahedron Lett.*, 1994; 12029 Rossi R., *Synlett.*, 1995; 344
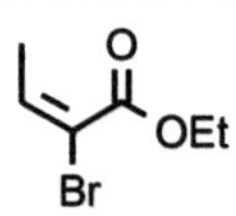	Br	R	Zn	Rossi R., *Synlett.*, 1995; 344
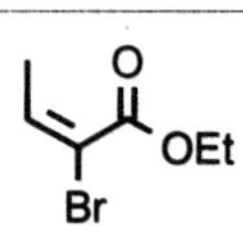	Br	Ar	Zn	Rossi R., *Tetrahedron Lett.*, 1994; 6913
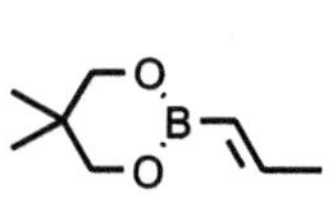	Br	C=C	Zn	Vaultier M., *Tetrahedron Lett.*, 1994; 3089

R	X	R'	M	References
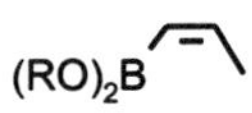	Br	C≡C	Zn	Wang K.K., *Tetrahedron Lett.*, 1994; 1829
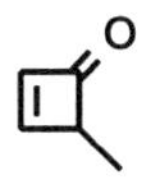	Cl	Het	Sn	Liebeskind L.S., *J. Org. Chem.*, 1993; 3550 Liebeskind L.S., *J. Am. Chem. Soc.*, 1993; 9868 Liebeskind L.S., *J. Org. Chem.*, 1994; 7737
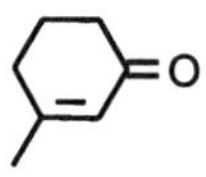	I	C=C	B	Blart E. Thesis, *Université Paris VI*, 1993
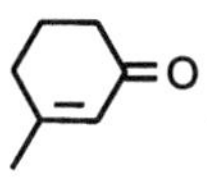	Br	Ar	B	Marson C. M., *J. Chem. Soc., Chem. Commun.*, 1994; 2305
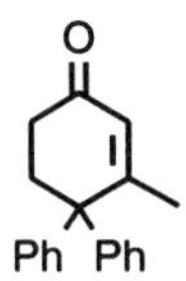	OSO_2CF_3	CH_2=CH-	Sn	Malleron J.L., *Synth. Commun.*, 1995; 2355
	Br	Ar	B	Marson C. M., *J. Chem. Soc., Chem. Commun.*, 1994; 2305
	I	CH_2=CH-	B	Genêt J-P., *Tetrahedron Lett.*, 1995; 1443
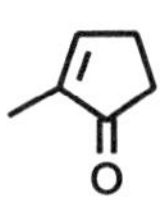	I	R'	B	Johnson C.R., *J. Am. Chem. Soc.*, 1993; 11014
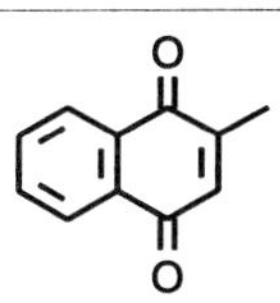	Br	R	Sn	Echavarren A.M., *J. Org. Chem.*, 1994; 6075
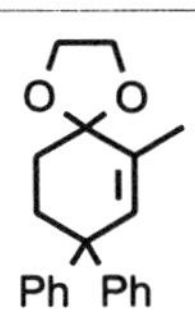	I	CH_2=CH-	Sn	Malleron J.L., *Synth. Commun.*, 1995; 2355
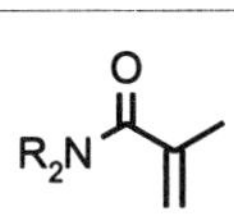	Br	R'	Sn	Meyers A.I., *Synlett.*, 1993; 573
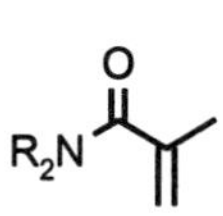	Br	R'	B	Meyers A.I., *Synlett.*, 1993; 573
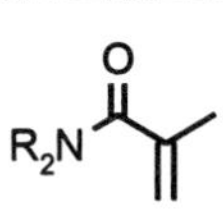	I	R'	Sn	Meyers A.I., *Synlett.*, 1993; 573
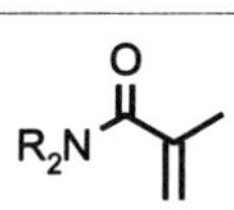	I	R'	B	Meyers A.I., *Synlett.*, 1993; 573

RXN1

R	X	R'	M	References
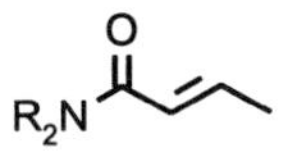	Br	R'	B	Kaga H., *Synlett.*, 1994; 607
$C=C-CO_2R$	I	Ar	Zn	Cossi P., *Tetrahedron*, 1992; 8801
$C=C-CO_2R$	Br	C=C	Sn	Kosugi M., *Bull. Chem. Soc. Jpn.*, 1993; 3058
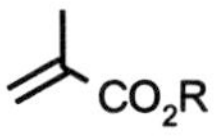	Br	C=C	B	Urdaneta N., *J. Organomet. Chem.*, 1994; 464, C33
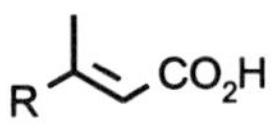	I	Ph	Zn	Duchêne A., *Tetrahedron Lett.*, 1995; 2469
$C=C(CH_2CO_2H)-$	I	R'	Zn	Duchêne A., *Synlett.*, 1994; 524 Duchêne A., *Tetrahedron Lett.*, 1995; 2469
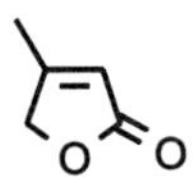	OSO_2CF_3	R'	B	Grigg R., *Tetrahedron*, 1994; 5489
$C=C-C=N-NR_2$	Br	Ar	Zn	Gilchrist T.L., *Tetrahedron*, 1993; 2543
C=C(OEt)-	$OPO(OEt)_2$	Het	Sn	Kosugi M., *Bull. Chem. Soc. Jpn.*, 1989; 3383
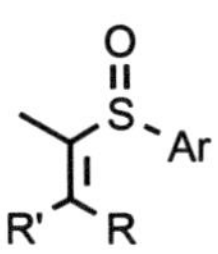	I	$CH_2=CH$	Sn	Paley R.S., *Tetrahedron Lett.*, 1995; 3605
$C=C-C=C-CO_2R$	Br	C=C	Sn	Crombie L., *Tetrahedron Lett.*, 1989; 4299
C=C-CH	X	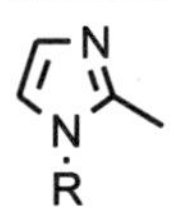	Sn	Kurusawa H., *J. Am. Chem. Soc.*, 1990; 2813
C=C-CH	X	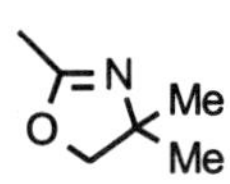	Sn	Kosugi M., *Chem. Lett.*, 1988; 1351 Beletskaya I.P., *J. Organomet. Chem.*, 1983; 250, 551
C=C-CH	Br	C=C	Sn	Adam W., *J. Org. Chem.*, 1994; 2695 Adam W., *Synthesis*, 1994; 557
Ar-C=C-CH	Br	Ar'	B	Moreno-Manas M., *J. Org. Chem.*, 1995; 2396
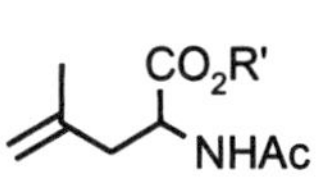	I	R	Sn	Crisp G.T., *Tetrahedron*, 1994; 2623
$C=C=C(CO_2Me)-$	I	Ar	B	Gillmann T., *Synlett.*, 1994; 649
Alkyne	Br	R'	B	Suzuki A., *Pure Appl. Chem.*, 1985; 1749
Alkyne	Br	C=C	B	Quintard J.P., *Tetrahedron Lett.*, 1992; 3647

R	X	R'	M	References
C_6H_5	Br		Sn	Stork G., *J. Am. Chem. Soc.*, 1990; 7399, (cine / ipso-substitutions)
Ar	X	R'	Mg	Ishikawa N., *J. Organomet. Chem.*,1976; 118, 349 Amatore C., *J. Organomet. Chem.*, 1990; 390, 389
Ar	X	R'	Zn	Amatore C., *J. Organomet. Chem.*, 1990; 390, 389
Ar	X	R'	B	Genêt J.P., *Synlett.*, 1992; 715
Ar	I	R'	B	Commerçon A., *Bioorg. Med. Chem. Lett.*, 1995; 809
Ar	Br, I	R'	Sn	Collum D.B., *Tetrahedron Lett.*, 1995; 3111
Ar	Br, I	C_6H_5	B	Miura Y., *Synthesis*, 1995; 1419
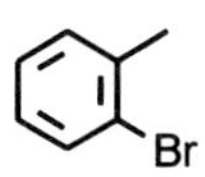	I	C=C	Zn	Wender P.A., *Synlett.*, 1995; 516
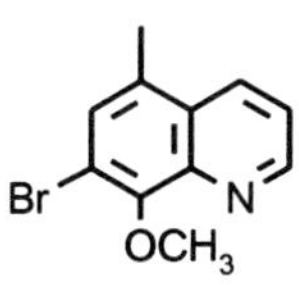	Br	Ar'	B	Queguiner G., *J. Heterocycl. Chem.*, 1995; 1261
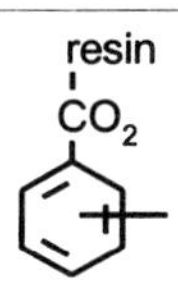	X	Ar	B	Friesen R. W., *Tetrahedron Lett.*, 1994; 9177
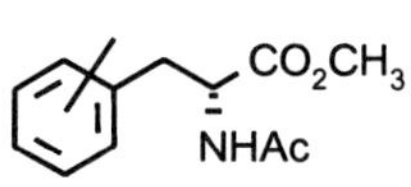	Br	Ar	B	Burk M. J., *J. Am. Chem. Soc.*, 1994; 10847
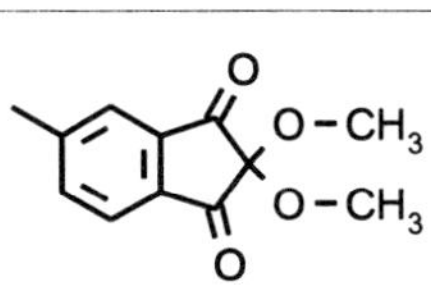	Br	Ar	M	Joullié M. M., *Tetrahedron Lett.*, 1994; 7719
Ar-$Cr(CO)_3$	OSO_2CF_3	R'	Sn	Wulff W.D., *J. Am. Chem. Soc.*, 1994; 7449
Ar-$Cr(CO)_3$	Br	Ar	B	Uemura M., *J. Chem. Soc., Chem. Commun.*, 1994; 2697 Uemura M., *Tetrahedron Lett.*, 1995; 6695
Ar-Ar'	X	C=C-CH	Zn	Katz A., *Tetrahedron Lett.*, 1993; 3551
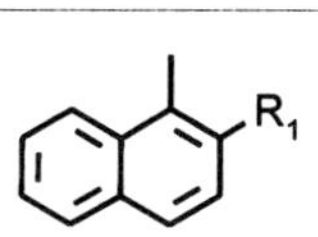	Br	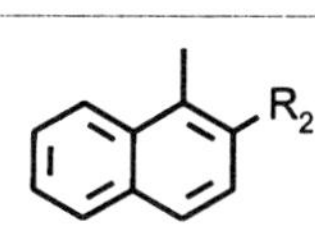	Mg	Frejd T., *Acta Chem. Scand.*, 1989; 43, 670
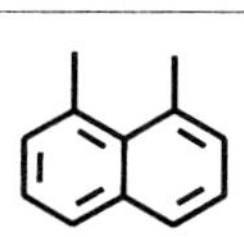	I	Ar	B	Grahn W., *Angew. Chem. Int. Ed. Engl.*, 1995; 1485

R	X	R'	M	References
Het	X	R'	M	Gronowitz S., *J. Heterocycl. Chem.*, 1989; 865 Gronowitz S., *J. Heterocycl. Chem.*, 1995; 751 Gronowitz S., *J. Heterocycl. Chem.*, 1994; 11 Yamanaka H., *Synthesis*, 1992; 552
Het	X	R'	M	Kalinin V.N., *Synthesis*, 1992; 413
Het	X	Ar, Het'	Sn	Achab S., *Tetrahedron Lett.*, 1995; 2615
Het	X	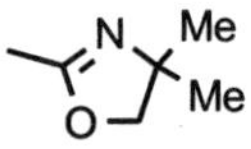	Sn	Dondoni A., *Synthesis*, 1987; 693
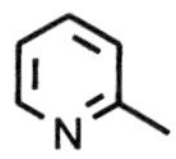	Br	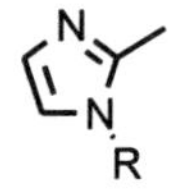	Zn	Bell A.S., *Tetrahedron Lett.*, 1988; 5013
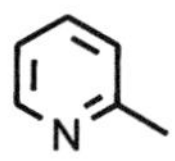	X	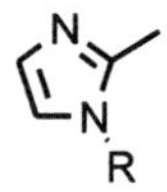	Sn	Gronowitz S., *Tetrahedron Lett.*, 1992; 2199
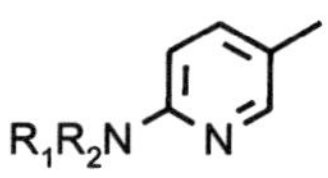	Br	Ar	Zn	Kuroboshi M., *Tetrahedron Lett.*, 1995; 563
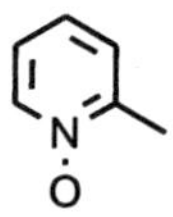	Cl, Br	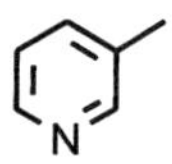	B	Zoltewicz J.A., *Tetrahedron*, 1995; 3103
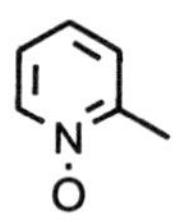	Cl, Br	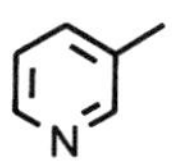	B	Zoltewicz J. A., *J. Org. Chem.*, 1995; 264
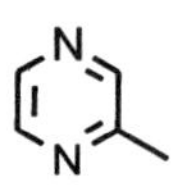	Cl		Sn	Ohta A., *Heterocycles*, 1992; 33, 257
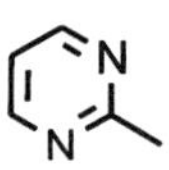	OSO_2CF_3	R'	Zn	Undheim K., *Heterocycles*, 1994; 37, 501
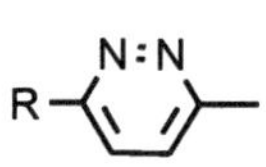	I	Ph	B	Bailey T.R., *J. Org. Chem.*, 1995; 748
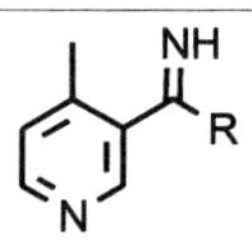	I	Ar	B	Queguiner G., *Tetrahedron Lett.*, 1994; 6489
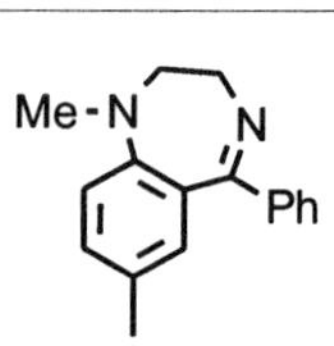	OSO_2CF_3	CN	K	Kraus G. A.,*Tetrahedron Lett.*, 1994; 9189
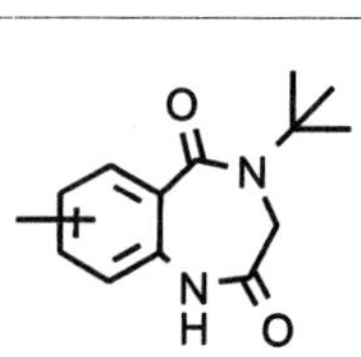	Br, I	Het	Sn	Karp G.M., *J. Org. Chem.*, 1995; 5814

R	X	R'	M	References
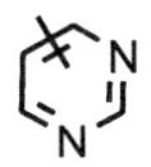	OSO_2CF_3	R'	Sn	Undheim K., *Heterocycles*, 1994; 37, 501
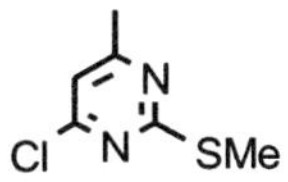	I	Ar	M	Queguiner G., *Tetrahedron*, 1994; 10299
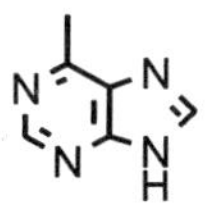	Cl	R'	Sn	Gundersen L.L., *Tetrahedron Lett.*, 1994; 3155
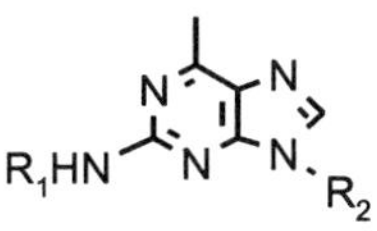	OSO_2CF_3, OTs	CH_2=CH-	Sn	Sasaki S., *Tetrahedron Lett.*, 1995; 421
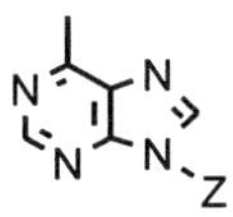	Cl	R'	Sn	Gundersen L.L., *Tetrahedron*, 1994; 9743
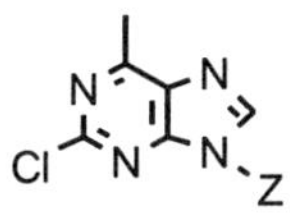	Cl	R'	Sn, Zn	Gundersen L.L., *Tetrahedron Lett.*,1995; 1945
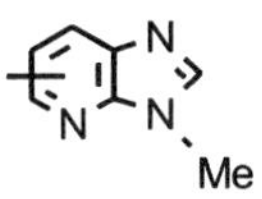	Br	C_6H_5	B	Grivas S., *J. Heterocycl. Chem.*, 1995; 467
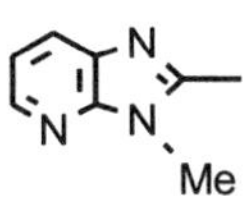	Cl	C_6H_5	B	Grivas S., *J. Heterocycl. Chem.*, 1995; 467
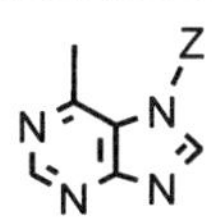	Cl	R'	Sn	Gundersen L.L., *Tetrahedron*, 1994; 9743
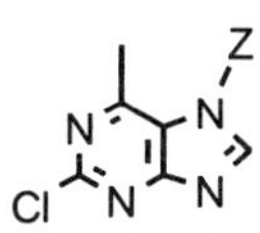	Cl	R'	M	Gundersen L.L., *Tetrahedron Lett.*, 1995; 1945
	Br	R'	M	Mathey F., *Synthesis*, 1995; 717
	X	R'	Sn	Mathey F., *J. Am. Chem. Soc.*, 1993; 10665
	X	H	Sn	Mathey F., *J. Am. Chem. Soc.*, 1993; 10665
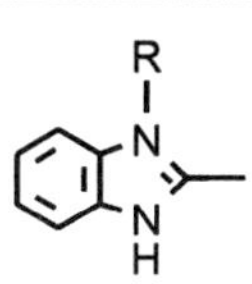	Br	C=C	Sn	Somei M., *Heterocycles*, 1988; 27, 1585
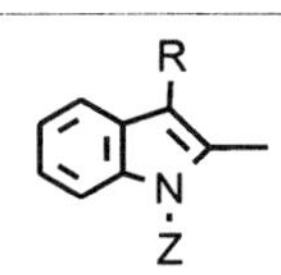	OSO_2CF_3	CH_2=CH-	Sn	Becalli E.M.,*Tetrahedron*, 1995; 2353

R	X	R'	M	References
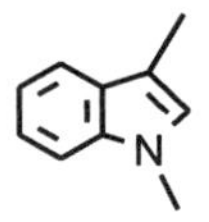	I	R'	B	Martin A.R., *Synth. Commun.*, 1992; 1757
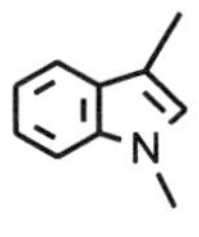	I	R'	Sn	Martin A.R., *Synth. Commun.*, 1992; 1757
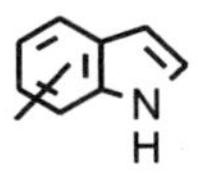	Br	Ar	B	Carrera G.M.Jr., *Synlett*., 1994; 93
	I	CH_2=CH	Sn	Scott A.I., *Tetrahedron Lett.*, 1995; 7043
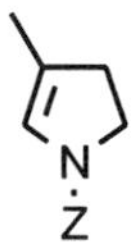	OSO_2CF_3	Ar	B	Baldwin J.E., *Tetrahedron Lett.*, 1995; 4869
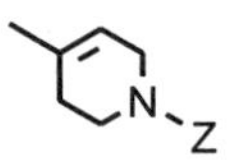	OSO_2CF_3	Ar	B	Wustron D.J., *Synthesis*, 1991; 993 Martin A.R., *Tetrahedron Lett.*, 1993; 2235
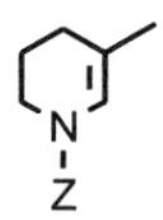	OSO_2CF_3	CH_2=CH-	Sn	Passarella D., *Heterocycles*, 1995; 41, 973
$ArCH_2$	X	Alkyl	Zn	Stille J.K., *J. Am. Chem. Soc.*, 1979; 4992 Srebnik M., *J. Org. Chem.*, 1993; 6908
$ArCH_2$	X	C=C	B	Suzuki A., *Pure Appl. Chem.*, 1985; 1749
RCH_2CH_2	I	R'	Mg	Castle P.L., *Tetrahedron Lett.*, 1986; 6013
RCH_2CH_2	X	R'	Mg	Yuan K., *Tetrahedron Lett.*, 1989; 4779
RCH_2CH_2	X	R'	B	Suzuki A., *Chem. Lett.*, 1992; 691
CO_2RCH_2	Br	Ar	B	Suzuki A., *Chem. Lett.*, 1989; 1405
$ROCH_2$-	X	C=C	Sn	Falck J.R., *Tetrahedron Lett.*, 1992; 4885
$RSCH_2$-	X	Alkyne	Sn	Falck J.R., *Tetrahedron Lett.*, 1992; 4885
$RSCH_2$-	X	Ar	Sn	Falck J.R., *Tetrahedron Lett.*, 1992; 4885
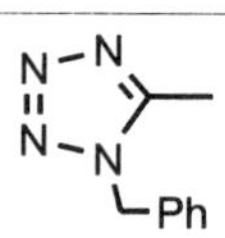	Br	R'	B	Yi K.Y., *Tetrahedron Lett.*, 1995; 1679
Ph-N=C-SPh	I	Ar	Zn	To Y., *Chem. Lett.*, 1989; 1261

R	X	R'	M	References
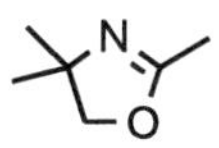	SCH_3	Ar	Mg	Pridgen L.N., *J. Org. Chem.*, 1981; 5402
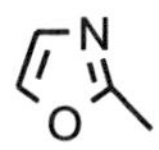	I	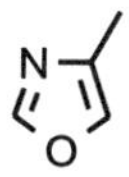	Sn	Barrett A.G.M., *Synlett.*, 1995; 415
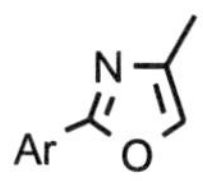	OSO_2CF_3	Het	Sn	Kelly T.R., *Tetrahedron Lett.*, 1995; 5319
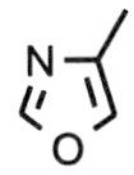	OSO_2CF_3	Sn	Sn	Barrett A.G.M., *Synlett.*, 1995; 415
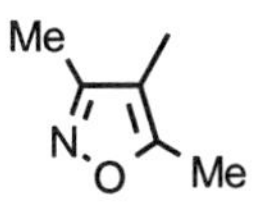	I	Ar	B, Sn	Labadie S.S., *Synth. Commun.*, 1994; 709
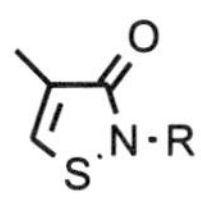	Br	R'	Sn	Fishwick C.W.G., *Tetrahedron Lett.*, 1994; 6551
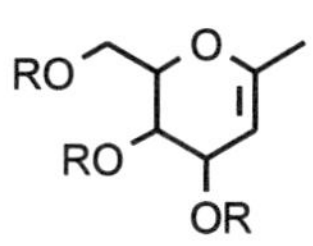	I	R'	M	Friesen R.W., *Can. J. Chem.*, 1994; 1262
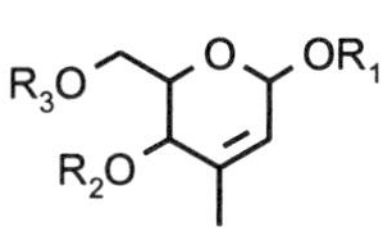	OSO_2CF_3	R'	Sn	Voelter W., *Angew. Chem. Int. Ed. Engl.*, 1994; 1499
Nucleoside	I	R'	Sn	Nair V., *J. Org. Chem.*, 1988; 3051
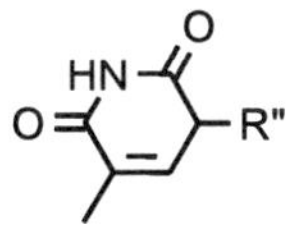	I	Het	Sn	Froehler B.C., *J. Am. Chem. Soc.*, 1994; 5540
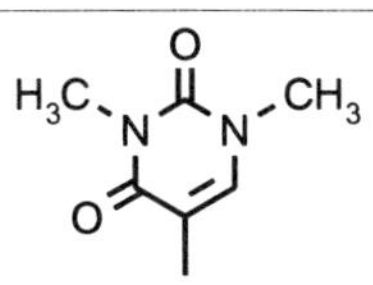	I	R	Sn	Crisp G.T., *Synth. Commun.*, 1992; 683
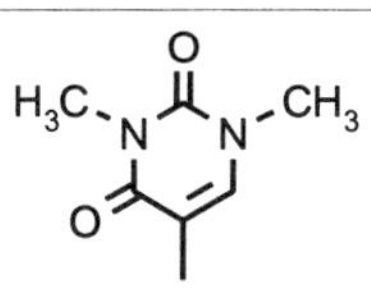	I	R	Zn	Koroniak H., *J. Fluorine Chem.*, 1995; 135
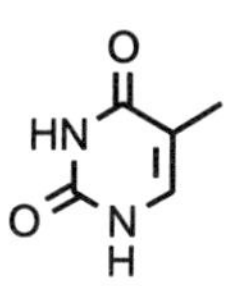	I	C=C	Sn	Farina V., *Synlett.*, 1991; 157
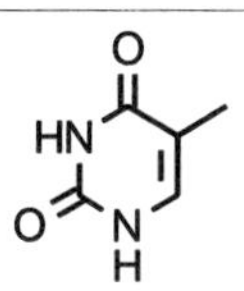	I	Alkyne	Cu	Cacchi S., *Tetrahedron Lett.*, 1989; 2581

R	X	R'	M	References
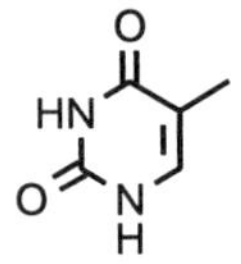	OSO_2CF_3	Alkyne	Cu	Cacchi S., *Tetrahedron Lett.*, 1989; 2581
Porphyrin	Br	Ar	B	Chan K.S., *Tetrahedron*, 1995; 3129
Porphyrin	OSO_2CF_3	Ar	B	Chan K.S., *J. Org. Chem.*, 1994; 6084
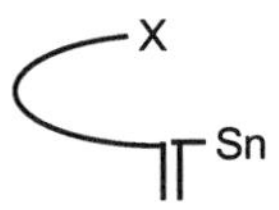	---	---	---	Finch H., *Tetrahedron Lett.*, 1993; 8353 (intramolecular)
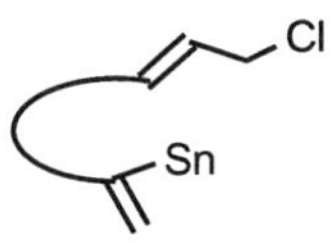	---	---	---	Pattenden G., *Synlett.*, 1994; 181 (intramolecular)

R	X	R'	M	References
C=C	Br	Het	B	Casalnuovo A.L., *J. Am. Chem. Soc.*, 1990; 4324 Terashima M., *Heterocycles*, 1984; 22, 2475
C=C	Br	Het	Sn	Liebeskind L.S., *Tetrahedron Lett.*, 1995; 2191
Ar	Cl	C=C	B	Mitchell M.B., *Tetrahedron Lett.*, 1991; 2273 Mc Killop A., *Tetrahedron*, 1992; 8117
Ar	Cl	Ar	B	Mc Killop A., *Tetrahedron*, 1992; 8117
Ar	Cl	C=C	Sn	Mitchell T.N., *Tetrahedron*, 1989; 969
$Ar-Cr(CO)_3$	Cl	C=C	Sn	Scott W.J., *J. Chem. Soc., Chem. Commun.*, 1987; 1755
$Ar-Cr(CO)_3$	Cl	Alkyl	Sn	Scott W.J., *J. Chem. Soc., Chem. Commun.*, 1987; 1755
Ar	I	R'	Mg	Ishikawa N., *J. Organomet. Chem.*, 1976; 349 Casalnuovo A.L., *J. Am. Chem. Soc.*, 1990; 4324
Ar	I	R'	Sn	Liebeskind L.S., *Tetrahedron Lett.*, 1995; 2191
C_6H_5	I	$CH_2=C-$	Sn	Liebeskind L.S., *J. Org. Chem.*, 1994; 5905
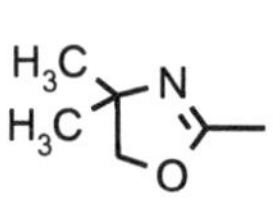	SR	Ar	Mg	Pridgen L.N., *J. Org. Chem.*, 1981; 5402
C=C	SeR	Me_3SiCH_2	Mg	Hevesi L., *Tetrahedron Lett.*, 1994; 6729
C=C-CH	OAc	C=C	Mg	Buono G., *Tetrahedron Lett.*, 1990; 77

R	X	R'	M	References
C=C-CH	OAc	C=C	Sn	Hegedus L.S., *J. Org. Chem.*, 1990; 3019
C=C-CH	OAc	Ar	Sn	Hegedus L.S., *J. Org. Chem.*, 1990; 3019
Ar	OSO_2R	H	Sn	Cabri W., *J. Org. Chem.*, 1990; 350
Het	OTs	CH_2=CH-	Sn	Sasaki S., *Tetrahedron Lett.*, 1995; 421
C=C	OTs	R	Cu	Tingoli M., *Tetrahedron*, 1995; 4691
C=C	OSO_2R	C=C	Sn	Scott W.J., *J. Org. Chem.*, 1991; 1489
C=C	OSO_2R	C=C-CH-	Sn	Scott W.J., *J. Org. Chem.*, 1991; 1489
Het	OSO_2F	Het	Zn	Roth G.P., *J. Org. Chem.*, 1991; 3493
Ar	OSO_2F	Ar	Zn	Roth G.P., *J. Org. Chem.*, 1991; 3493
Ar	OSO_2F	C=C	Sn	Roth G.P., *J. Org. Chem.*, 1991; 3493
Het	OSO_2CF_3	R'	Sn	Farina V., *Tetrahedron Lett.*, 1988; 6043
Ar	OSO_2CF_3	H	Sn	Cacchi S., *Tetrahedron Lett.*, 1986; 5541
Ar	OSO_2CF_3	R'	Sn	Stille J.K., *J. Am. Chem. Soc.*, 1987; 5478 Stille J.K., *J. Am. Chem. Soc.*, 1986; 3033
Ar	OSO_2CF_3	Alkyne	Cu	Scott W.J., *Acc. Chem. Res.*, 1988; 47
C=C	OSO_2CF_3	C=C	Sn	Stille J.K., *Org. Synth.*, 1990; 68, 116 Liebeskind L.S., *Tetrahedron Lett.*, 1995; 2191
Ar	OSO_2Ar'	R'	Sn	Badone D., *J. Org. Chem.*, 1992; 6321
C=C	$OPO(OEt)_2$	R'	Mn	Oshima K., *Chem. Lett.*, 1987; 2203
Ar	N_2^+	R'	Sn	Kikukawa K., *J. Org. Chem.*, 1983; 1333
C=C	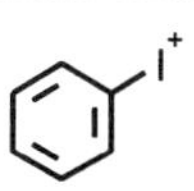	Alkyne	Sn	Stang P.J., *J. Am. Chem. Soc.*, 1993; 11626

R	X	R'	M	References
R	OSO_2CF_3	H	Si	Kotsuki H., *Synthesis*, 1995; 1348
C=C	I	H	Sn	Utimoto K., *Bull. Chem. Soc. Jpn.*, 1991; 2593

R	X	R'	M	References
C=C	OSO_2CF_3	H	Sn	Cacchi S., *Org. Synth.*, 1990; 68, 138 Chen Q.Y., *Synthesis*, 1988; 896
Ar	Br	HCO_2	Na	Helquist P., *Tetrahedron Lett.*, 1978; 1913
Ar	OSO_2CF_3	H	Sn	Saa J.M., *J. Org. Chem.*, 1990; 991 Farina V., *Tetrahedron Lett.*, 1988; 6043
$ArCH_2$	X	Alkyl	Sn	Stille J.K., *J. Am. Chem. Soc.*, 1979; 4992
Ar	X	Alkyl	Mg	Ishikawa N., *J. Organomet. Chem.*, 1976; 118, 349
C=C	X	Alkyl	M	Murahashi S.I., *J. Org. Chem.*, 1979; 2408
Ar	I	Alkyl	M	Hamon D.P.G., *Tetrahedron Lett.*, 1993; 5333
Ar	X	Alkyl	B	Soderquist J.A., *Tetrahedron Lett.*, 1990; 5541 Ellman J.A., *J. Am. Chem. Soc.*, 1994; 11171
Ar	I	Alkyl	B	Ames M.M., *Synthesis*, 1994; 1433
C=C	X	Alkyl	B	Suzuki A., *Chem. Lett.*, 1989; 1405
$HOCH_2$-C=C-	I	6	Zn	Neghishi E.I., *Tetrahedron Lett.*, 1993; 1437
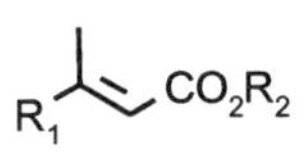	OSO_2CF_3	CH_3	B	Gibbs R.A., *Tetrahedron Lett.*, 1995; 5669
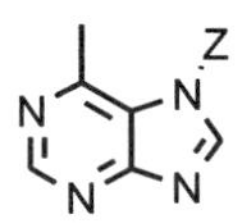	Cl	CH_3	Zn	Gundersen L.L., *Tetrahedron*, 1994; 9743
C=C	Cl	CH_3	Mg	Ogasawara K., *Synlett.*, 1994; 665
Ar	I	CH_3	Sn	Collum D.B., *Tetrahedron Lett.*, 1995; 3111
Ar	OSO_2CF_3	CH_3	Sn	Johansson A.M., *J. Med. Chem.*, 1995; 647
C=C	Br	C_6H_5-CH-CH_3	Mg	Richards C.J., *Tetrahedron Lett.*, 1995; 3745 (asymmetric catalysis)
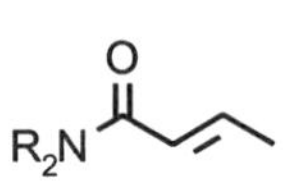	Br	R-$(CH_2)_4$-	B	Kaga H., *Synlett.*, 1994; 607
Ar	X	Z-$(CH_2)_n$-	Si	Hiyama T., *Tetrahedron Lett.*, 1994; 6507
Ar	I	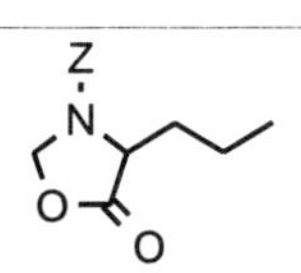	Zn	Jackson R.F.W., *Synlett.*, 1994; 379
Ar	I	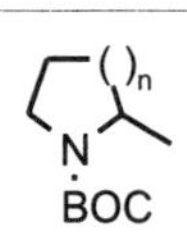	Li	Dieter R.K., *Tetrahedron Lett.*, 1995; 3613

R	X	R'	M	References
Si-C=C-C-	Cl	C=C	Al	Oehlschlager A.C., *J. Org. Chem.*, 1995; 4595
R	X	C=C	B	De Lera A.R., *Synthesis*, 1995; 285
Ph	I	C=C	Si	Hiyama T., *Tetrahedron Lett.*, 1993; 8263
C=C	I	H_2C=CH-	Sn	Linstrumelle G., *Tetrahedron Lett.*, 1995; 3687
C=C	I	C=C	Mg/ Cu	Linstrumelle G., *Tetrahedron Lett.*, 1978; 191
C=C	I	C=C	Cu	Linstrumelle G., *Tetrahedron Lett.*, 1978; 191
Het	X	C=C	Sn	Nair V., *J. Org. Chem.*, 1988; 3051
Ar	X	C=C	Sn	Farina V., *J. Am. Chem. Soc.*, 1991; 9585
Ar	I	C=C	Sn	Haack R.A., *Tetrahedron Lett.*, 1988; 2783 Deshpande M.S., *Tetrahedron Lett.*, 1994; 5613 (catal. on solid support)
Ar	Br	C=C	Sn	Haack R.A., *Tetrahedron Lett.*, 1988; 2783
Ar	I	C=C	Zn	Wender P.A., *Synlett.*, 1995; 516
R	I	C=C	Zn	Sato F., *J. Org. Chem.*, 1995; 290
C=C	I	C=C	Zn	Mascarenas J.L., *Tetrahedron Lett.*, 1995; 5413
C=C	I	$(CH=CH)_n$-C=C-	B	De Lera A.R., *Tetrahedron*, 1995; 2435
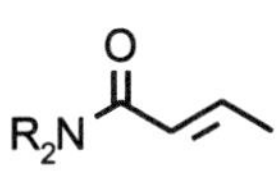	Br	R-CH=CH-	B	Kaga H., *Synlett.*, 1994; 607
R	Br, I	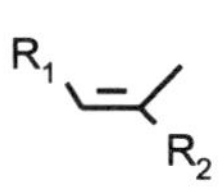	B	Srebnik H., *J. Org. Chem.*, 1995; 3276
C=C	OSO_2CF_3	C=C(OEt)-	Sn	Stille J.K., *J. Org. Chem.*, 1990; 3114
C=C	Br	C=C(OEt)-	Zn	Suzuki A., *Chem. Lett.*, 1989; 1959
Ar	Br	C=C(OEt)-	Zn	Farina V., *Tetrahedron Lett.*, 1988; 6043
Het	Br	C=C(OEt)-	Sn	Kelly T.R., *Tetrahedron Lett.*, 1995; 5319
Ar	X	-C=C-OEt	Sn	Yamanaka H., *Heterocycles*, 1990; 31, 219
Ar	Br	-C=C-OEt	Sn	Sakamoto T., *Tetrahedron*, 1991; 1877

R	X	R'	M	References
Ar	I	-C=C-OEt	B	Kumar K., *J. Org. Chem.*, 1992; 6995
R	X	C=C(OR)-	Sn	Kocienski P., *J. Chem. Soc., Perkin Trans.1*, 1994; 1187
R	X	C=C(OR)-	Zn	Kocienski P., *J. Chem. Soc., Perkin Trans.1*, 1994; 1187
Het	I	C=COAc-	Sn	Kosugi M., *Chem. Lett.*, 1982; 939
Ar	Br	C=COAc-	Sn	Migita T., *J. Chem. Soc., Chem. Commun.*, 1983; 344
Ar	I	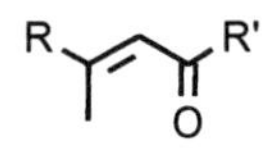	Sn	Takeda T., *Tetrahedron*, 1995; 2515
Ar	I	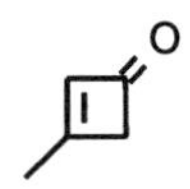	Sn	Liebeskind L.S., *J. Org. Chem.*, 1994; 7917
Ar	I	Ar-N=CEt-	Zn	Murakami M., *Chem. Lett.*, 1989; 1603
Ar	Br	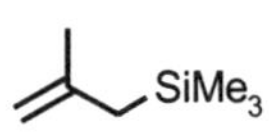	Sn	Kang K.T., *Tetrahedron Lett.*, 1991; 4341
C=C	X	Me_3Si-C=C-	Sn	Stille J.K., *J. Am. Chem. Soc.*, 1987; 817
C=C	Br	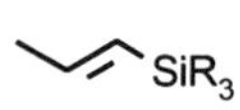	B	Soderquist J.A., *Tetrahedron Lett.*, 1994; 27
R	X	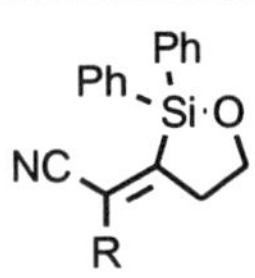	-	Ito Y., *Tetrahedron Lett.*, 1994; 8635 (intermolecular)
Ar	X	H_2C=C(F)-	Sn	Mc Carthy J.R., *Tetrahedron Lett.*, 1994; 5177
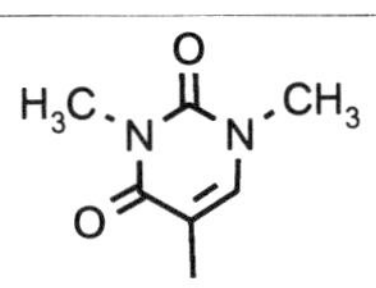	I	CF_2=CF-	Zn	Koroniak H., *J. Fluorine Chem.*, 1995; 135
Ar	I	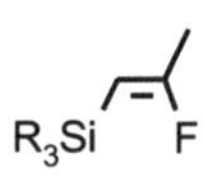	Sn	Mc Carthy J.R., *Tetrahedron Lett.*, 1994; 1027
R	I , OSO_2CF_3	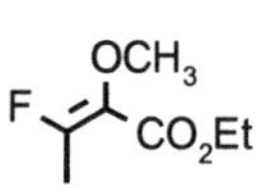	Sn	Shi G.Q., *J. Org. Chem.*, 1995; 6608
R	Br, I	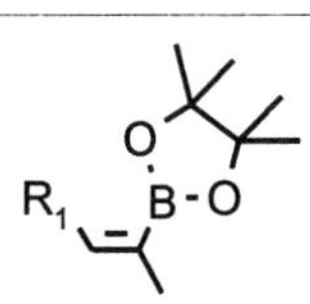	Zr	Srebnik M., *J. Org. Chem.*, 1995; 3276
C=C-CH-	X	C=C-CH-	Sn	Stille J.K., *Tetrahedron Lett.*, 1980; 2599

R	X	R'	M	References
Ar	OSO_2CF_3	C=C-CH-	Si	Hatanaka Y., *Tetrahedron Lett.*, 1994; 1279
Ar	X	C=C-CH-	SiF_3	Hatanaka Y., *Tetrahedron Lett.*, 1994; 6511
Ar	I , OSO_2CF_3	$Me_3Si\text{-}CH_2\text{-}CH{=}CH\text{-}CH_2\text{-}$	Sn	Tsuji Y., *J. Org. Chem.*, 1995; 4647
C=C-CH	Br	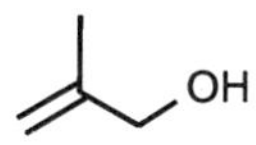	Sn	Adam W., *J. Org. Chem.*, 1994; 2695
Ar	I	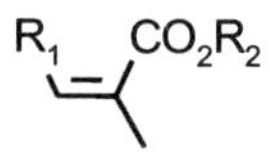	Sn	Rossi R., *Tetrahedron Lett.*, 1994; 12029
C=C	Br	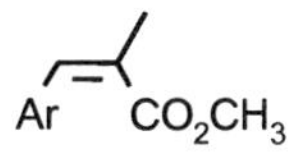	Sn	Horikawa H., *Synthesis*, 1995; 582
Ar	I	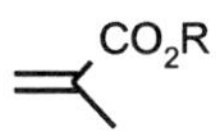	Sn	Levin J.I., *Tetrahedron Lett.*, 1993; 6211
C_6H_5	I	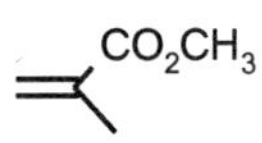	Sn	Busacca C.A., *J. Org. Chem.*, 1994; 7553 (anomalous Stille reaction)
Ar	X	$H_2N\text{-}C\text{-}C{=}C\text{-}$	Sn	Corriu R.J.P., *J. Org. Chem.*, 1993; 1443
C=C	X	$H_2N\text{-}C\text{-}C{=}C\text{-}$	Sn	Corriu R.J.P., *J. Org. Chem.*, 1993; 1443
C=C	I	BOCHN-C-C=C-	Sn	Degl'Innocenti A., *Tetrahedron Lett.*, 1994; 2081
Ar	Br	$(SiMe_3)_2N\text{-}C\text{-}C{=}C\text{-}$	Sn	Corriu R.J.P., *Tetrahedron Lett.*, 1991; 4121
C=C	X	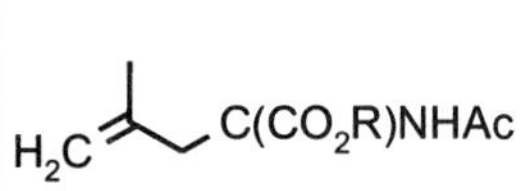	Sn	Crisp G.T., *Tetrahedron*, 1994; 3213
C=C	OSO_2CF_3	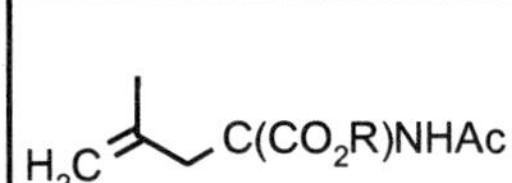	Sn	Crisp G.T., *Tetrahedron*, 1994; 3213 (ipso/cine-substitu.)
Ar	X	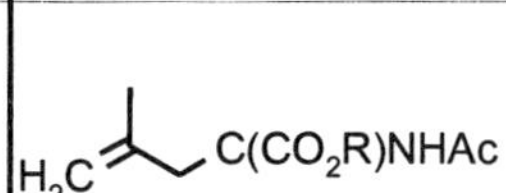	Sn	Crisp G.T., *Tetrahedron*, 1994; 3213 (ipso/ cine-substitu.)
Ar	I	$F_2C{=}C\text{-}R$	B	Ichikawa J., *Chem. Lett.*, 1991; 961
Ar	X	$F_2C{=}C\text{-}R$	B	Ichikawa J., *Tetrahedron Lett.*, 1992; 3779
C=C	X	$F_2C{=}C\text{-}R$	B	Minami T., *Synlett.*, 1992; 739
C=C	X	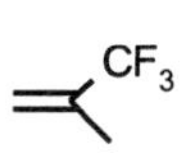	Zn	Jiang B., *Tetrahedron Lett.*, 1992; 511

R	X	R'	M	References
Alkyne	Br	$(OEt)_2HC\text{-}(CH_2)_n\text{-}C{=}C\text{-}$	Sn	Quintard J.P., *Tetrahedron Lett.*, 1992; 3647
Ar	I	C=C=C	Sn	Braslau R., *Synth. Commun.*, 1994; 789
Ar	OSO_2CF_3	C=C=C	Sn	Badone D., *Tetrahedron Lett.*, 1994; 5477
C=C	I	C=C=C	Zn	Wang K.K., *Tetrahedron Lett.*, 1994; 1829
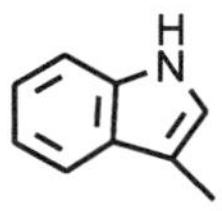	I	$C{=}C{=}C(OCH_3)\text{-}$	Zn	Hegedus L.S., *J. Org. Chem.*, 1989; 4141
Ar	I	$C{=}C{=}C(OCH_3)\text{-}$	Zn	Hegedus L.S., *J. Am. Chem. Soc.*, 1983; 943
C=C	Br	$C{=}C{=}C(OCH_3)\text{-}$	Zn	Hegedus L.S., *J. Am. Chem. Soc.*, 1983; 943
R	X		Zn Sn	De Meijere A., *Chem. Ber.*, 1994; 1511
C=C	I	HC≡C–	Sn	Pancrazi A., *Synlett.*, 1994; 998
R	Br	R_1–C≡C–	B	Soderquist J.A., *Tetrahedron Lett.*, 1995; 6847
C=C	OSO_2CF_3	R_1–C≡C–	Sn	Farina V., *Tetrahedron Lett.*, 1988; 6043
Ar	OSO_2CF_3	R_1–C≡C–	Sn	Stille J.K., *J. Org. Chem.*, 1989; 5856
Ar	X	–C≡C–CO_2Et	Zn	Yamanaka H., *Synthesis*, 1992; 746
Ar	I	–C≡C–OEt	Sn	Yamanaka H., *Synlett.*, 1992; 502 Sakamoto T., *Chem. Pharm. Bull.*, 1994; 2032
C=C	I	–C≡C–$N(CH_3)_2$	Si	Neuenschwander M., *Chimia*, 1995; 72
C=C	I	–C≡C–$N(CH_3)_2$	Sn	Neuenschwander M., *Chimia*, 1995; 72
Ar	I	–C≡C–	Sn	Cummins C.H., *Tetrahedron Lett.*, 1994; 857 (disubstitu.)
R	OSO_2CF_3	C_6H_5	Sn	Garcia Martinez A., *Synlett.*, 1994; 1047
C=C	OSO_2CF_3	C_6H_5	Zn	Stork G., *J. Am. Chem. Soc.*, 1990; 7399 (cine / ipso-substitutions)
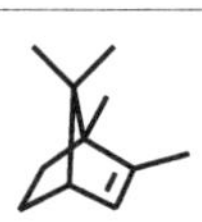	OSO_2CF_3	Ar	B	Myers A.G., *J. Am. Chem. Soc.*, 1994; 11556
C=C	X	Ar	B	Terashima M., *Heterocycles*, 1984; 22, 2475
Ar'	X	Ar	B	Haber S., *Patent*, DE 4414499-A1

R	X	R'	M	References
C=C	OSO_2CF_3	Ar	Sn	Farina V., *J. Org. Chem.*, 1993; 5434 Farina V., *Tetrahedron Lett.*, 1988; 6043
$HOCH_2$-C=C-	I	Ar	Zn	Neghishi E.I., *Tetrahedron Lett.*, 1993; 1437
Ar'	OSO_2CF_3	Ar	Zn	Koch K., *Synlett.*, 1994; 347
Ar'	X	Ar	Zn	Koch K., *Synlett.*, 1994; 347
Ar	Br,I	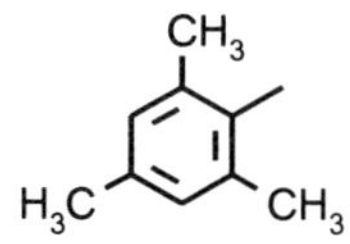	B	Anderson J.C., *Synlett.*, 1995; 765
Ar	Br	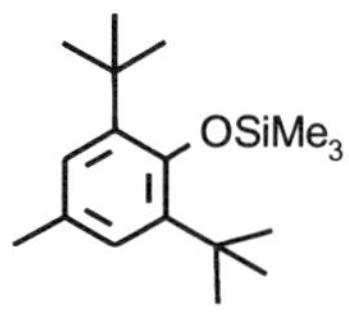	B	Satoh Y., *Synthesis*, 1994; 1146
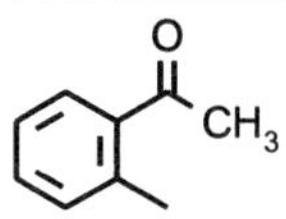	Br	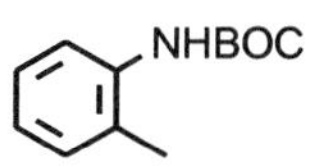	B	Tour J.M., *J. Am. Chem. Soc.*, 1994; 11723
Het	X	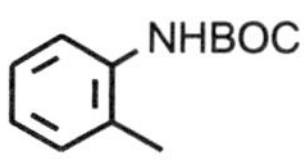	B	Snieckus V., *J. Org. Chem.*, 1995; 292
Ar	Br	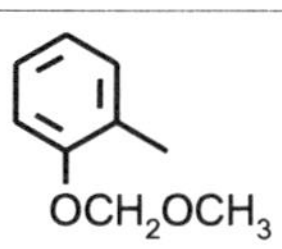	B	Roques B.P., *J. Med. Chem.*, 1995; 4693
Ar'	X	$ArCH_2$	B	Soderquist J.A., *Tetrahedron Lett.*, 1990; 5541
C=C	X	Heterocycle	Mg	Minato A., *Tetrahedron Lett.*, 1984; 83
C=C	X	Heterocycle	Zn	Minato A., *Tetrahedron Lett.*, 1984; 83
Ar	X	Heterocycle	Sn	Bailey T.R., *Tetrahedron Lett.*, 1986; 4407 Gronowitz S., *J. Organomet. Chem.*, 1993; 460, 127
Het	X	Heterocycle	Sn	Gronowitz S., *J. Heterocycl. Chem.*, 1995; 751 Gronowitz S., *J. Heterocycl. Chem.*, 1994; 641
Ar	I	Heterocycle	Sn	Armitage M.A., *Tetrahedron Lett.*, 1995; 775
Ar	X	Heterocycle	Mg	Kumada M., *Tetrahedron Lett.*, 1981; 5319
Het	X	Heterocycle	Mg	Kumada M., *Tetrahedron Lett.*, 1981; 5319
Het	X	Heterocycle	Sn	Gronowitz S., *J. Organomet. Chem.*, 1993; 460, 127
Ar	Br	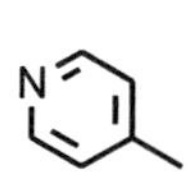	Sn	Fugita M., *Tetrahedron Lett.*, 1995; 5247

R	X	R'	M	References
Ar	X	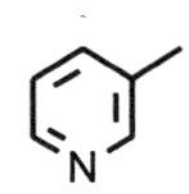	B	Sonesson C., *Tetrahedron Lett.*, 1994; 9063
Ar	I	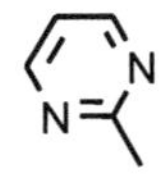	Sn	Undheim K., *Tetrahedron*, 1994; 275
Het	I	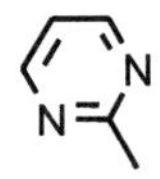	Sn	Undheim K., *Tetrahedron*, 1994; 275
Het	Br	—HetN—O	Sn	Zoltewicz J.A., *J. Org. Chem.*, 1995; 3487
C=C	OSO_2CF_3	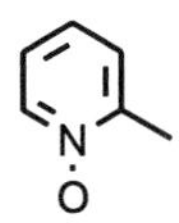	Zn	Koga H., *Heterocycles*, 1995; 41, 2405
Het	Br	—HetN$^+$—CH_3	Sn	Zoltewicz J.A., *J. Org. Chem.*, 1995; 3487
R	X	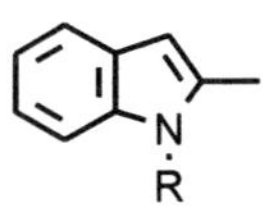	Sn	Labadie S.S., *J. Org. Chem.*, 1994; 4250
R	X	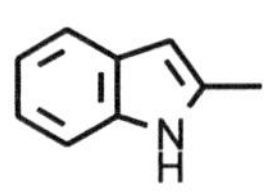	Sn	Fukuyama T., *J. Am. Chem. Soc.*, 1994; 3127
C=C	X	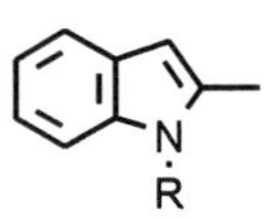	Mg, Sn	Kumada M., *Tetrahedron Lett.*, 1981; 5319 Palmisano G., *Helv. Chim. Acta*, 1993; 2356
Het	X	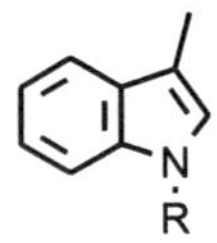	Zn	Amat M., *Tetrahedron Lett.*, 1994; 793
R	X	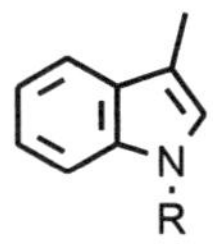	Sn	Ortar G., *Tetrahedron Lett.*, 1994; 2405
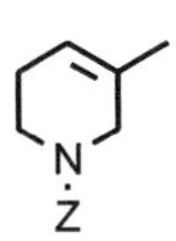	OSO_2CF_3	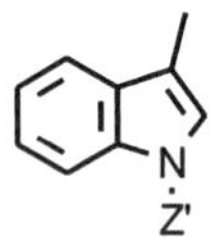	B	Martin A.R., *Heterocycles*, 1994; 37, 1761
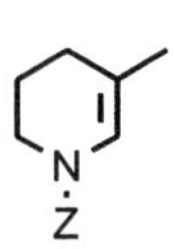	OSO_2CF_3	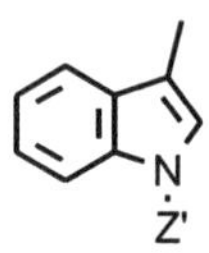	B	Martin A.R., *Heterocycles*, 1994; 37, 1761
Ar	Br	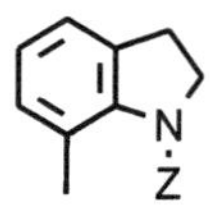	Sn	Iwao M., *Heterocycles*, 1994; 38, 1717
Het	Br	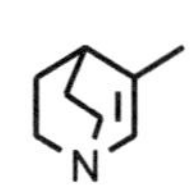	Sn	Hacksell U., *Bioorg. Med. Chem. Lett.*, 1994; 2837

R	X	R'	M	References
Het	X	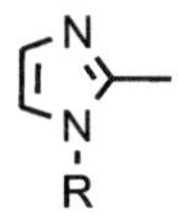	Zn	Bell A.S., *Tetrahedron Lett.*, 1988; 5013
Het	I	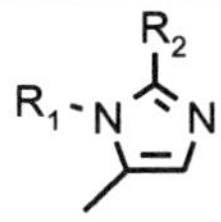	Sn	Achab S., *Tetrahedron Lett.*, 1995; 2615
Het	I		Sn	Hibino S., *J. Org. Chem.*, 1995; 5899
Ar	X	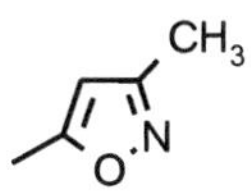	Sn	Yamanaka H., *Tetrahedron Lett.*, 1989; 4249
Ar	X	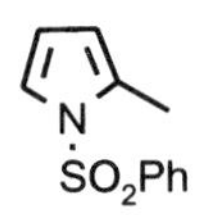	B	Ketcha D.M., *Synth. Commun.*, 1995; 2145
C=C	OSO_2CF_3	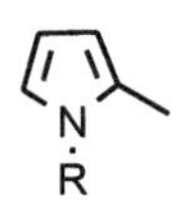	Sn	Farina V., *J. Org. Chem.*, 1990; 5833
Ar	X		B	Muchowski J.M., *J. Org. Chem.*, 1992; 1653
C=C	I	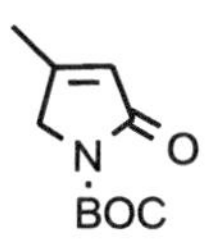	Sn	Reginato G., *Tetrahedron*, 1995; 2129
R	X	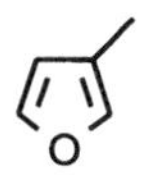	Sn	Wong H.N.C., *Tetrahedron*, 1994; 9583
C=C	Br	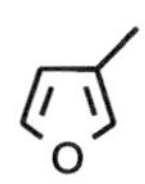	B	Keay B.A., *Synlett.*, 1994; 625
C=C	I	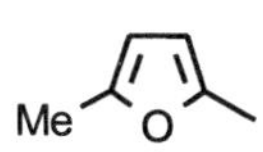	B (in situ)	Keay B.A., *J. Org. Chem.*, 1994; 6501
Ar	X	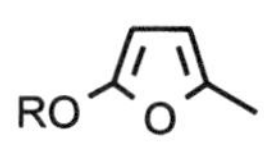	Sn	Pearce B.C., *Synth. Commun.*, 1992; 1627
Ar	X	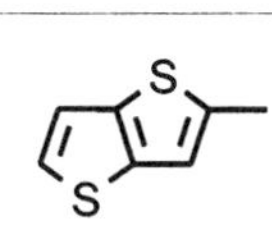	B, Sn, Zn	Kirsch G., *J. Chem. Soc., Perkin Trans. 1*, 1994; 2603
Ar	OSO_2CF_3, I	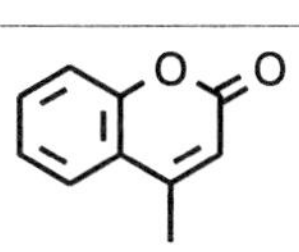	Sn	Ortar G., *Synth. Commun.*, 1995; 2883
Ar	I	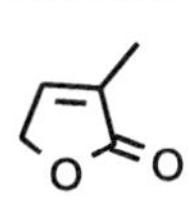	Sn	Sweeney J.B., *Tetrahedron Lett.*, 1992; 7049

R	X	R'	M	References
Ar	I	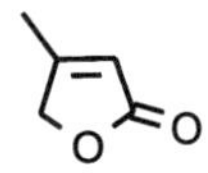	Sn	Sweeney J.B., *Tetrahedron Lett.*, 1992; 7049
C=C	I	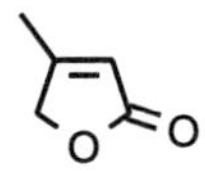	Sn	Reginato G., *Tetrahedron*, 1995; 2129
Het	X	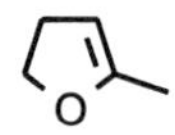	Sn	Doyledaves G. Jr., *Tetrahedron Lett.*, 1993; 1571
C=C-CH	Br	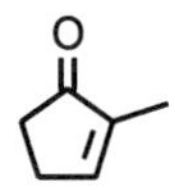	Sn	Adam W., *Synthesis*, 1994; 557
Ar	I	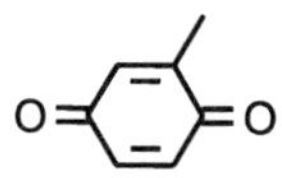	Sn	Liebeskind L., *J. Org. Chem.*, 1993; 408
Het	I	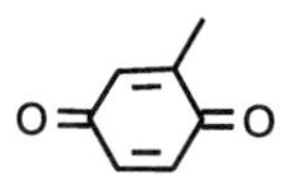	Sn	Liebeskind L., *J. Org. Chem.*, 1993; 408
Ph	Br	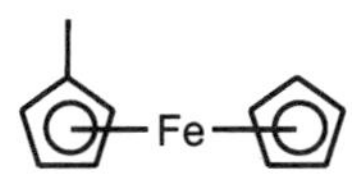	Sn	Kagan H.B., *J. Org. Chem.*, 1995; 2502
Het	X	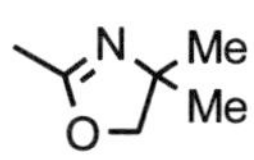	Sn	Dondoni A., *Tetrahedron Lett.*, 1986; 5269 Dondoni A., *Synthesis*, 1987; 693 Kosugi M., *Chem. Lett.*, 1988; 1351
C=C	X	CN	K	Sekiya A., *Chem. Lett.*, 1975; 277 Yamamura K., *Tetrahedron Lett.*, 1977; 4429
Ar	X	CN	K	Takagi K., *Bull. Chem. Soc. Jpn.*, 1976; 3177
Ar	I	CN	Si	Chatani N., *J. Org. Chem.*, 1986; 4714
C=C	OSO_2CF_3	CN	Li	Piers E., *Can. J. Chem.*, 1993; 1867
Ar	X	CN	Zn	Desmond R., *Synth. Commun.*, 1994; 887
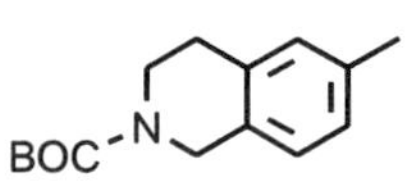	OSO_2CF_3	CN	Zn	Selnick H.G., *Synth. Commun.*, 1995; 3255
C=C	X	RO_2CCH_2-	Zn	Fauvarque J.F., *J. Organomet. Chem.*, 1981; 209, 109
Ar	I	$RO_2C(CH_2)_2$-	Zn (in situ)	Yoshida Z., *Tetrahedron Lett.*, 1986; 955
C=C	I, OSO_2CF_3	$RO_2C(CH_2)_2$-	Zn (in situ)	Yoshida Z., *Tetrahedron Lett.*, 1986; 955
Ar	I	$RO_2C(CH_2)_3$-	Zn (in situ)	Yoshida Z., *Tetrahedron Lett.*, 1986; 955
Ar	X	$RO_2C(CH_2)_3$-	Zn	Knochel P., *Tetrahedron Lett.*, 1994; 1177

R	X	R'	M	References
C=C	I, OSO_2CF_3	$RO_2C(CH_2)_3$-	Zn (in situ)	Yoshida Z., *Tetrahedron Lett.*, 1986; 955
Het	I	CH_2CN	Sn	Nair V., *J. Org. Chem.*, 1988; 3051
Ar	OSO_2CF_3	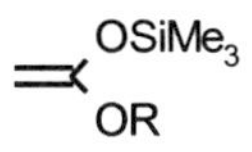	Si	Carfagna C., *J. Org. Chem.*, 1991; 261
Ar	Br	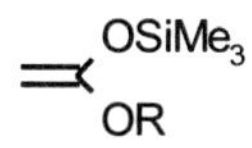	Si	Carfagna C., *J. Org. Chem.*, 1991; 261
Ar	X	Z-CH-CN	Na	Uno M., *J. Chem. Soc., Chem. Commun.*, 1984; 932 Uno M., *Synthesis*, 1985; 506
Ar	X	Z-CH-CN	Na	Yamanaka H., *Synthesis*, 1992; 552 Sakamoto J., *Chem. Pharm. Bull.*, 1990; 1513
Ar	X	Z-CH-CN	Na	Ciufolini M.A., *J. Org. Chem.*, 1988; 4149
Ar	I	$CH(CN)_2$	Na	Zauhar J., *Synthesis*, 1995; 703
C=C	I	$-CO-CH_2-$	K	Piers E., *J. Org. Chem.*, 1990; 3454
Ar	Br	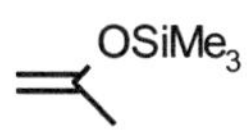	Sn	Kuwajima I., *J. Am. Chem. Soc.*, 1982; 6831
Ar	X	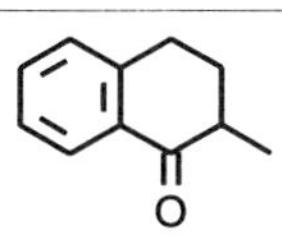	Sn	Donelly D.M.X., *Heterocycles*, 1989; 28, 15
Het	X	CH_2OCH_3	Sn	Undheim K., *Tetrahedron*, 1989; 993
Het	X	CH_2OSiMe_3	Sn	Undheim K., *Tetrahedron*, 1989; 993
Ar	OSO_2CF_3	CH_2SiMe_3	Al	Saulnier M.G., *J. Am. Chem. Soc.*, 1989; 8320
C=C	OSO_2CF_3	CH_2SiMe_3	Al	Saulnier M.G., *J. Am. Chem. Soc.*, 1989; 8320
C=C	I	CH_2SiMe_3	Mg, Zn	Negishi E., *Tetrahedron Lett.*, 1982; 27
C=C	X	$CH_2B(OR)_2$	Zn	Miyaura N., *J. Organomet. Chem.*, 1993; 444, C1 Miyaura N., *J. C. S., Chem. Commun.*, 1994; 467
Ar	X	$(CH_2)_nSnR_3$	B	Ishiyama T., *Synlett.*, 1991; 687
C=C	X	$(CH_2)_nSnR_3$	B	Ishiyama T., *Synlett.*, 1991; 687
Ar	X	>N	M	Migita T., *Chem. Lett.*, 1983; 927
Ar	Br	>N	Sn	Hartwig J.F., *J. Am. Chem. Soc.*, 1994; 5969

R	X	R'	M	References
Ar	Br	>N	Na	Buchwald S.L., *Angew. Chem. Int. Ed. Engl.*, 1995; 1348 Beller M., *Angew. Chem. Int. Ed. Engl.*, 1995; 1316
Ar	Br	>N	Sn	Buchwald S.L., *J. Am. Chem. Soc.*, 1994; 7901
Ar	Br	>N	---	Hartwig J.F., *Tetrahedron Lett.*, 1995; 3609
Ar	X	SR	Sn	Crisp G.T., *Synth. Commun.*, 1992; 683
Het	X	SR	Sn	Crisp G.T., *Synth. Commun.*, 1992; 683
Ar	X	SR	Na	Murahashi S.I., *J. Org. Chem.*, 1979; 2408 Kosugi M., *Chem. Lett.*, 1978; 13
C=C	OSO_2CF_3	SR	Li	Garcia Martinez A., *Synlett.*, 1994; 561
C=C	X	SR	Sn	Rossi R., *Tetrahedron Lett.*, 1989; 2699
R	X	S-SiR_3	K	Soderquist J.A., *Tetrahedron Lett.*, 1994; 3225
Ar	I	EtO_2CCH_2S-	Sn	Dickens M., *J. C. S., Perkin.Trans. 1*, 1992; 323
Ar	I	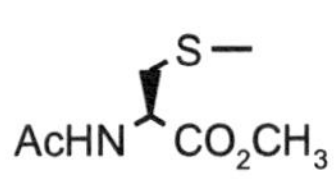	---	Ortar G., *Tetrahedron Lett.*, 1995; 4133
Ar	I	H(R'O)OP-		Schwabacher A.W., *J. Org. Chem.*, 1994; 4206
Ar	I	R(R'O)OP-	---	Schwabacher A.W., *Synthesis*, 1992; 1255
Ar	Br	R(R'O)OP-	---	Xu Y., *Tetrahedron Lett.*, 1988; 1955
Ar	Br	[CH(EtO)$_2$](EtO)OP-	---	Hall R.G., *J. Chem. Soc., Perkin Trans. 1*, 1995; 1145
C=C	Br	(RO)$_2$OP-	---	Hirao T., *Bull. Chem. Soc. Jpn.*, 1982; 909
C=C	OSO_2CF_3	(RO)$_2$OP-	---	Holt D.A., *Tetrahedron Lett.*, 1989; 5393
Ar	Br	(RO)$_2$OP-	---	Hirao T., *Bull. Chem. Soc. Jpn.*, 1982; 909
Het	Br	(RO)$_2$OP-	---	Thompson C.M., *J. Heterocycl. Chem.*, 1994; 1701
C=C	Br	R(R'O)OP-	---	Xu Y., *Tetrahedron Lett.*, 1989; 949
Ar	OSO_2CF_3	>POH	---	Kurz L., *Tetrahedron Lett.*, 1990; 6321
Ar	X	>PH	---	Beletskaya I.P., *Bull. Acad. Sci. USSR*, 1992; 1272
C=C	OSO_2CF_3	PPh_3	---	Stang P.J., *J. Org. Chem.*, 1989; 2783 Stang P.J., *J. Org. Chem.*, 1990; 5033

R	X	R'	M	References
Ar	Br, I	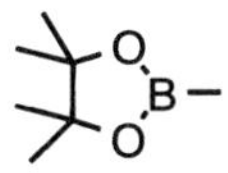	B	Miyaura N., *J. Org. Chem.*, 1995; 7508
Ar	Br	Me_3Si	Si	Matsumoto H., *J. Organomet. Chem.*, 1975; 85, C1 Dunogues J., *J. Organomet. Chem.*, 1993; 446, 135
C=C-CH-	Cl	Me_3Si	Si	Hayashi T., *Organometallics*, 1993; 4051
Ar	Br	$R(R'O)_2Si$	Si	Dunogues J., *J. Organomet. Chem.*, 1994; 468, 69
Ar	X	SnR_3	Sn	Azizian H., *J. Organomet. Chem.*, 1981; 215, 49 Kung H.F., *J. Med. Chem.*, 1990; 171 Mori M., *J. Org. Chem.*, 1991; 3486 Guiles J. W., *Synlett.*, 1995; 165
C=C	X	SnR_3	Sn	Farina V., *J. Org. Chem.*, 1991; 4317
Het	OSO_2CF_3	Me_3Sn	Sn	Barrett A.G.M., *Synlett.*, 1995; 415
C=C	OSO_2CF_3	Me_3Sn	Sn	Paquette L.A., *J. Org. Chem.*, 1995; 764
Ar	SO_2Cl	I	Zn	Miura M., *Bull. Chem. Soc. Jpn.*, 1993; 2121

R	X	R'	M	References
C=C	I	C=C	Al	Negishi E.I., *J. Am. Chem. Soc.*, 1976; 6729
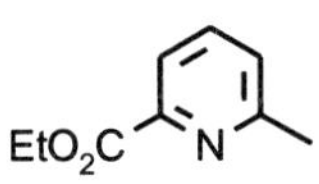	I	C=C	Al	Chandraratna R.A.S., *J. Med. Chem.*, 1995; 2820
C=C	X	C=C	B	Suzuki A., *Pure Appl. Chem.*, 1985; 1749 Suzuki A., *J. Org. Chem.*, 1993; 2201
C=C	Br	C=C	B	William J.S., *Org. Synth.*, 1990; 68, 130
Ar	OSO_2CF_3	C=C	B	Suzuki A., *J. Am. Chem. Soc.*, 1989; 314 Suzuki A., *J. Org. Chem.*, 1993; 2201
Ar	X	C=C	B	Suzuki A., *Pure Appl. Chem.*, 1985; 1749 Beletskaya I.P., *Dokl. Chem.*,1991; 354
Ar	X	Ar	B	Buchecker R., *Tetrahedron Lett.*, 1994; 3277 Novak B. M., *J. Org. Chem.*, 1994; 5034 Wright S. W., *J. Org. Chem.*, 1994; 6095
Ar	OSO_2CF_3	Ar'	B	Ortar G., *Tetrahedron Lett.*, 1992; 4815 Buchecker R., *Tetrahetron Lett.*, 1994; 3277
Ar	X	Z-CH-CN	B	Uno M., *Synthesis*, 1985; 506

R	X	R'	M	References
Ar	Br	R'	B (in situ)	Keay B.A., *J. Org. Chem.*, 1994; 6501
C=C	X	R'	Cu	Normant J.F., *Tetrahedron Lett.*, 1981; 959
C=C	OTs	R	Cu	Tingoli M., *Tetrahedron*, 1995; 4691
R	X	R'	Cu	Beletskaya I.P., *J. Organomet. Chem.*, 1983; 250, 551
Het	X	Ar	Cu	Shimizu N., *Tetrahedron Lett.*, 1993; 3421
Ar	I	R'	Hg	Beletskaya I.P., *J. Organomet. Chem.*, 1983; 250, 551
Ar	I	CN	K	Takagi K., *Bull. Chem. Soc. Jpn.*, 1991; 1118
C=C	OSO_2CF_3	CN	Li	Piers E., *Can. J. Chem.*, 1993; 1867
C=C	X	R'	Mg	Linstrumelle G., *Tetrahedron Lett.*, 1978; 191
C=C	Cl	R'	Mg	Linstrumelle G., *Tetrahedron Lett.*, 1985; 2575
C=C	Cl	CH_3	Mg	Ogasawara K., *Synlett.*, 1994; 665
C=C	$OP(OR)_2$	R'	Mn	Oshima K., *Chem. Lett.*, 1987; 2203
C=C	$OCOCF_3$	R'	Mn	Oshima K., *Chem. Lett.*, 1987; 2203
C=C	OSO_2CF_3	Ar	Na	Ortar G., *Tetrahedron Lett.*, 1992; 4815
Ar	X	Z-CH-CN	Na	Yamanaka H., *Synthesis*, 1992; 552
R	X	R'	Si	Hiyama T., *J. Org. Chem.*, 1988; 918
Ar	N_2^+	C=C	Si	Kenaga K., *Chem. Lett.*, 1988; 873
Ar	X	C=C	Si	Hiyama T., *J. Org. Chem.*, 1988; 918
C=C	X	C=C	Si	Hiyama T., *J. Org. Chem.*, 1988; 918
C=C	Br	C=C	$(RO)_2Si$	Ito Y., *Tetrahedron Lett.*, 1989; 6051
Ar	I	C=C	$(RO)_2Si$	Ito Y., *Tetrahedron Lett.*, 1989; 6051
Het	I	C=C	$(RO)_2Si$	Ito Y., *Tetrahedron Lett.*, 1989; 6051
Ar	I	C=C	F_2Si-	Hiyama T., *Chem. Lett.*, 1989; 1711

R	X	R'	M	References
C=C	Br	C=C	F_2Si-	Hiyama T., *Heterocycles*, 1990; 30, 303
C=C	OSO_2CF_3	C=C	F_2Si-	Hiyama T., *Tetrahedron Lett.*, 1990; 2719
Ar	I	C=C	R_2FSi-	Hatanaka Y., *J. Organomet. Chem.*, 1994; 465, 97 (ipso/ cine-substitutions)
Ar	X	Ar'	X_2RSi	Hatanaka Y., *Tetrahedron*, 1994; 8301
R	I	R'	Sn	Stille J.K., *J. Am. Chem. Soc.*, 1987; 813 Beletskaya I.P., *J. Organomet. Chem.*, 1983; 250, 551 Farina V., *Tetrahedron Lett.*, 1988; 6043
Ar	I	C=C	Sn	Stille J.K., *J. Org. Chem.*, 1987; 422
C=C	X	C=C	Ti	Mori M., *Heterocycles*, 1992; 33, 819
Ar	X	C=C	Ti	Mori M., *Heterocycles*, 1992; 33, 819
Het	Br	Het	Zn	Bell A.S., *Tetrahedron Lett.*, 1988; 5013 Erdik E., *Tetrahedron*, 1992; 9577
C=C	OSO_2CF_3	Ar	Zn	Keenan R.M., *Synth. Commun.*, 1989; 793
Het	Br	Ar	Zn	Vorbruggen H., *Heterocycles*, 1988; 27, 2659
R	I	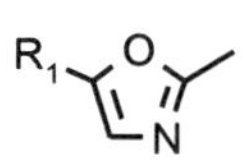	Zn	Hughes M.J., *Tetrahedron*, 1995; 8889
Het	Cl	Het'	Zn (in situ)	Bracher F., *Tetrahedron*, 1994; 12329
C=C	Br	C=C	Zr	Negishi E.I., *Tetrahedron Lett.*, 1978; 1027
C=C	I	C=C	Zr	Negishi E.I., *Tetrahedron Lett.*, 1978; 1027

RXN2 Cross-Coupling of Organometallics with RCOX Derivatives

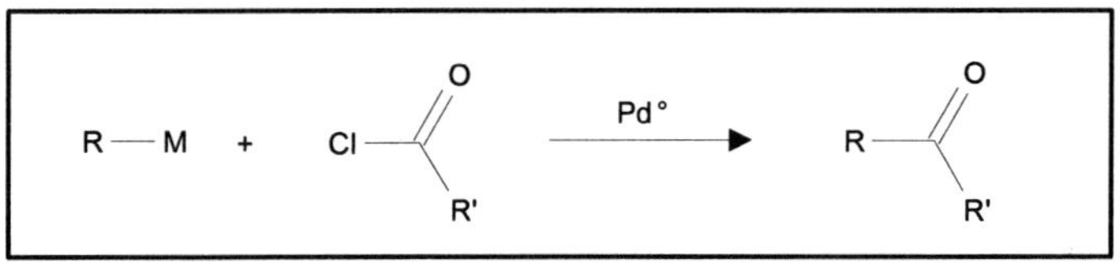

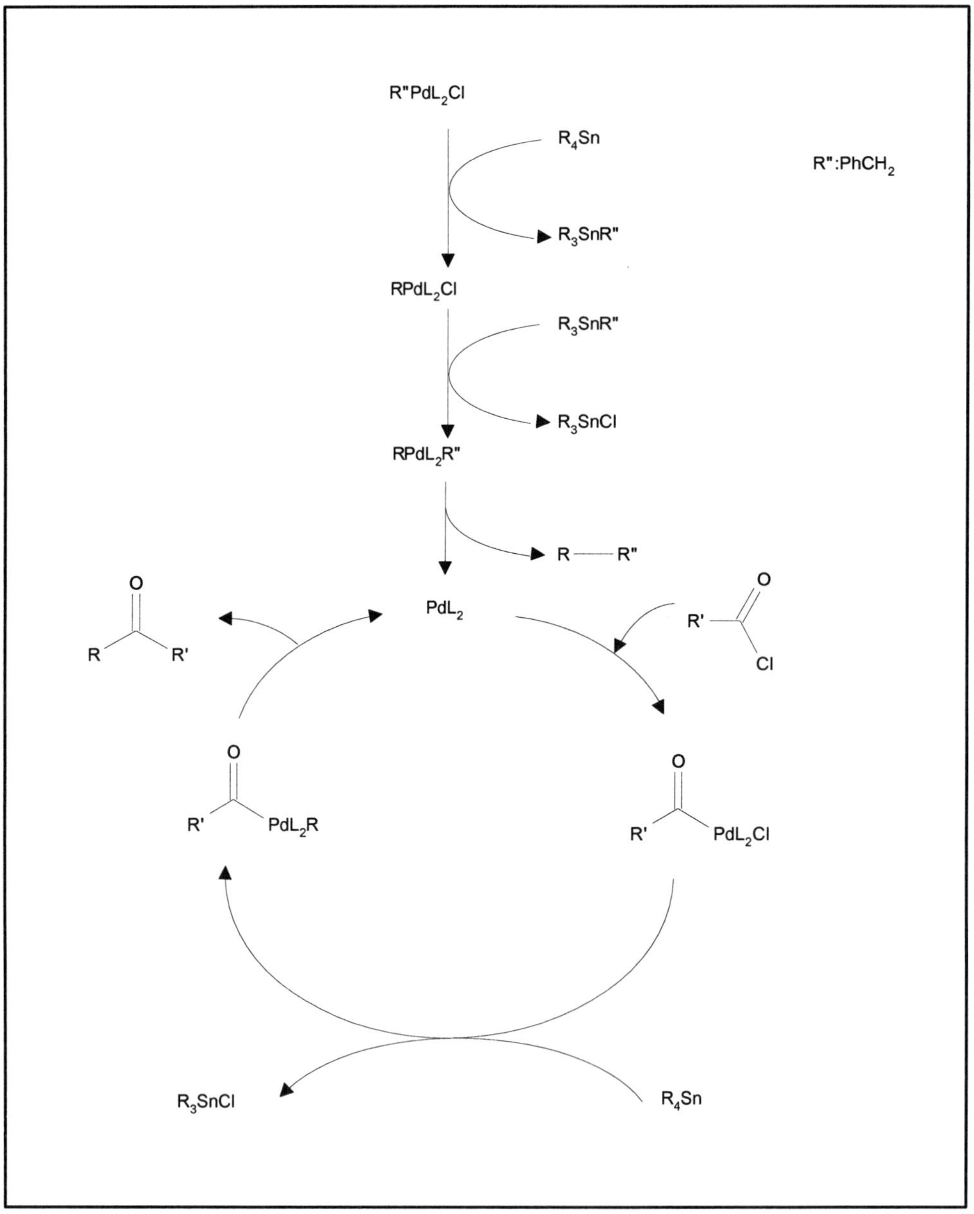

R	References
Ar	Negishi E.I., *Tetrahedron Lett.*, 1983; 5181 Stille J.K., *J. Org. Chem.*, 1979; 1613 Barton D.H.R., *Tetrahedron*, 1988; 5661 Yu K-L., *Tetrahedron Lett.*, 1994; 8955 Evans P.A., *J. Org. Chem.*, 1995; 2298
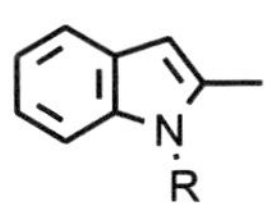	Labadie S.S., *J. Org. Chem.*, 1994; 4250
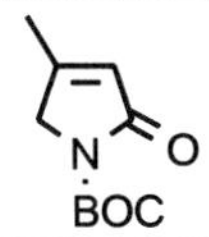	Reginato G., *Tetrahedron*, 1995; 2129
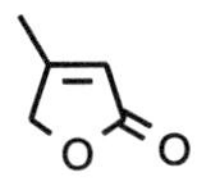	Reginato G., *Tetrahedron*, 1995; 2129
	Wong H.N.C., *Tetrahedron*, 1994; 9583
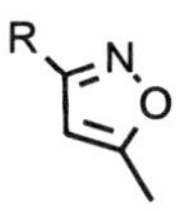	Yamanaka H., *Tetrahedron Lett.*, 1989; 4249
$ArCH_2$-	Fujisawa T., *Chem. Lett.*, 1981; 1135
C=C-	Negishi E.I., *Tetrahedron Lett.*, 1983; 5181
R_1OCO-C=C-	Stille J.K., *Org. Synth.*, 1989; 67, 86
R_1-CO-C=C-	Echavarren A.M., *J. Org. Chem.*, 1992; 5047 Echavarren A.M., *J. Org. Chem.*, 1994; 4179 Echavarren A.M., *Tetrahedron Lett.*, 1994; 7435
$(EtO)_2$CH-C=C-	Quintard J.P., *Tetrahedron Lett.*, 1993; 5445
R_1-C(OAc)-C=C-	Knochel P., *Angew. Chem. Int. Ed. Engl.*, 1993; 582
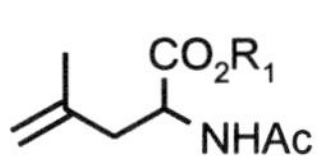	Crisp G.T., *Tetrahedron*, 1994; 3213
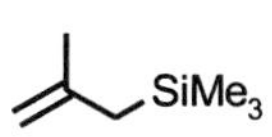	Kang K.T., *Tetrahedron Lett.*, 1991; 4341
Bu_3Sn-C=C-	Echavarren A.M., *J. Org. Chem.*, 1992; 5047
C=C(F)-	Mc Carthy J.R., *Tetrahedron Lett.*, 1994; 5177
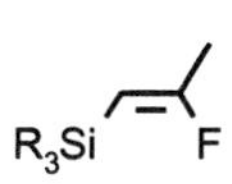	Mc Carthy J.R., *Tetrahedron Lett.*, 1994; 1027

R	References
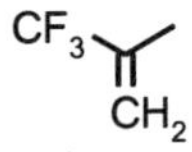	Xu Y., *J. Org. Chem.*, 1994; 2638
C≡C−	Negishi E.I., *Tetrahedron Lett.*, 1983; 5181 Crisp G.T., *Synth. Commun.*, 1989; 1745
Alkyl	Negishi E.I., *Tetrahedron Lett.*, 1983; 5181 Grey R.A:., *J. Org. Chem.*, 1984; 2288 Stille J.K., *J. Org. Chem.*, 1979; 1613
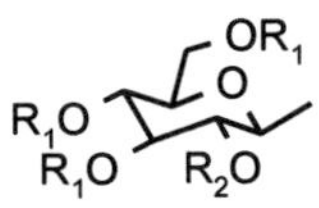	Falck J.R., *Tetrahedron Lett.*, 1995; 5881
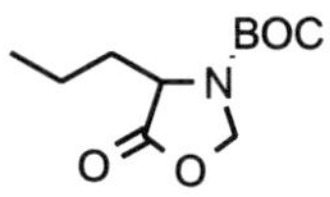	Jackson R.F.W., *Synlett.*, 1995; 819
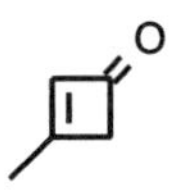	Liebeskind L.S., *J. Org. Chem.*, 1994; 7917
R_2O-$(R_1)C$-	Falck J.R., *J. Am. Chem. Soc.*, 1994; 1
	Fetizon M., *Synthesis*, 1990; 755
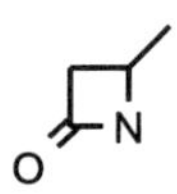	Ricci A., *Tetrahedron Lett.*, 1990; 2637
Me_3Sn-	Mitchell T.N., *Synthesis*, 1990; 1001
H	Lee K.S., *Makromol. Chem. Macro. Chem. Phy.*, 1989; 1547 Guibé F., *J. Org. Chem.*, 1981; 4439 Sonoda N., *Tetrahedron Lett.*, 1993; 2491

M	References
Al	Oshima K., *Bull. Chem. Soc. Jpn.*, 1985; 2425
Bi	Barton D.H.R., *Tetrahedron*, 1988; 5661
Hg	Hegedus L.S., *J. Org. Chem.*, 1989; 4141
Na	Uemura S., *J. Organomet. Chem.*, 1993; 443, 253
Pb	Yamamoto Y., *J. Chem. Soc., Chem. Commun.*, 1987; 1302

M	References
Sn	Stille J.K., *J. Org. Chem.*, 1979; 1613 Falck J.R., *Tetrahedron Lett.*, 1993; 8007
Zn	Negishi E.I., *Tetrahedron Lett.*, 1983; 5181 Grey R.A., *J. Org. Chem.*, 1984; 2288 Sato T., *Chem. Lett.*, 1981; 1135 Eddik E., *Tetrahedron*, 1992; 9577 Jackson R.F.W., *Synlett.*, 1995; 819

R'	References
Ar	Negishi E.I., *Tetrahedron Lett.*, 1983; 5181 Grey R.A., *J. Org. Chem.*, 1984; 2288 Barton D.H.R., *Tetrahedron*, 1988; 5661
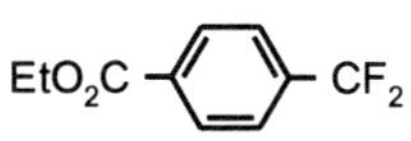	Yu K-L., *Tetrahedron Lett.*, 1994; 8955
Heterocycle	Stille J.K., *J. Org. Chem.*, 1979; 1613 Sato T., *Chem. Lett.*, 1981; 1135
C=C	Negishi E.I., *Tetrahedron Lett.*, 1983; 5181 Stille J.K., *J. Org. Chem.*, 1979; 1613
C=C-C=C-	Duchêne A., *Tetrahedron Lett.*, 1995; 2469
CH_3	Boireau G., *Synth. Commun.*, 1995; 149
Alkyl	Negishi E.I., *Tetrahedron Lett.*, 1983; 5181 Grey R.A., *J. Org. Chem.*, 1984; 2288 Barton D.H.R., *Tetrahedron*, 1988; 5661
CF_3	Guiles J.W., *Synlett.*, 1995; 165
OR_1	Negishi E.I., *Tetrahedron Lett.*, 1983; 5181 Jousseaume B., *Organometallics*, 1991; 366 Jousseaume B., *Synlett.*, 1993; 117 Baldwin J.E., *J. Chem. Soc., Perkin Trans. 1*, 1994; 1697 (intramolecular reaction)
R_1-N-R_2	Jousseaume B., *Organometallics*, 1991; 366 Jousseaume B., *Synlett.*, 1993; 117
SEt	Falck J.R., *Tetrahedron Lett.*, 1995; 5881

R'X	References
PhO-CS-Cl	Falck J.R., *Tetrahedron Lett.*, 1995; 5881

R'X	References
R_1-CO-SAr	Sonoda N., *Tetrahedron Lett.*, 1993; 2491
R_1SO_2Cl	Labadie S.L., *J. Org. Chem.*, 1989; 2496
R_1-CO-SeAr	Sonoda N., *Organometallics*, 1992; 3937
CF_3CO-O-$COCF_3$	Guiles J.W., *Synlett.*, 1995; 165

RXN3 Cross-Coupling of Siloxycyclopropanes with RX and RCOX Derivatives

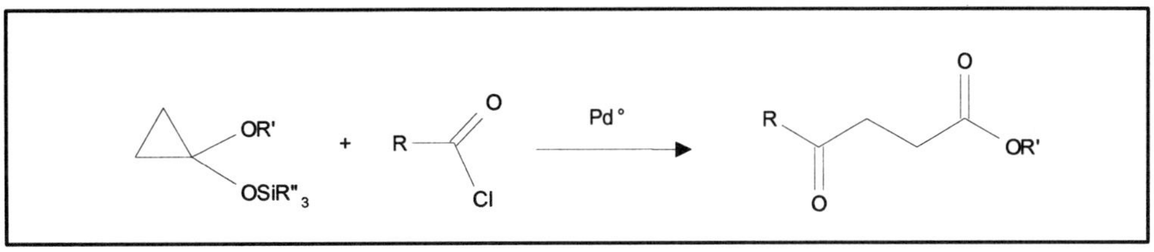

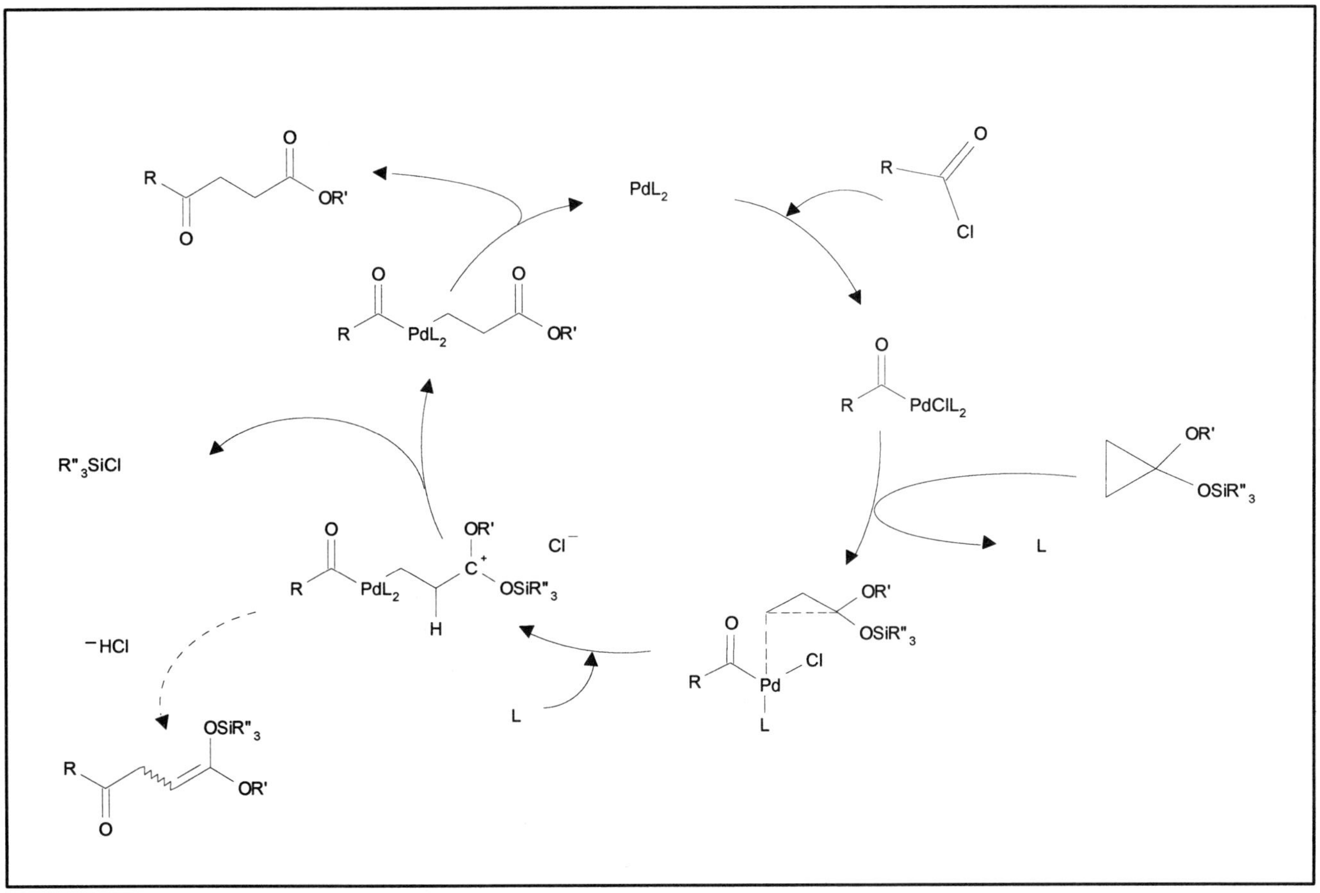

	RX	Product	References
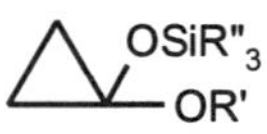	R-CO-Cl	R-CO-C-C-CO_2R'	Nakamura E., *J. Org. Chem.*, 1991; 2809 Nakamura E., *Tetrahedron Lett.*, 1989; 6541
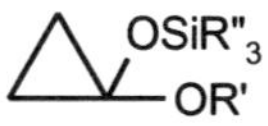	C=C-O-Tf	Ar'-C-C-CO-Ar	Nakamura E., *J. Am. Chem. Soc.*, 1988; 3296
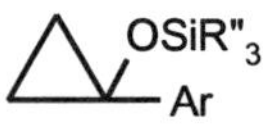	R-CO-Cl	R-CO-C-C-CO-Ar	Nakamura E., *J. Org. Chem.*, 1991; 2809
	Ar'-O-Tf	Ar'-C-C-CO-Ar	Nakamura E., *J. Am. Chem. Soc.*, 1988; 3296

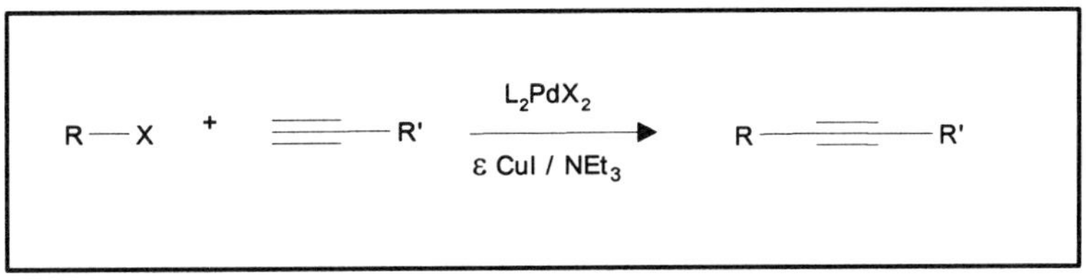
R—X
+
R'
L_2PdX_2
ε CuI / NEt_3
R'

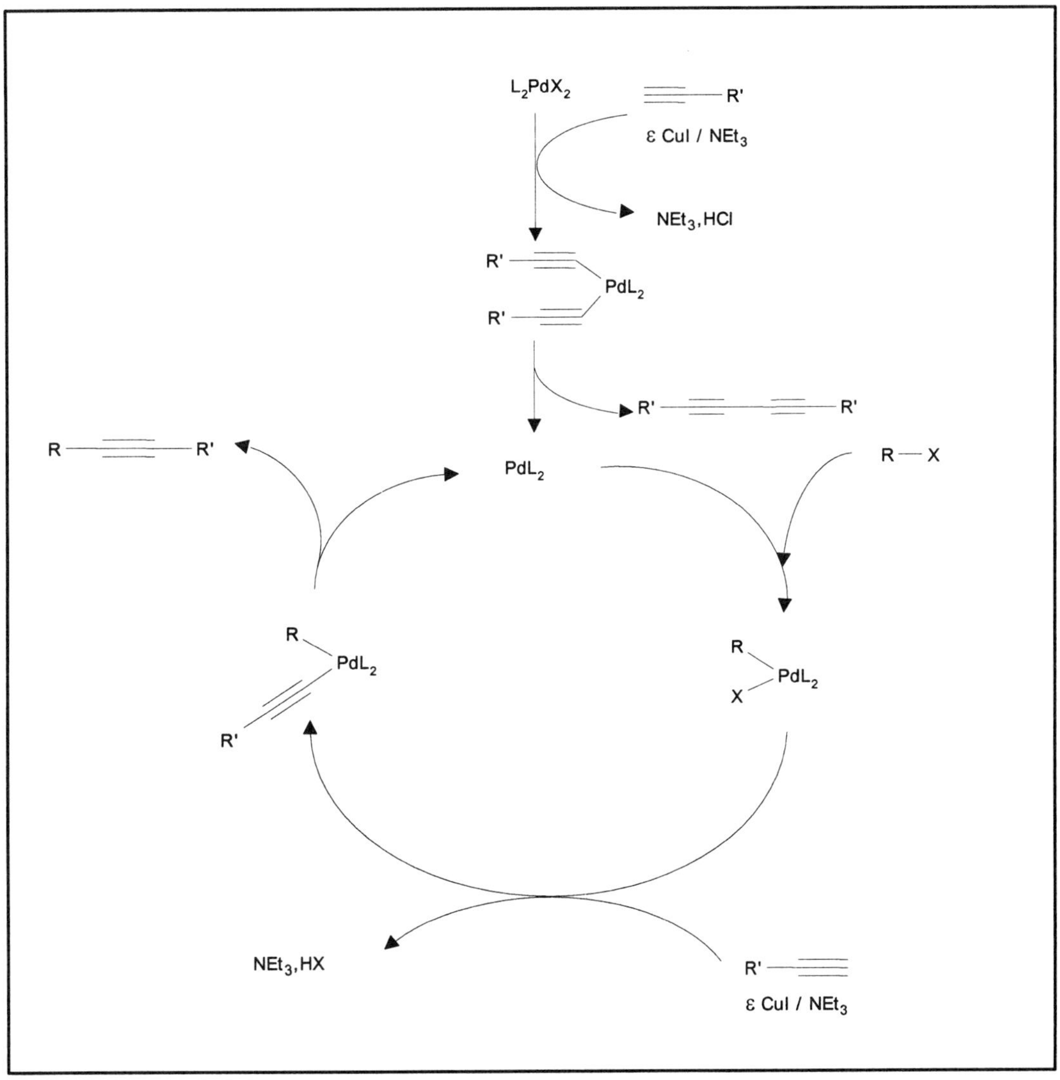
L_2PdX_2
R'
ε CuI / NEt_3
NEt_3,HCl
R'
PdL_2
R'
R'
R'
R
R'
PdL_2
R—X
R
PdL_2
R'
R
PdL_2
X
NEt_3,HX
R'
ε CuI / NEt_3

R / X	References
Ar / I	Sonogashira K., *Tetrahedron Lett.*, 1975; 4467
Ar / I	Heck F.R., *J. Organomet. Chem.*, 1975; 93, 259 (without CuI)
Ar / I	Sita L. R., *Tetrahedron Lett.*, 1995; 4525
Ar, Het / Br, I	Cosford N.D.P., *Synlett.*, 1995; 1115
Ar, Het / I	Genêt J.P., *Synlett.*, 1992; 715 (without CuI)
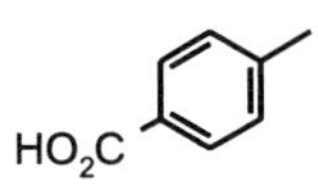 / I	Schmidtchen F.P., *Tetrahedron*, 1995; 2325
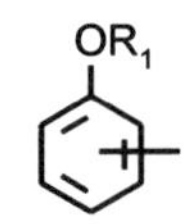 / I	Bates R.W., *Tetrahedron*, 1995; 8199
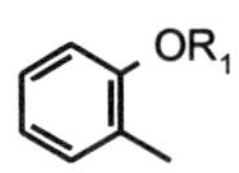 / X	Tsukayama M., *Heterocycles*, 1994; 38, 1487
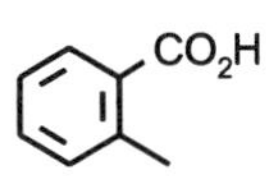 / I	Cheng C.H., *J. Org. Chem.*, 1995; 3711
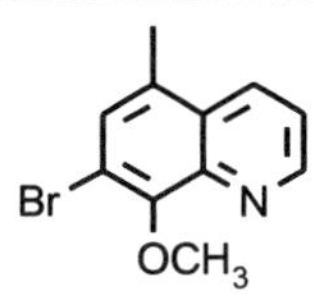 / Br	Queguiner G., *J. Heterocycl. Chem.*, 1995; 1261
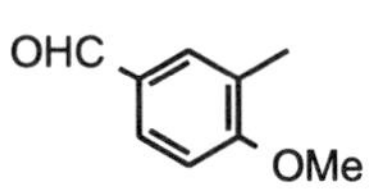 / OTf	Ogasawara K., *Heterocycles*, 1994; 38, 2463
Aromatic(F) / I	Burton D.J., *J. Org. Chem.*, 1993; 7368
Heterocycle(N) / Br	Sonogashira K., *Tetrahedron Lett.*, 1975; 4467
Heterocycle(N) / OTf	Isobe M., *Synlett.*, 1994; 589 Isobe M., *Tetrahedron*, 1995; 3737
Heterocycle(N) / X	Vorbrüggen H., *Heterocycles*, 1988; 27, 2659 Kalinin V.N., *Synthesis*, 1992; 413
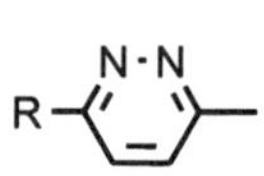 / I	Bailey T.R., *J. Org. Chem.*,1995; 748
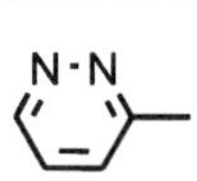 / OTf	Suffert J., *Heterocycles*, 1994; 38, 1273

R / X	References
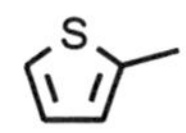 / I	Merriman G., *Bioorg. Med. Chem. Lett.*, 1995; 2483
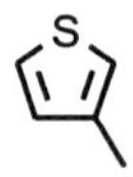 / Br	Merriman G., *Bioorg. Med. Chem. Lett.*, 1995; 2483
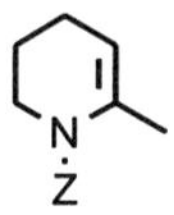 / OTf	Isobe M., *Synlett.*, 1994; 589 Isobe M., *Tetrahedron*, 1995; 3737
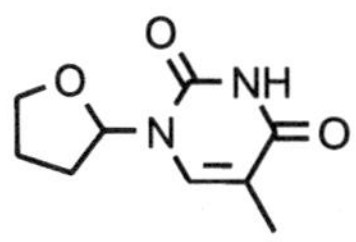 / OTf	Crisp G.T., *J. Org. Chem.*, 1993; 6614
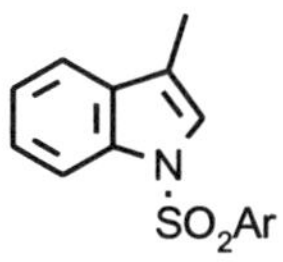 / OTf	Gribble G.W., *Synth. Commun.*, 1992; 2129
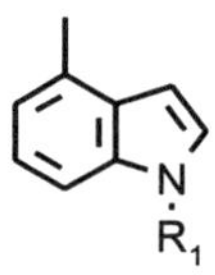 / I	Galambos G., *Heterocycles*, 1993; 36, 2241
Het-CO- / Cl	Hagihara N., *Synthesis*, 1977; 777 Crisp G.T., *Synth. Commun.*, 1989; 1745
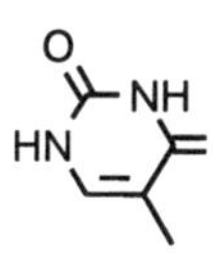 / I	Cacchi S., *Tetrahedron Lett.*, 1989; 2581
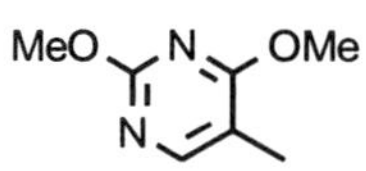 / I	Kundu N.G., *J. Chem. Soc., Perkin Trans. 1*, 1993; 2657
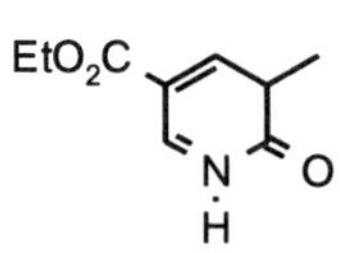 / I	Houpis I. N., *Tetrahedron lett.*, 1994; 9355
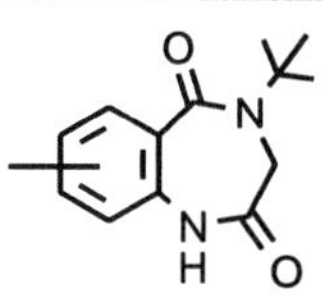 / Br, I	Karp G.M., *J. Org. Chem.*, 1995; 5814
Porphyrin / X	Van Lier J. E., *Tetrahedron*, 1994; 11933
Porphyrin / I	Dolphin D., *J. Chem. Soc., Chem. Commun.*, 1995; 527
Porphyrin / OSO_2CF_3	Chan K.S., *Synth. Commun.*, 1993; 1489
C=C- / X	Heck F.R., *J. Organomet. Chem.*, 1975; 93, 259 (without CuI)

R / X	References
C=C- / Br	Sonogashira K., *Tetrahedron Lett.*, 1975; 4467
C=C- / I	Takahashi T., *Synlett.*, 1995; 513
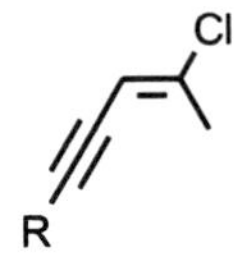 / I	Linstrumelle G., *Tetrahedron Lett.*, 1995; 3687
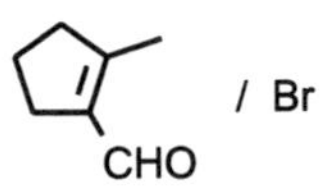 / Br	Wang K.K., *Tetrahedron Lett.*, 1995; 3785
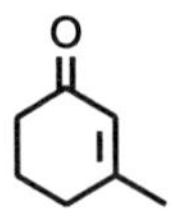 / I	Genêt J.P., *Tetrahedron Lett.*, 1995; 5891 (without CuI)
Cl-C=C- / Cl	Linstrumelle G., *Tetrahedron*, 1994; 5335 Linstrumelle G., *Tetrahedron*, 1995; 4245
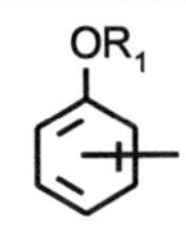 / Cl	Linstrumelle G., *Tetrahedron Lett.*, 1994; 5335
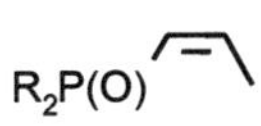 / I	Lu X., *Synthesis*, 1995; 769
F-C=C(F)- / I	Burton D.J., *Tetrahedron Lett.*, 1990; 1369
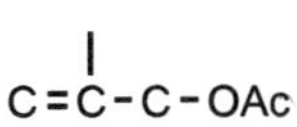 / Br	Nwokogu G.C., *Tetrahedron Lett.*, 1984; 3263 Nwokogu G.C., *J. Org. Chem.*, 1985; 3900
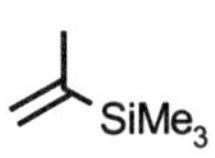 / Br	Schinzer D., *Synthesis*, 1995; 299
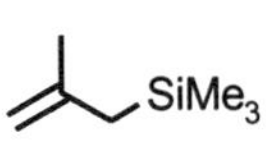 / Br	Schinzer D., *Synthesis*, 1995; 299
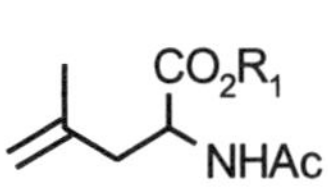 / I	Crisp G.T., *Tetrahedron*, 1994; 2623
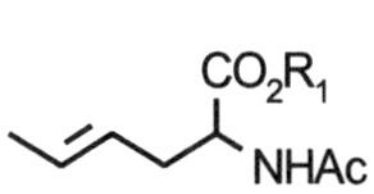 / I	Crisp G.T., *Tetrahedron*, 1994; 2623
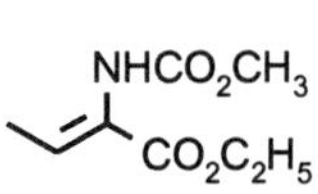 / Br	Danion D., *Synthesis*, 1994; 1171
C=C-C=C- / Cl	Linstrumelle G., *Synth. Commun.*, 1994; 2273

R / X	References
C=C=C- / Br	Linstrumelle G., *Synthesis*, 1983; 32
C=C=C-C(SPh)- / OAc	Padwa A., *Synlett.*, 1992; 869
C≡C− / I	Genêt J.P., *J. Org. Chem.*, 1995; 6829 (without CuI)
C≡C-C− / OCO_2CH_3	Tsuji J., *Tetrahedron Lett.*, 1990; 7179 Tsuji J., *J. Organomet. Chem.*, 1991; 417, 305
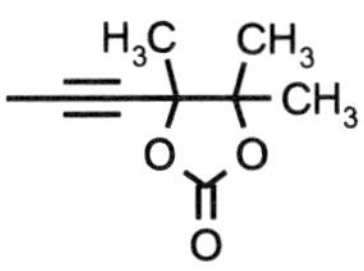	Dixneuf P.H., *J. Chem. Soc., Chem. Commun.*, 1994; 1845
C≡C-C− / X	Linstrumelle G., *Tetrahedron Lett.*, 1993; 3853
C≡C-C− / OAc	Linstrumelle G., *Tetrahedron Lett.*, 1993; 3853
R_1R_2N=C- / Cl	Lin S.Y., *Synthesis*, 1991; 235
$R_1N=(CF_3)C$- / I	Uneyama K., *Tetrahedron Lett.*, 1991; 1459

X	References
Halogen	Sonogashira K., *Tetrahedron Lett.*, 1975; 4467 Casalnuovo A.L., *J. Am. Chem. Soc.*, 1990; 4324
Br	Guzman A., *Synth. Commun.*, 1990; 2059
I	Commerçon A., *Bioorg. Med. Chem. Lett.*, 1995; 809
OSO_2CF_3	Chen Q.Y., *Tetrahedron Lett.*, 1986; 1171 Suffert J., *Tetrahedron Lett.*, 1991; 1453 Cacchi S., *Tetrahedron*, 1993; 4955
$OSO_2CF_2OR_1$	Chen Q.Y., *Tetrahedron Lett.*, 1986; 1171
OCO_2CH_3	Tsuji J.,*Tetrahedron Lett.*, 1990; 7179
OAc	Padwa A., *Synlett.*, 1992; 869

R'	References
Ar	Sonogashira K., *Tetrahedron Lett.*, 1975; 4467 Chen Q.Y., *Tetrahedron Lett.*, 1986; 1171
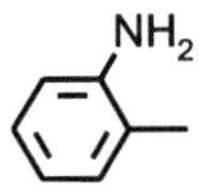	Cacchi S., *Tetrahedron Lett.*, 1989; 2581 Cacchi S., *J. Organomet. Chem.*, 1994; 475, 289
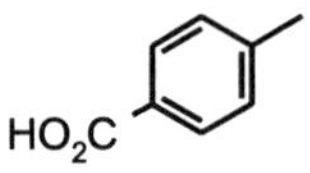	Schmidtchen F.P., *Tetrahedron*, 1995; 2325
C=C-	Mourino A., *Tetrahedron*, 1991; 3485 Genêt J.P., *Tetrahedron Lett.*, 1995; 5891 (without CuI)
Me_3Si	Hagihara N., *Synthesis*, 1980; 627 Lau K.S.Y., *J. Org. Chem.*, 1981; 2280 Genêt J.P., *Synlett.*, 1992; 715 (without CuI) Blart E., *Thesis, Université Paris VI*, 1993
$(EtO)_3C$	Yamanaka H., *Synthesis*, 1992; 746
$MeOCH_2$	Chen Q.Y., *Synthesis*, 1988; 896
$EtO_2C\text{-}CH_2\text{-}CO_2\text{-}C(R_1)(R_2)\text{-}$	Cacchi S., *Synlett.*, 1993; 65
$F_3CONH\text{-}CH_2\text{-}$	Robins M.J., *Tetrahedron Lett.*, 1990; 3731
$HO\text{-}CHR_1\text{-}$	Kundu N.G., *J. Chem. Res.*, 1995; 4
$HO\text{-}C(CH_3)_2\text{-}$	Swindell C. S., *Tetrahedron Lett.*, 1994; 4959 Litt M.H., *J. Org. Chem.*, 1994; 5818
$HOCH_2\text{-}$	Genêt J.P., *Synlett.*, 1992; 715 (without CuI) Beletskaya I.P., *Synthesis*, 1984; 728
$HO(CH_2)_2\text{-}$	Linstrumelle G., *Tetrahedron Lett.*, 1993; 6403 (without CuI)
$-CH_2-CH(NHAc)(CO_2Et)$	Crisp G.T., *Tetrahedron*, 1992; 3239
$HO_2C\text{-}CH_2\text{-}CH_2\text{-}$	Cacchi S., *Synlett.*, 1991; 409

RXN5 Intermolecular HECK Reaction

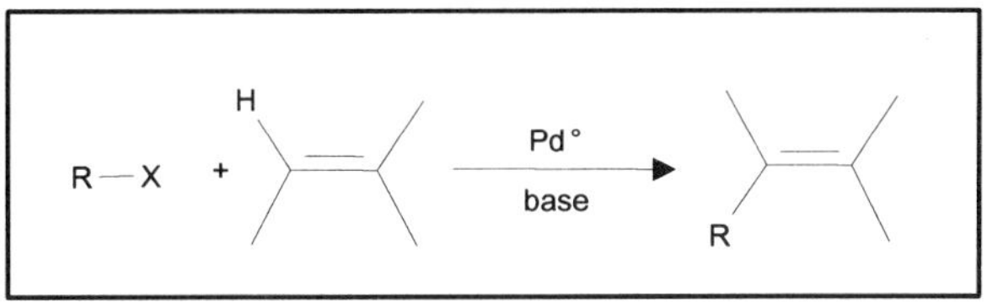

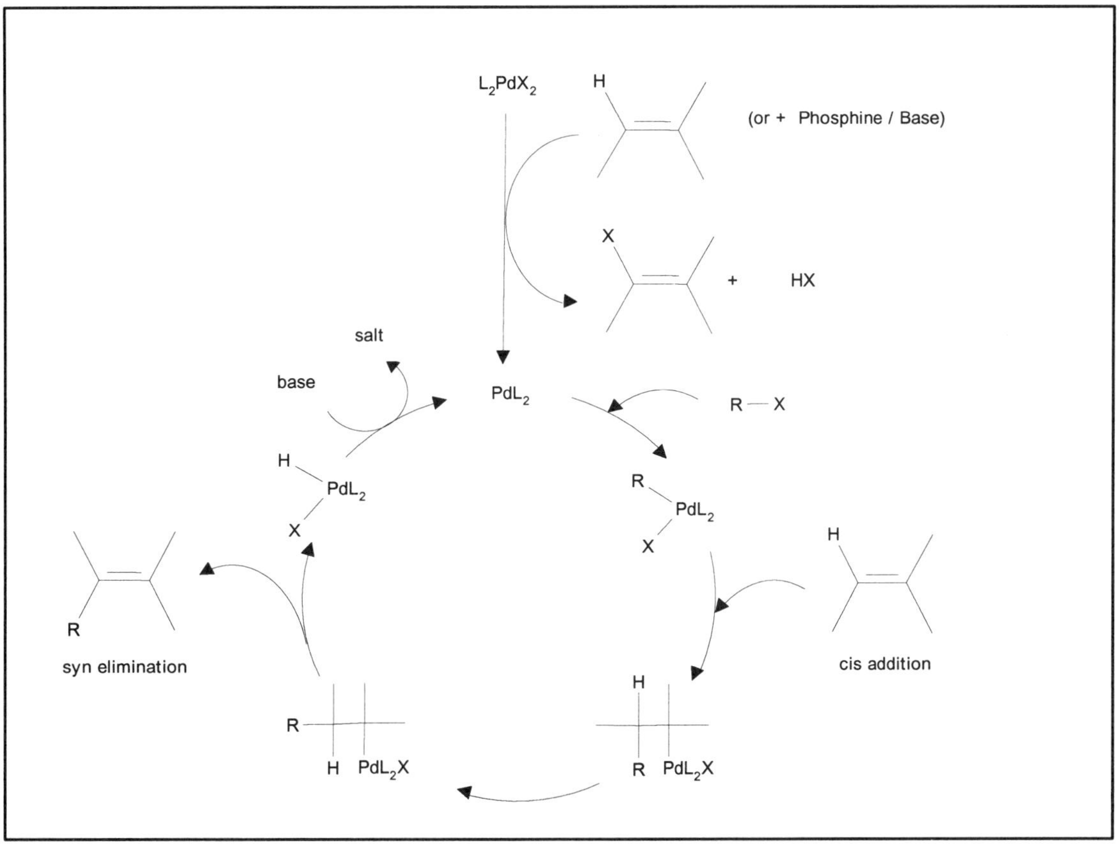

RXN5 Intermolecular HECK Reaction

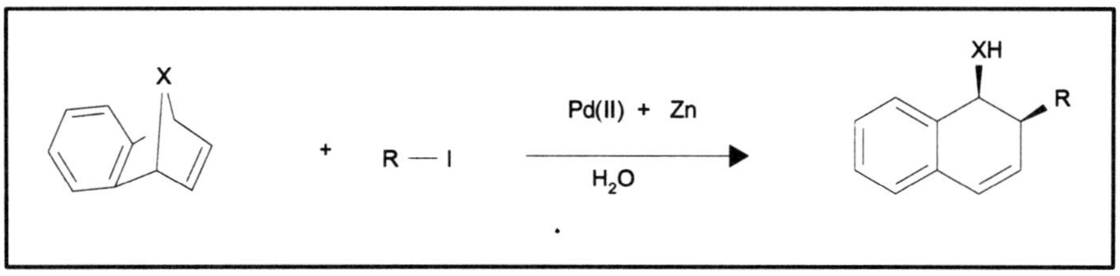

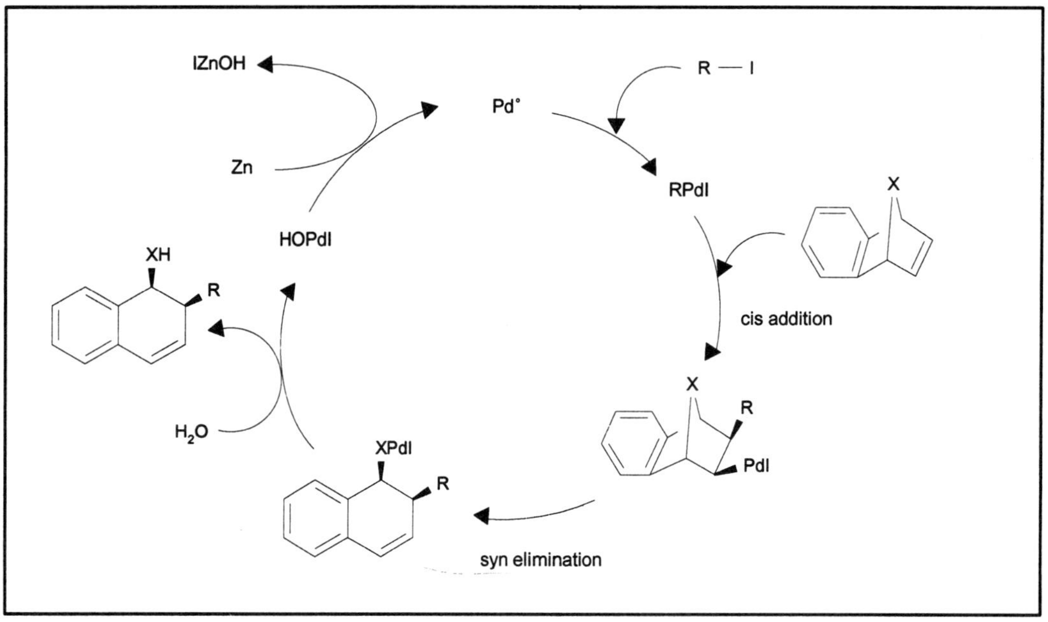

R	X	H-C=C	References
Ar	Br	$H_2C=CH_2$	Heck R.F., *J. Org. Chem.*, 1978; 2454 De Vries R.A., *Organometallics*, 1994; 2405
Ar	N_2^+ (in situ)	$CH_2=CH_2$	Beller M., *Tetrahedron Lett.*, 1994; 8773
Ar	X	C=C	Heck R.F., *Org. Reactions*, 1982; 27, 345
Ar	I	C=C	Jeffery T., *J. Chem. Soc., Chem. Commun.*, 1984; 1287
Ar	I	C=C	Zhou P., *Tetrahedron Lett.*, 1995; 4567 (solid phase)
Ar	OSO_2CF_3	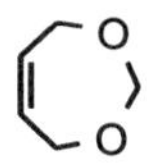	Shibasaki M., *Tetrahedron Lett.*, 1994; 1227 (asym. catal.)
Ph	I	C=C	Jeffery T., *Tetrahedron Lett.*, 1994; 4103
Het	X	C=C	Heck R.F., *Org. Reactions*, 1982; 27, 345 Kalinin V.N., *Synthesis*, 1992; 413
Het	Br	C=C	Heck R.F., *J. Org. Chem.*, 1978; 2947 Yoshida Z.I., *Tetrahedron Lett.*, 1978; 919
Het (N)	X	C=C	Vorbrüggen H., *Heterocycles*, 1988; 27, 2659
Het (N)	Cl		Ohta A., *Heterocycles*, 1992; 33, 257
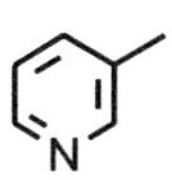	OSO_2CF_3	C=C	Bailey T.R., *Synlett.*, 1995; 157
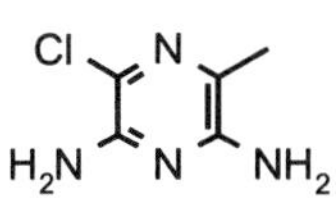	I	C=C-O	Townsend L.B., *Tetrahedron Lett.*, 1995; 8363
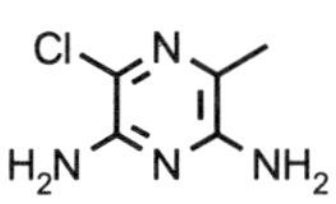	I	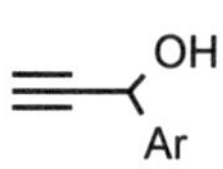	Kundu N. G., *J. Chem. Soc., Chem. Commun.*, 1995; 99
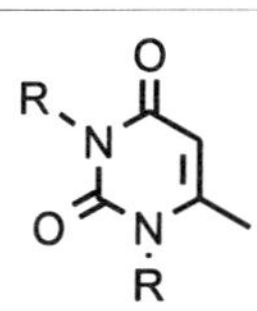	I	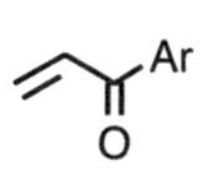	Kundu N.G., *Bioorg. and Med. Chem. Lett.*, 1995; 1627
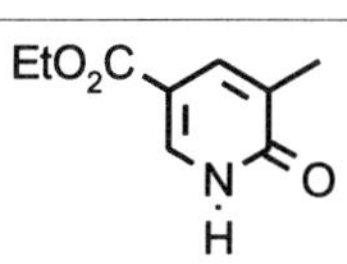	I	$CH_2=CH_2$	Choi W.B., *Tetrahedron Lett.*, 1995; 4571
$ArCH_2$	Cl	C=C	Heck R.F., *J. Org. Chem.*, 1972; 2320 Zhanggu Z., *Synth. Commun.*, 1992; 2019
C=C	X	C=C	Heck R.F., *Org. Reactions*, 1982; 27, 345 De Meijere A., *Synlett.*, 1994; 189

R	X	H-C=C	References
C=C	X	C=C	Heck R.F., *Org. Reactions*, 1982; 27, 345 De Meijere A., *Synlett.*, 1994; 189
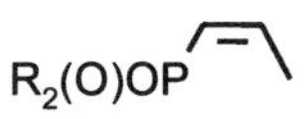	I	C=C	Lu X., *Synthesis*, 1995; 769
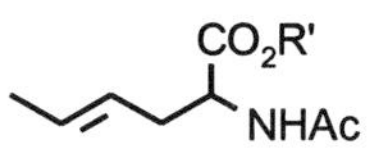	I	C=C	Crisp G.T., *Tetrahedron*, 1994; 2623
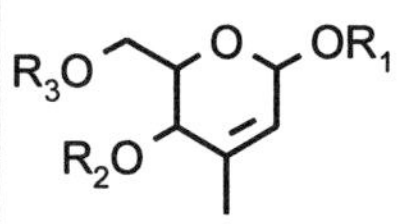	OSO_2CF_3	C=C	Voelter W., *Angew. Chem. Int. Ed. Engl.*, 1994; 1499
C≡C—	I	C=C	Jeffery T., *Synthesis*, 1987; 70
C≡C-C—	OCO_2CH_3	C=C	Mandai T., *Tetrahedron Lett.*, 1991; 3397
$RN=C(CF_3)$-	I	C=C	Uneyama K., *Tetrahedron Lett.*, 1991; 1459

R	X	H-C=C	References
C=C	Br	C=C	Heck R.F., *Org. Reactions*, 1982; 27, 345
Ar	Br, Cl, I	C=C	Julia M., *Bull. Soc. Chim. Fr.*, 1973; 2790
Ar	Br	$C=C-CO_2Et$	Hanaoka T.A., *J. Mol. Catal.*, 1994; 88, L113
Ar	Cl	C=C	Davison J.B., *J. Mol. Catal.*, 1984; 22, 349 Milstein D., *Organometallics*, 1993; 4734
Ar	Cl ----> I	C=C	Bozell J.J., *J. Am. Chem. Soc.*, 1988; 2655
Ar	I	C=C	Casalnuovo A.L., *J. Am. Chem. Soc.*, 1990; 4324
Ar	I	$H_2C=CH-Z$	Jeffery T., *Tetrahedron Lett.*, 1994; 3051
$ArCH_2$	$N(+)Bu_3$ Br(-)	C=C	Zhuangyu Z., *Synthesis*, 1995; 245
Ar	N_2+ (in situ)	$CH_2=CH_2$	Beller M., *Tetrahedron Lett.*, 1994; 8773
Ar	N_2+ (in situ)	C=C	Sengupta S., *Tetrahedron Lett.*, 1995; 4475 Matsuda T., *J. Org. Chem.*, 1981; 4885 Sengupta S., *J. Chem. Soc., Perkin Trans. 1*, 1993; 1943

R	X	H-C=C	References
Ar	N_2^+	C=C	Kikukawa K., *J. Chem. Soc., Perkin Trans. 1*, 1986; 1959 (-PdSi) Matsuda T., *Tetrahedron*, 1981; 31 Ikenaga K., *Chem. Lett.*, 1988; 873 (-PdSi) Beller M., *Synlett.*, 1995; 441
Ar	$N{=}N{-}SO_2Ar'$	C=C	Kamigata N., *Bull. Chem. Soc. Jpn.*, 1988; 3575
Ar	N=N-NRR'	C=C	Sengupta S., *Synth. Commun.*, 1995; 651
Ar	$ON{=}NSO_2Ar'$	C=C	Kamigata N., *Bull. Chem. Soc. Jpn.*, 1990; 2118
C=C	$Ph{-}I^+$	C=C	Moriarty R.M., *J. Am. Chem. Soc.*, 1991; 6315
C=C	OSO_2CF_3	C=C	Stille J.K., *J. Org. Chem.*, 1985; 2302 Ortar G., *Tetrahedron Lett.*, 1984; 2271
Ar	OSO_2CF_3	C=C	Cabri W., *Synlett.*, 1992, 871 Cabri W., *J. Org. Chem.*, 1993; 7421
Ar	OSO_2CF_2OR	C=C	Chen Q.Y., *Tetrahedron Lett.*, 1986; 1171 Chen Q.Y., *Synthesis*, 1988; 896
Ar	CO-Cl	C=C	Spencer A., *J. Organomet. Chem.*, 1982; 233, 267 Spencer A., *J. Organomet. Chem.*, 1984; 265, 323 Cabri W., *Tetrahedron Lett.*, 1991; 1753
Het	CO-Cl	C=C	Kasahara A., *J. Heterocycl. Chem.*, 1989; 597
Ar	SO_2Cl	C=C	Miura M., *Tetrahedron Lett.*, 1989; 975 Miura M., *J. Chem. Soc., Perkin Trans. 1*, 1990; 2207
Ar	$B(OH)_2$	C=C	Uemura S., *J. Organomet. Chem.*, 1994; 465, 85 Uemura S., *J. Org. Chem.*, 1995; 883
Ph	BPh_2	C=C	Uemura S., *J. Organomet. Chem.*, 1994; 465, 85 Uemura S., *Tetrahedron Lett.*, 1994; 1739 Uemura S., *J. Org. Chem.*, 1995; 883

R	X	H-C=C	References
Ar	X	C=C	Heck R.F., *Org. Reactions*, 1982; 27, 345
C=C	X	C=C	Heck R.F., *Org. Reactions*, 1982; 27, 345
Ar	Br,I	$CH_2{=}CH_2$	Kiji J., *J. Mol. Catal.*, 1995; 97, 73
Ar	Br	$CH_2{=}CH_2$	Kasahara A., *J. Heterocycl. Chem.*, 1989; 1405
Ar	Br,I	C=C-Ph	Bumagin N.A., *J. Organomet. Chem.*, 1995; 486, 259

R	X	H-C=C	References
Het	OSO_2CF_3	C=C-Ph	Gribble G.W., *Synt. Commun.*, 1992; 2129
Ar	X		Prashad M., *Tetrahedron Lett.*, 1989; 2877
C=C	X	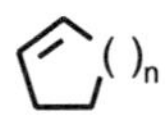	Larock R.C., *J. Org. Chem.*, 1989; 2047
Ar	I	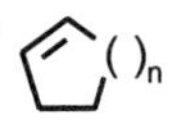	Larock R.C., *Tetrahedron Lett.*, 1988; 905
Ar	X	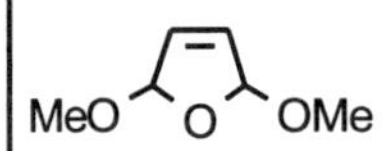	Ogasawara K., *Heterocycles*, 1994; 38, 1745 Ogasawara K., *Synlett.*, 1994; 499
C=C	I	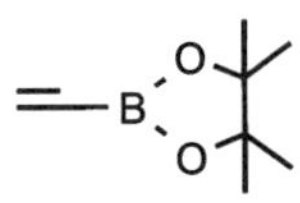	Whiting A., *Tetrahedron Lett.*, 1995; 3925
Ar	X	C=C-$SiMe_3$	Hallberg A., *J. Org. Chem.*, 1986; 5286
Ar	I	C=C-$SiMe_3$	Westerlund C., *Chem. Lett.*, 1982; 1993 (-PdSi)
C=C	X	C=C-$SiMe_3$	Hallberg A., *J. Org. Chem.*, 1986; 5286
C=C	OSO_2CF_3	C=C-$SiMe_3$	Hallberg A., J. Org. Chem., 1986; 5286
Ar	Br	C=C-$SiMeCl_2$	Tanaka M., *Chem. Lett.*, 1990; 2175
C=C	X	C=C-$SiMeCl_2$	Tanaka M., *Chem. Lett.*, 1990; 2175
Ar	I	Me_3SiC=$CSiMe_3$	Hallberg A., *J. Org. Chem.*, 1989; 1773 (-PdSi)
Ar	OSO_2CF_3	Me_3SiC=$CSiMe_3$	Hallberg A., *J. Org. Chem.*, 1989; 1773 (-PdSi)
Ar	N_2^+	C=C-$SnBu_3$	Kikukawa K., *J. Organomet. Chem.*, 1986; 311, C44 (-PdSn)
Ar	N_2^+	C=C-$GeMe_3$	Ikenaga K., *Chem. Lett.*, 1990; 185 (-PdGe)
Ar	Br	C=C-Z	Heck R.F., *J. Org. Chem.*, 1977; 3903
Ar	I	C=C-Z	Genêt J.P., *Synlett.*, 1992; 715 Sugihara T., *Tetrahedron Lett.*, 1995; 5547
Ar	X	C=C-Z	Blart E., *Thesis, Université Paris VI*, 1993
Ar	OSO_2CF_3	C=C-Z	Cabri W., *Synlett.*, 1992; 871

R	X	H-C=C	References
Ar	Br,I	$C=C-CO_2H$	Bumagin N.A., *J. Organomet. Chem.*, 1995; 486, 259
Ar	I	$C=C-CO_2H$	Zhuangyu Z., *Synth. Commun.*, 1995; 595
Ar	X	C=C-CN	Choudary B.M., *Tetrahedron*, 1992; 719
Ar	X	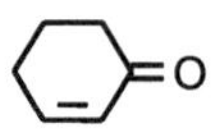	Blart E., *Thesis, Université Paris VI*, 1993
Ar	X	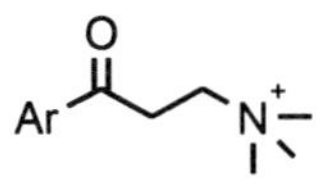	Blart E., *Thesis, Université Paris VI*, 1993
Ar	I	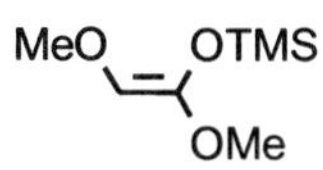	Sakamoto T., *J. Chem. Soc., Perkin Trans.1*, 1994; 235
Ar	I	$C=C(OEt)-CO_2Et$	Yamanaka H., *Heterocycles*, 1988; 27, 453 Yamanaka H., *Heterocycles*, 1988; 27, 257 Ortar G., *Tetrahedron Lett.*, 1987; 3039
C=C	OSO_2CF_3	$C=C(OEt)-CO_2Et$	Ortar G., *Tetrahedron Lett.*, 1987; 3039
Ar	I	$C=C(NR_2)-CO_2R'$	Frejd T., *Synthesis*, 1989; 414
Ar	Br	$C=C(NHAc)-CO_2H$	Naso F., *Tetrahedron Lett.*, 1983; 4603
Het(N)	Br	$C=C(NHAc)-CO_2R'$	Bozell J.J., *J. Org. Chem.*, 1991; 2584
Ar	I	$C=C(NHR)-CO_2R'$	Frejd T., *Acta Chem. Scand.*, 1992; 163
Het(N)	X	$EtO-C=C-CO_2Et$	Yamanaka H., *Heterocycles*, 1988; 27, 257
Ar	Br	$C=C-SO_2NH_2$	Kuroda S., *Bull. Chem. Soc. Jpn.*, 1991; 1431
C=C	OSO_2CF_3	$C=C-PO(OEt)_2$	Cacchi S., *Synlett.*, 1995; 677
Ar	X	$C=C-P(O)R_2$	Pietrusiewicz K.M., *Tetrahedron: Asymmetry*, 1993; 2143
Ar	X	C=C-OR	Andersson C.M., *J. Org. Chem.*, 1990; 5757
Ar	OSO_2CF_3	C=C-OR	Cabri W., *J. Org. Chem.*, 1990; 3654 Pendrak I., *J. Org. Chem.*, 1995; 3249
C=C	OSO_2CF_3	C=C-OR	Hallberg A., *J. Org. Chem.*, 1989; 1502
Ar	X	C=C-OR	Cabri W., *Tetrahedron Lett.*, 1991; 1753

R	X	H-C=C	References
Het(N)	I	C=C-OAc	Daves Jr. G.D., *J. Heterocycl. Chem.*, 1978; 351
C=C	OSO_2CF_3	C=C-OAc	Ortar G., *Tetrahedron Lett.*, 1991; 1579
Ar	I	C=C-OAc	Larock R.C., *Tetrahedron*, 1994; 305
Ar	I		Daves Jr. G.D., *J. Org. Chem.*, 1979; 21 Larock R.C., *Tetrahedron Lett.*, 1988; 905 Larock R.C., *Tetrahedron Lett.*, 1989; 2603 Larock R.C., *J. Org. Chem.*, 1990; 407 Daves Jr. G.D., *Organometallics*, 1990; 3151
Ar	I	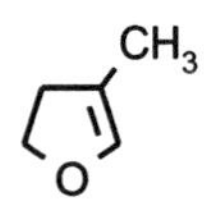	Reiser O., *Synlett.*, 1995; 153
Ar	OTf		Hayashi T., *Tetrahedron. Lett.*, 1992; 1485 (asymmetric catalysis)
C=C	OTf		Hayashi T., *Tetrahedron. Lett.*, 1993; 2505 (asymmetric catalysis)
Ar	X	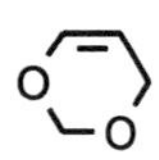	Yamanaka H., *Tetrahedron Lett.*, 1992; 6845
Ar	I	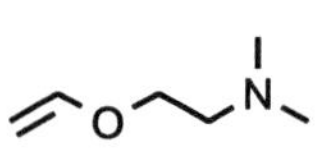	Hallberg A., *Acta Chem. Scand.*, 1993, 212
Ar	OSO_2CF_3	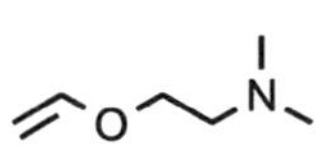	Hallberg A., *Tetrahedron*, 1994; 285
Ar	Br	$C{=}C{-}C(OR)_2$	Heck R.F., *J. Org. Chem.*, 1977; 3907
Ar	X	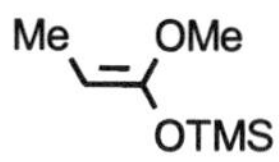	Sakamoto T., *Heterocycles*, 1993; 36, 2509
Ar	OSO_2CF_3	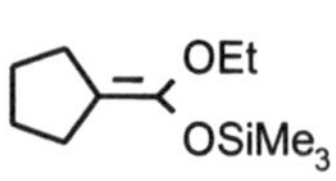	Koch K., *J. Med. Chem.*, 1994; 3197 (-PdSi)
Het(N)	Br	$C{=}C{-}CH_2OH$	Yoshida Z.I., *Chem. Lett.*, 1978; 975
Ar	X	$C{=}C{-}CH_2OH$	Heck R.F., *J. Org. Chem.*, 1976; 265
Ar	I	$C{=}C{-}CH_2OH$	Larock R.C., *Tetrahedron Lett.*, 1991; 569 Kang S.K., *Tetrahedron Lett.*, 1995; 6287
C=C	Br	$C{=}C{-}CH_2OH$	Yoshida Z.I., *Tetrahedron*, 1979; 329

R	X	H-C=C	References
C=C	I	$C{=}C{-}CH_2OH$	Jeffery T., *Tetrahedron Lett.*, 1990; 6641
C=C	X	$C{=}C{-}CH_2OH$	Jeffery T., *J. Chem. Soc., Chem. Commun.*, 1991; 324
Ar	I	$C{=}C{-}(CH_2)_n{-}OH$	Larock R.C., *Tetrahedron Lett.*, 1989; 6629
Ar	I	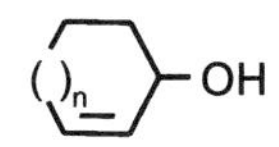	Larock R.C., *Tetrahedron*, 1994; 305
Ar	I		Mandai T., *Synlett*, 1990; 85
C=C	X		Mandai T., *Synlett*, 1990; 85
C=C	OSO_2CF_3		Cacchi S., *Tetrahedron*, 1991; 1525
Ar	I	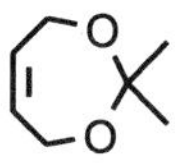	Sakamoto T., *Heterocycles*, 1993; 36, 2437
Ar	I	$C{=}C{-}CH_2{-}O{-}CO{-}NHR$	Tamaru Y., *Tetrahedron Lett.*, 1994; 4133
Ar	I	$C{=}C{-}CH_2{-}NRR'$	Filippini L., *Tetrahedron Lett.*, 1993; 1643
Ar	X	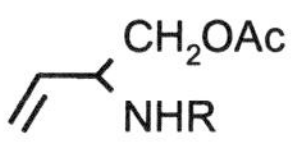	Crisp G.T., *Tetrahedron*, 1992; 3541
C=C	OSO_2CF_3	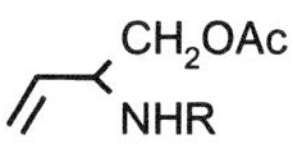	Crisp G.T., *Tetrahedron*, 1992; 3541
Ar	X	$C{=}C{-}CH_2NHCOMe$	Heck R.F., *Org. React.*, 1982; 27, 345
C=C	OSO_2CF_3	$C{=}C{-}CH_2N(BOC)_2$	Arcadi A., *Tetrahedron Lett.*, 1990; 2463
C=C	Br	$C{=}C{-}CH_2N(BOC)_2$	Helquist P., *Tetrahedron Lett.*, 1995; 3115
R	X	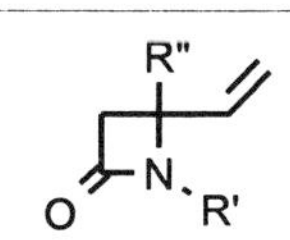	Shimizu I., *Synlett.*, 1994; 839
Ar	I	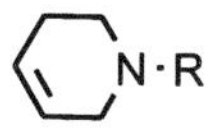	Hallberg A., *J. Org. Chem.*, 1992; 4015

R	X	H-C=C	References
Ph	I	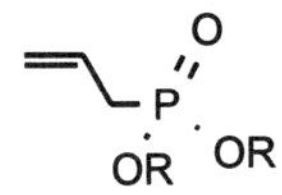	Demik N.N., *Zh. Org. Khim.*, 1994; 876
Ar	Br	$C=C-CH_2-SO_2NH_2$	Kuroda S., *Bull. Chem. Soc. Jpn.*, 1991; 1431
Ar	X	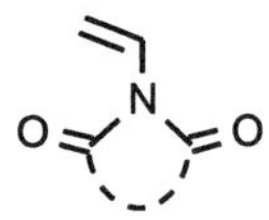	Heck R.F., *J. Org. Chem.*, 1978; 2949 Zhuangyu Z., *Synth. Commun.*, 1995; 595
Ar	I	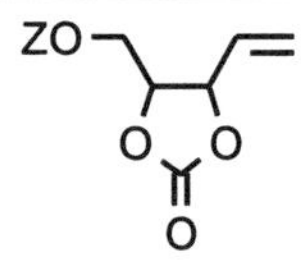	Kang S.K., *Tetrahedron Lett.*, 1995; 8047 (-PdOR)
Ar	I	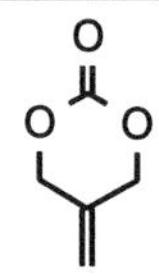	Kang S.K., *Tetrahedron Lett.*, 1995; 8047 (-PdOR)
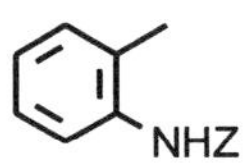	I, Br	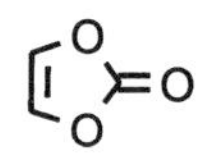	Ogasawara K., *Heterocycles*, 1995; 41, 1627 (-PdOR, cyclisation)
Ar	X	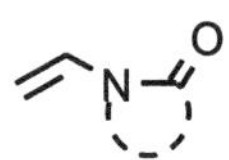	Heck R.F., *J. Org. Chem.*, 1978; 2949
Ar	I		Hallberg A., *J. Org. Chem.*, 1990; 2464
C=C	OTf	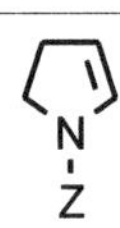	Hayashi T., *Tetrahedron. Lett.*, 1993; 2505 (asymmetric catalysis)
Ar	I	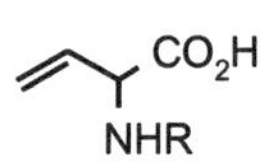	Itaya T., *Chem. Pharm. Bull.*, 1995; 398
C=C	OSO_2CF_3	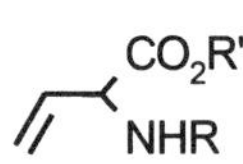	Crisp G.T., *Tetrahedron*, 1992; 3541
Ar	X	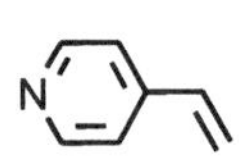	Heck R.F., *J. Org. Chem.*, 1978; 2947
Ar	X	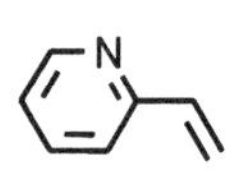	Heck R.F., *J. Org. Chem.*, 1978; 2947
Ar	I	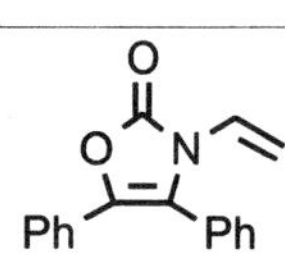	Busacca C.A., *J. Org. Chem.*, 1993; 3299
Ar	I	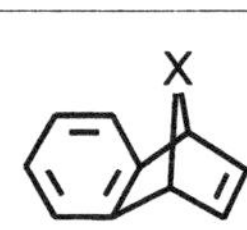	Cheng C.H., *Tetrahedron Lett.*, 1993; 4019 (-PdOR,-PdNR2)

R	X	H-C=C	References
Ar	Br		Ohta A., *Heterocycles*, 1990; 31, 1951
Ar	Br	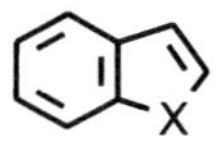	Ohta A., *Heterocycles*, 1990; 31, 1951
Het(N)	Cl	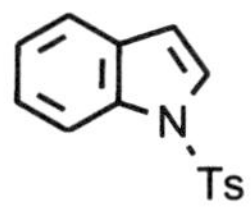	Ohta A., *Chem. Pharm. Bull.*, 1989; 1477
C=C	OSO_2CF_3	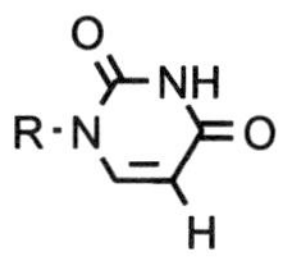	Hassan M.E., *Can. J. Chem.*, 1991; 198
Ar	Br	C=C-C=C	Heck R.F., *J. Org. Chem.*, 1978; 5018
Ar	I	C=C-C=C	Jeffery T., *Tetrahedron Lett.*, 1992; 1989
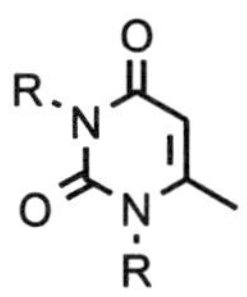	I	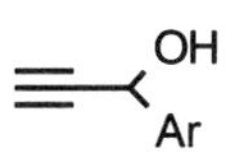	Kundu N. G., *J. Chem. Soc., Chem. Commun.*, 1995; 99

RXN6 Intramolecular HECK Reaction

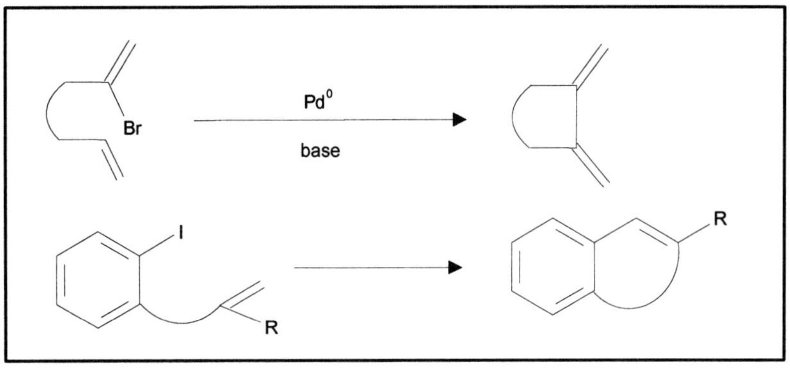

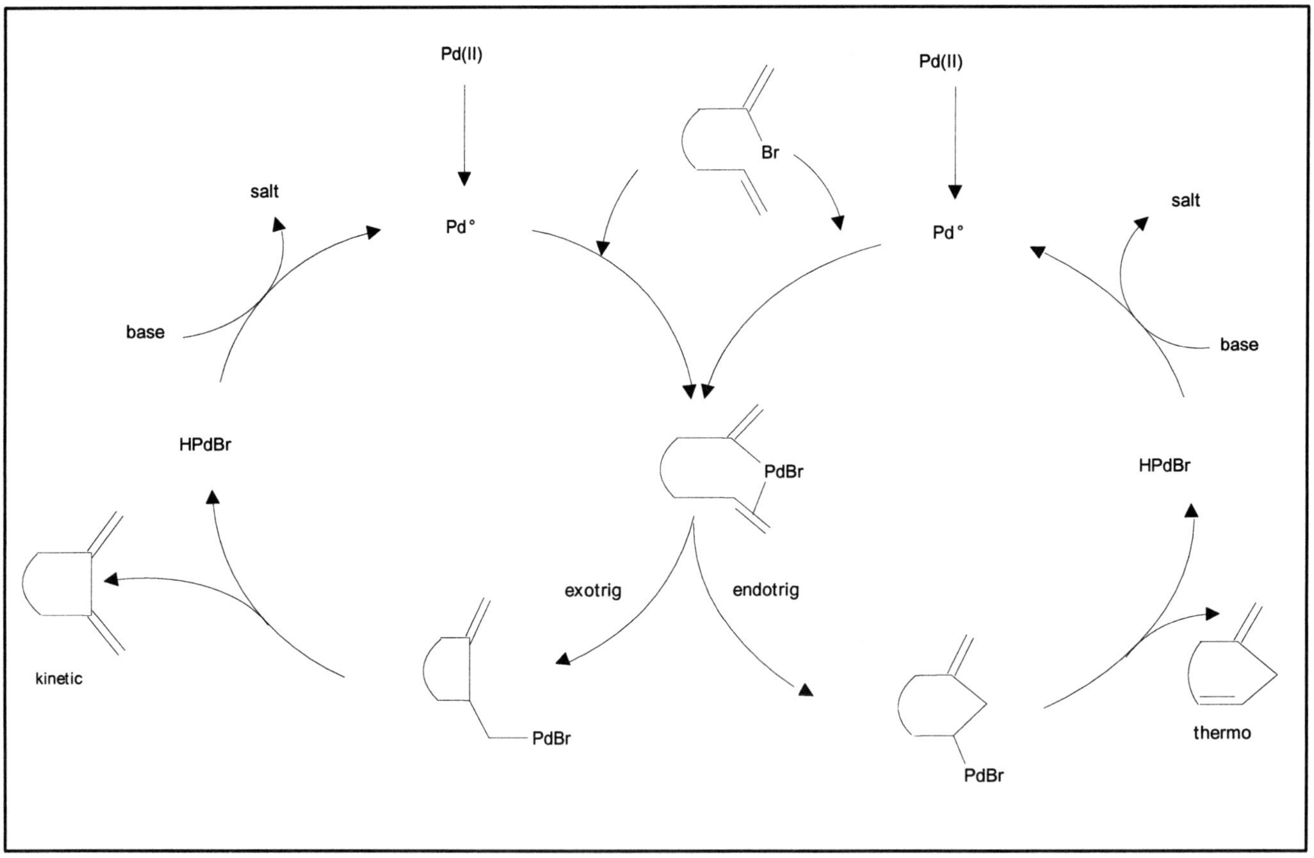

A ---> B	References
	Grigg R., *Tetrahedron*, 1988; 2033 Gaudin J.M., *Tetrahedron Lett.*, 1991; 6113 Fuchs P.L., *Tetrahedron Lett.*, 1993; 5209 De Meijere A., *Synlett.*, 1994; 191 Masters J.J., *Tetrahedron Lett.*, 1993; 7253 Grigg R., *J. Chem. Soc., Chem. Commun.*, 1984; 1073
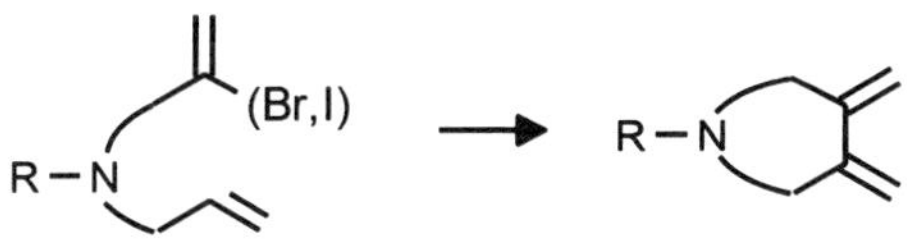	Grigg R., *Tetrahedron*, 1988; 2033 Blechert S. *Tetrahedron Lett.*, 1994; 9537
	Youg W.B., *J. Am. Chem. Soc.*, 1995; 5228
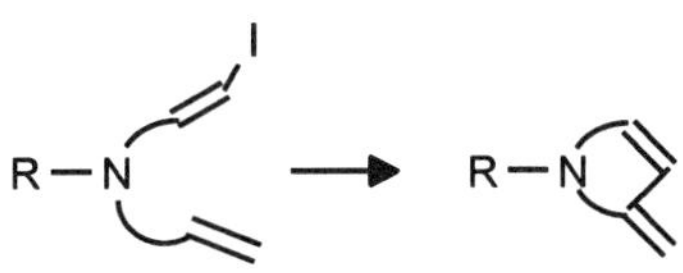	Shibasaki S., *Tetrahedron Lett.*, 1993; 4965
	Shibasaki M., *Tetrahedron*, 1994; 371 (asym. catalysis)
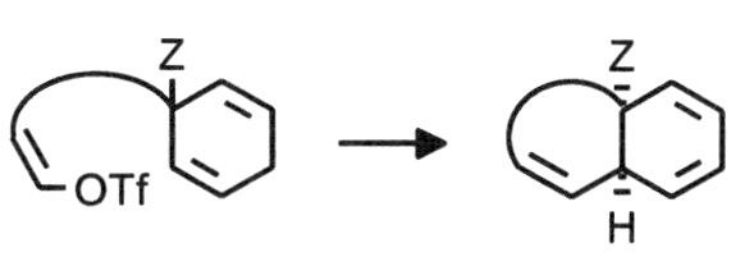	Shibasaki M., *J. Am. Chem. Soc.*, 1994; 11737 (asym. cat.)
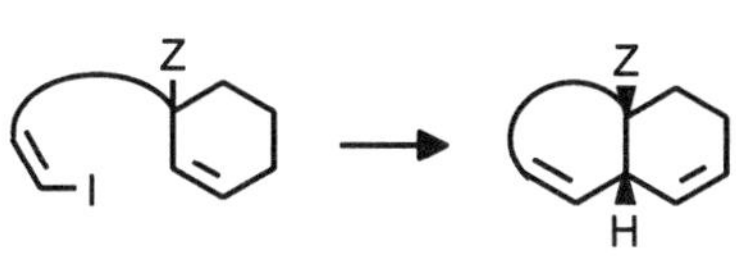	Larock R.C., *Tetrahedron Lett.*, 1988; 2919
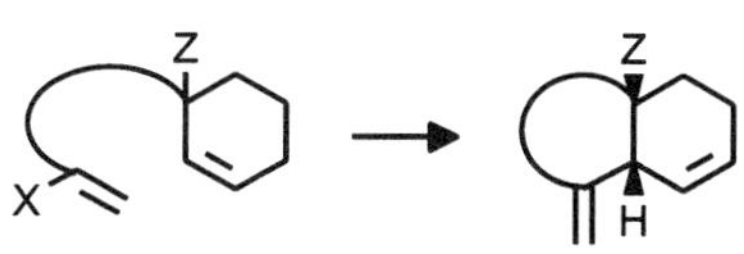	Larock R.C., *Tetrahedron Lett.*, 1988; 2919
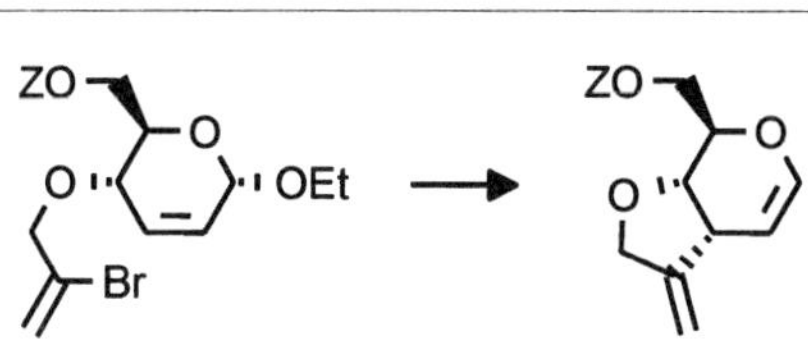	Sinou D., *J. Chem. Soc., Chem. Commun.*, 1995; 1893 (-PdOR)

A ---> B	References
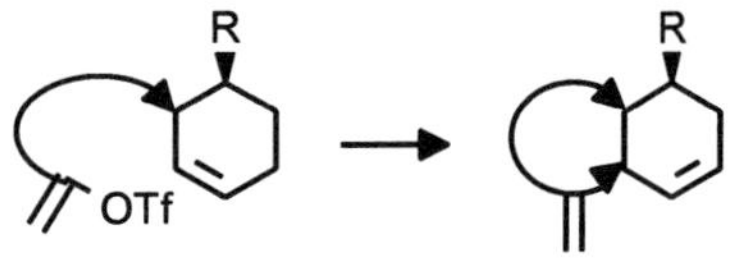	Fukumoto K., *Synlett.*, 1995; 761
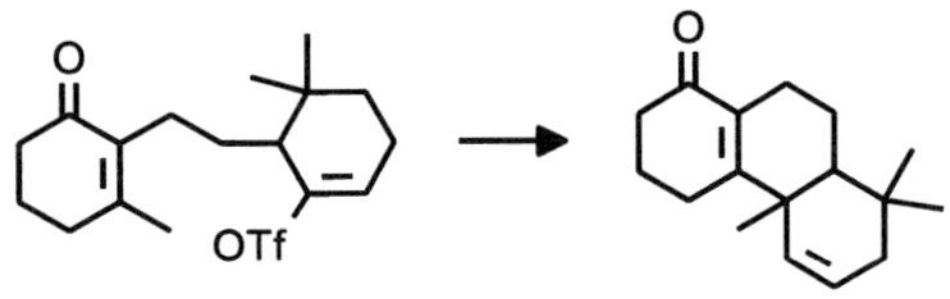	Overman L.E., *Tetrahedron*, 1994; 347
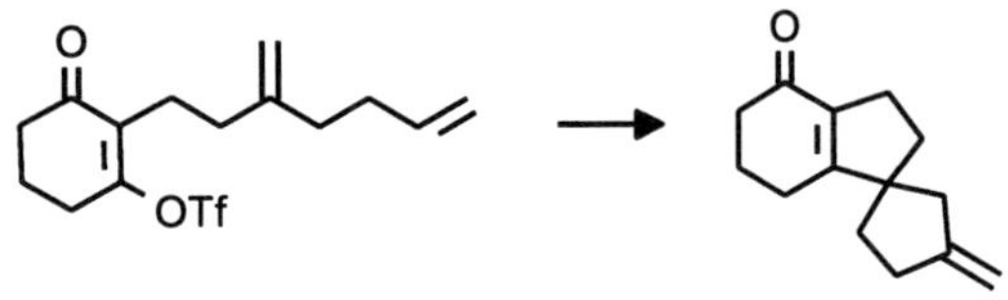	Overman L.E., *J. Org. Chem.*, 1989; 5846
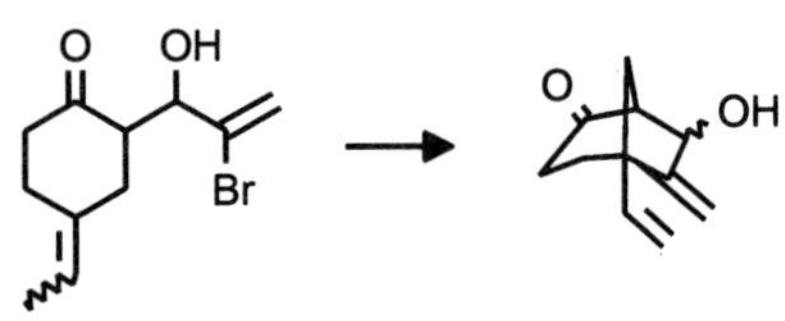	Fukumoto K., *Tetrahedron Lett.*, 1994; 6495
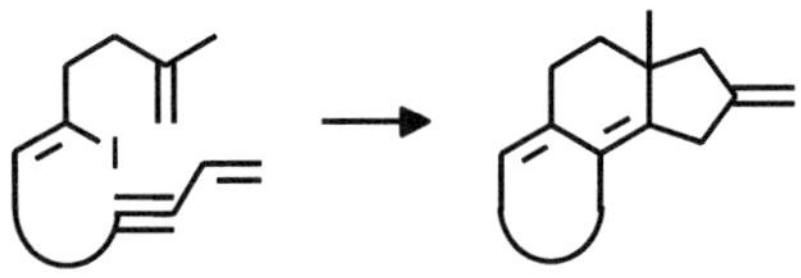	Negishi E.I., *J. Am. Chem. Soc.*, 1990; 8590
	De Meijere A., *J. Chem. Soc., Chem. Commun.*, 1992; 390
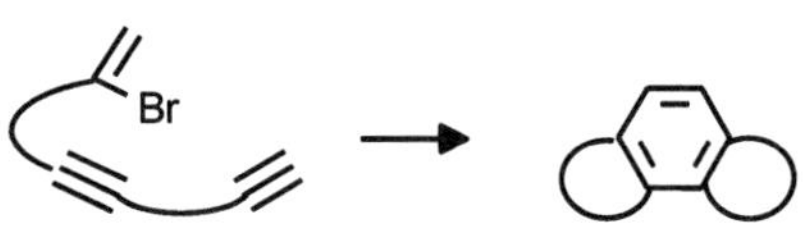	De Meijere A., *Synlett.*, 1991; 777
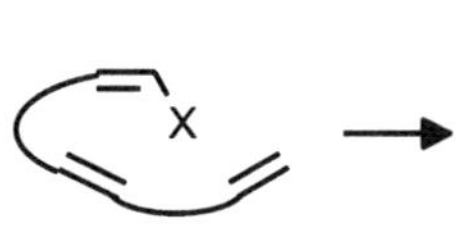 Polycyclisation	Overman L.E., *J. Org. Chem.*, 1993; 5304
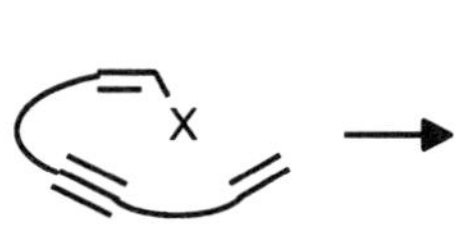 Polycyclisation	De Meijere A., *J. Org. Chem.*, 1991; 6487

A ---> B	References
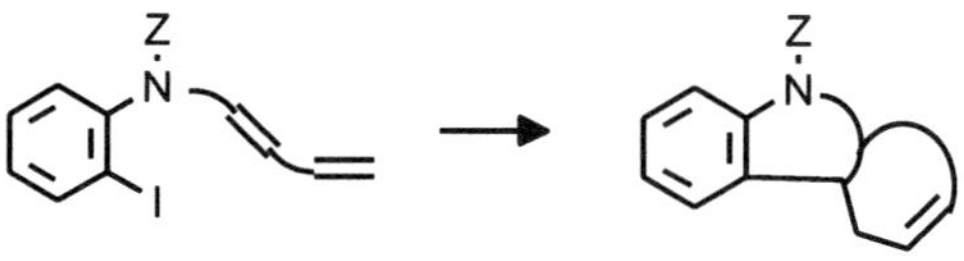	Grigg R., *Tetrahedron*, 1994; 359
	Grigg R., *Tetrahedron Lett.*, 1995; 8137
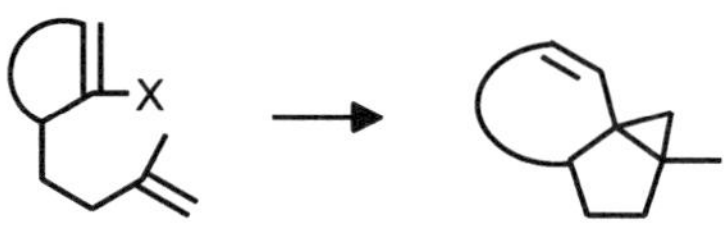	Grigg R., *Tetrahedron Lett.*, 1991; 3855 Grigg R., *Tetrahedron Lett.*, 1994; 2753
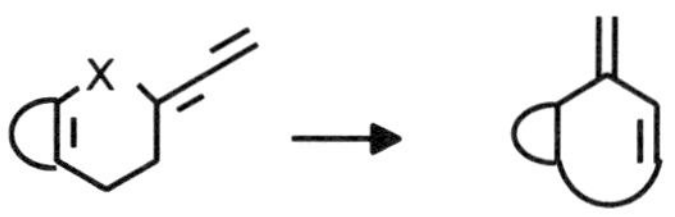	Negishi E.I., *J. Org. Chem.*, 1994; 4730
	Negishi E.I., *J. Am. Chem. Soc.*, 1992; 10091
	Mori M., *Tetrahedron*, 1985; 5465 (Pd° elimination) Mori M., *J. Chem. Soc., Chem. Commun.*, 1988; 12 (Pd° elimination)
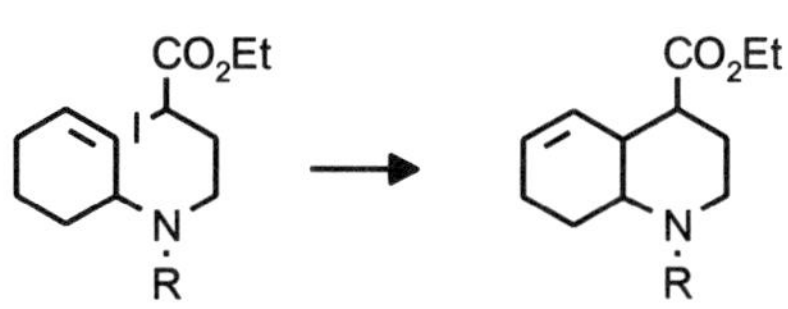	Mori M., *Tetrahedron Lett.*, 1985; 1519 (Pd° elimination)
	Negishi E.I., *Tetrahedron Lett.*, 1988; 2915 Fuchs P.L., *Tetrahedron Lett.*, 1993; 5205 Isobe M., *Tetrahedron*, 1994; 11143
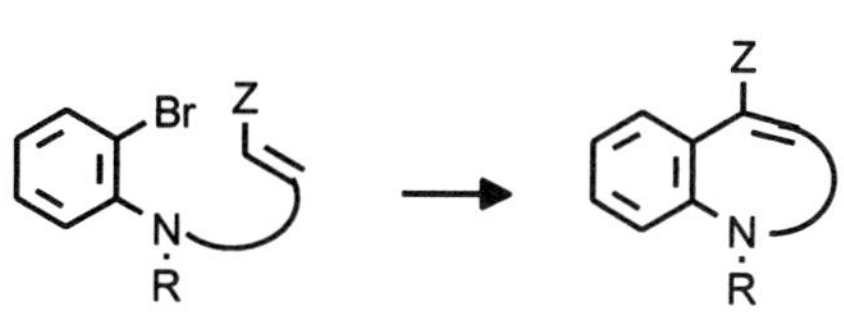	Michael J.P., *Tetrahedron Lett.*, 1993; 8365

A ---> B	References
	Gronowitz S., *Tetrahedron*, 1995; 10323
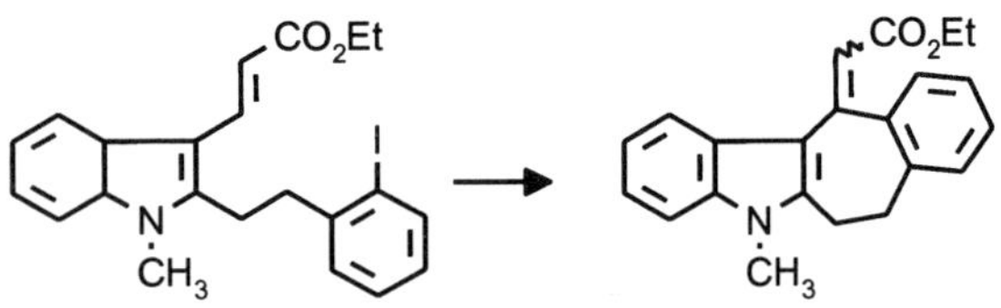	Merour J.Y., *Tetrahedron Lett.*, 1995; 8587
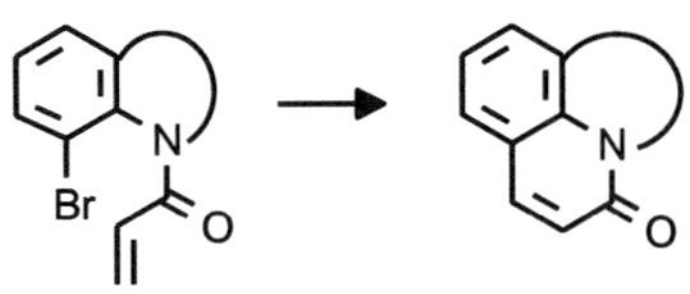	Dankwardt J.W., *J. Org. Chem.*, 1995; 2312
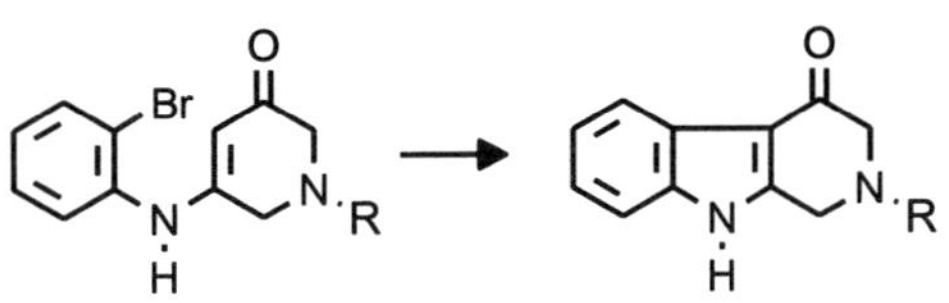	Chen L.C., *Synthesis*, 1995; 385
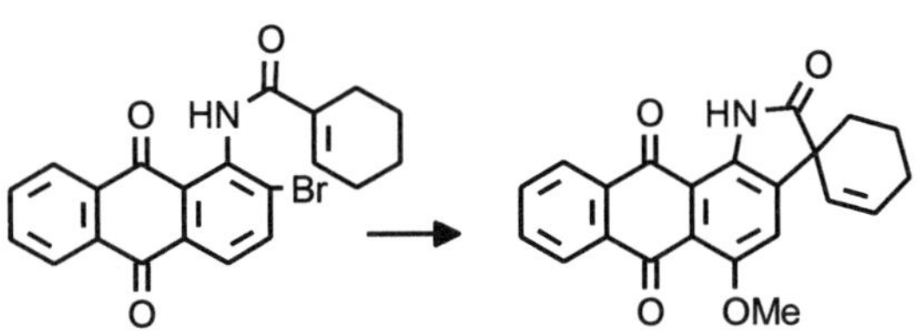	Isobe M., *Tetrahedron*, 1994; 11143
	Larock R.C., *Tetrahedron Lett.*, 1988; 4687
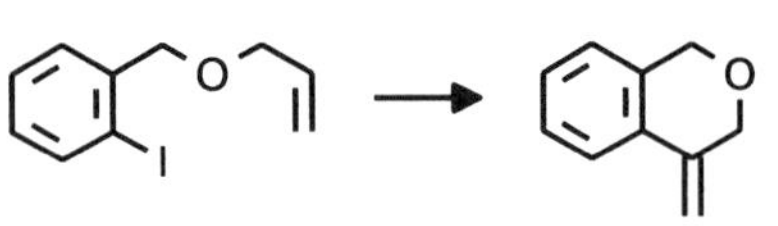	Overman L. E., *Heterocycles*, 1994; 39, 497
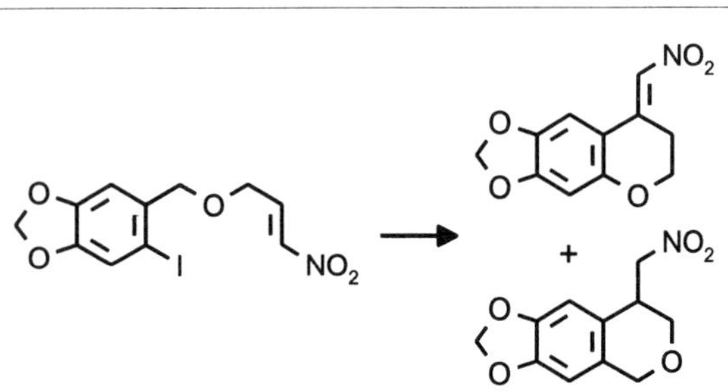	Denmark S.E., *J. Org. Chem.*, 1995; 1013

A ---> B	References
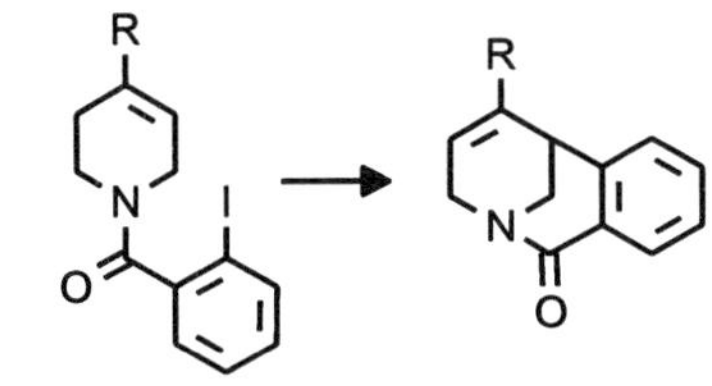	Grigg R., *J. Chem. Soc., Chem. Commun.*, 1986; 1697
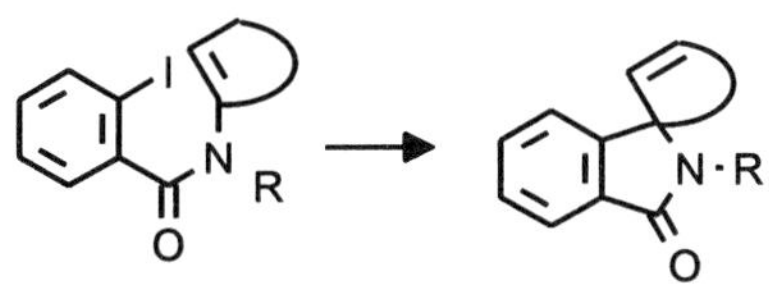	Grigg R., *J. Chem. Soc., Chem. Commun.*, 1986; 1697
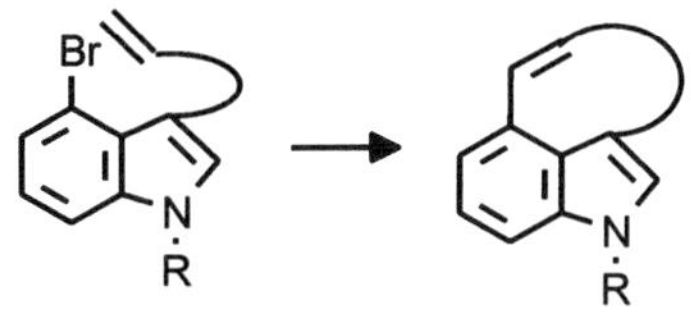	Hegedus L.S., *J. Org. Chem.*, 1989; 4141
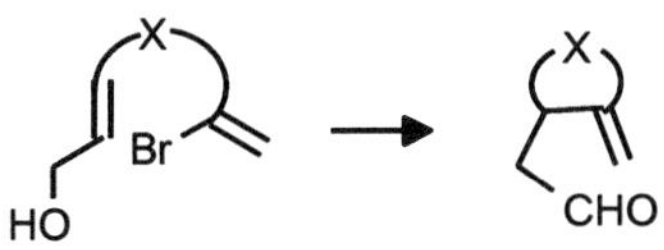	Gaudin J.M., *Tetrahedron Lett.*, 1991; 6113
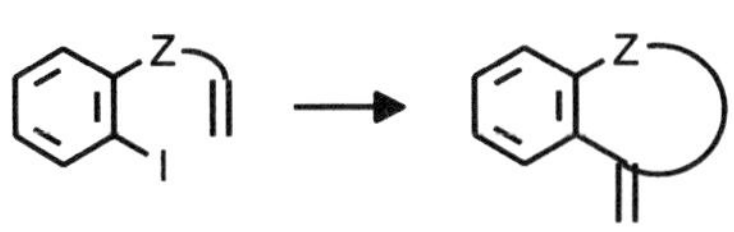	Larock R.C., *Tetrahedron Lett.*, 1987; 5291 Negishi E.I., *Tetrahedron Lett.*, 1988; 3903
	Stocks M.J., *Tetrahedron Lett.*, 1995; 6555 Goff D.A., *J. Org. Chem.*, 1995; 5748
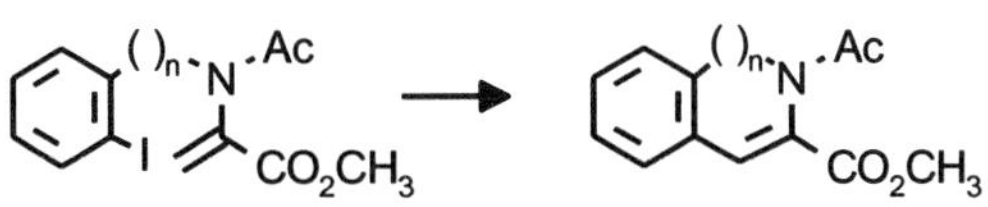	Gibson S.E., *J. Chem. Soc., Chem. Commun.*, 1995; 1743
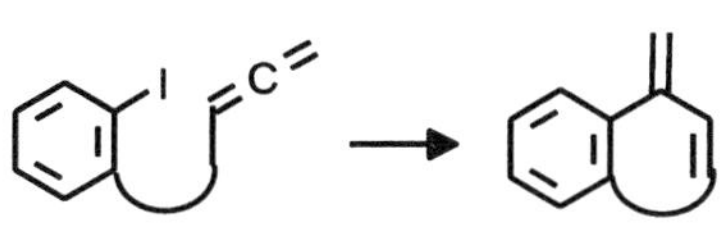	Negishi E-I., *J. Am. Chem. Soc.*, 1995; 6345
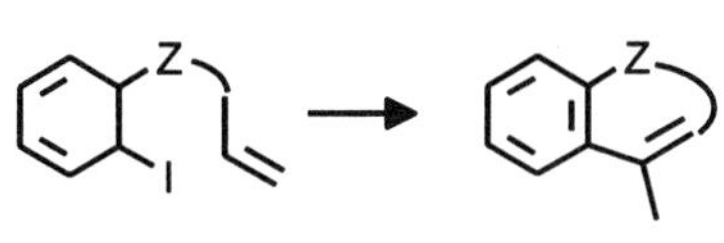	Catellani M., *Tetrahedron Lett.*, 1994; 5919
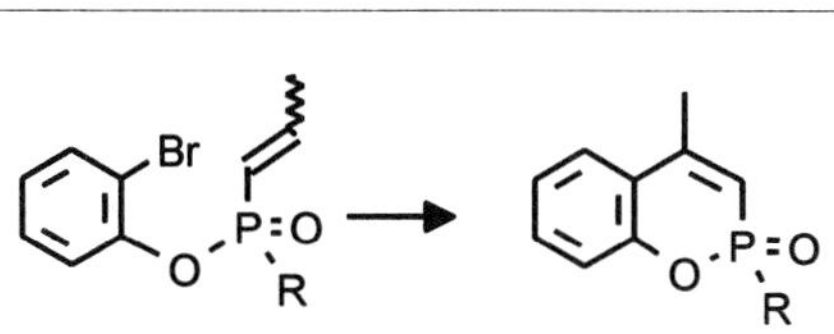	Xu Y., *J. Chem. Soc, Perkin Trans. 1*, 1994; 1665

A ---> B	References
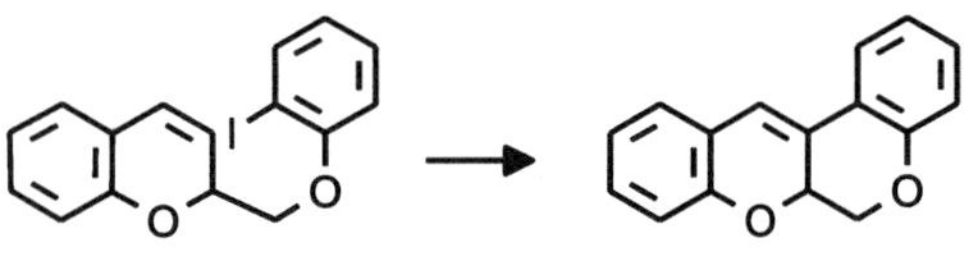	Whiting D.A., *J. Chem. Soc., Chem. Commun.*, 1987; 510
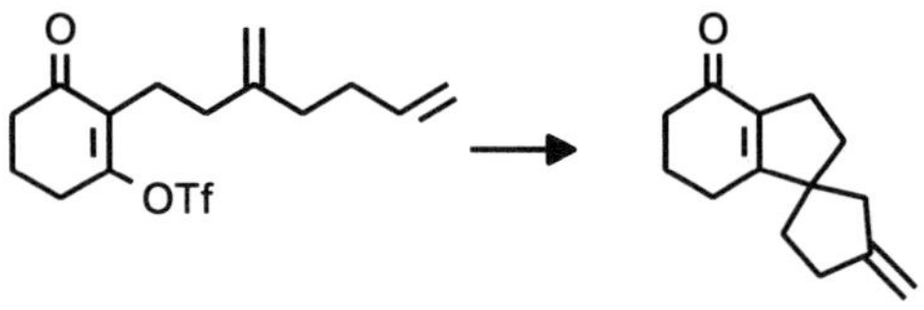	Grigg R., *Tetrahedron Lett.*, 1991; 687 Kozikowski A.P., *Tetrahedron Lett.*, 1991; 3317
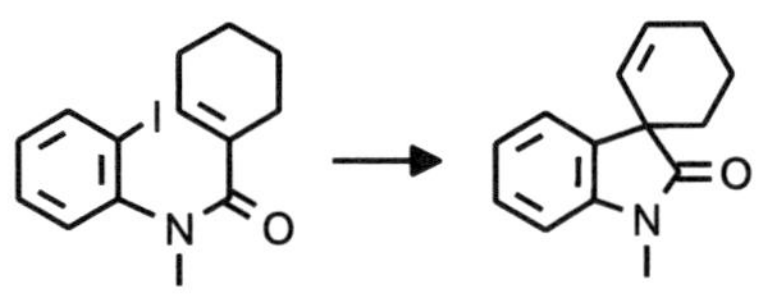	Overman L.E., *J. Org. Chem.*, 1992; 4571
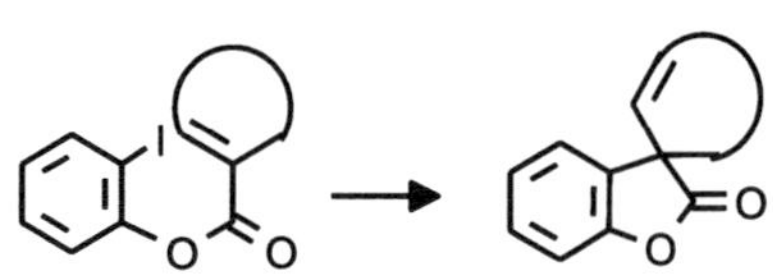	Marinelli F., *Synthesis*, 1995; 831
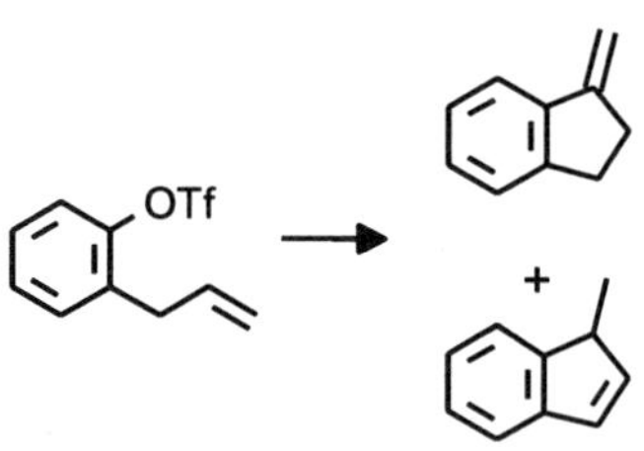	Blechert S., *Tetrahedron*, 1995; 1167
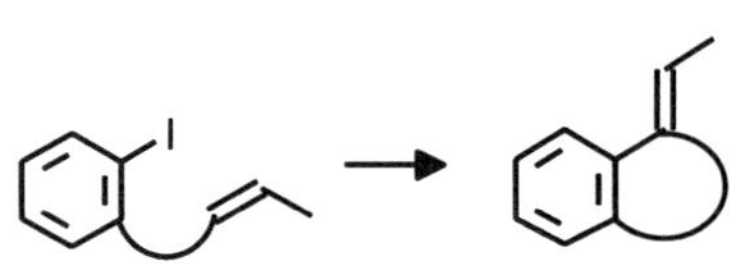	Danishefsky S.J., *J. Org. Chem.*, 1994; 355 Tietze L.F., *Synthesis*, 1993; 876
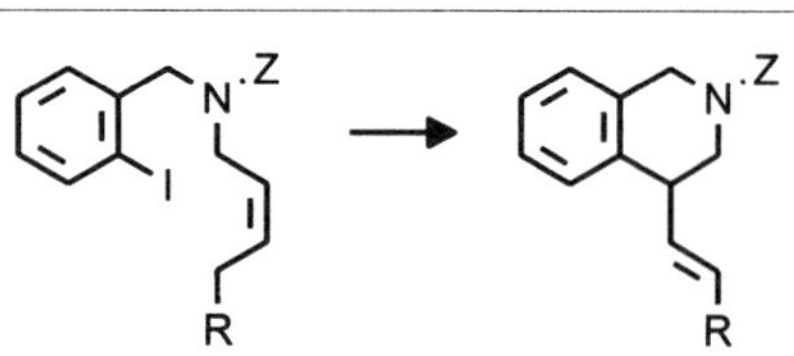	Tietze L.E., *Angew. Chem., Int. Ed. Engl.*, 1994; 1089
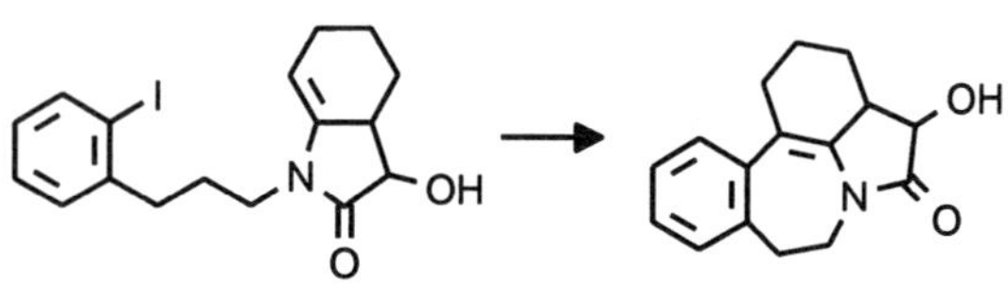	Rigby J.H., *J. Am. Chem. Soc.*, 1995; 7834

A ---> B	References
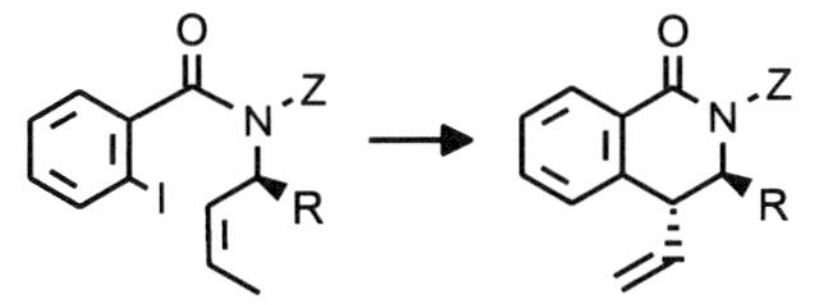	Tietze L.F., *Liebigs Ann.*, 1995; 1153
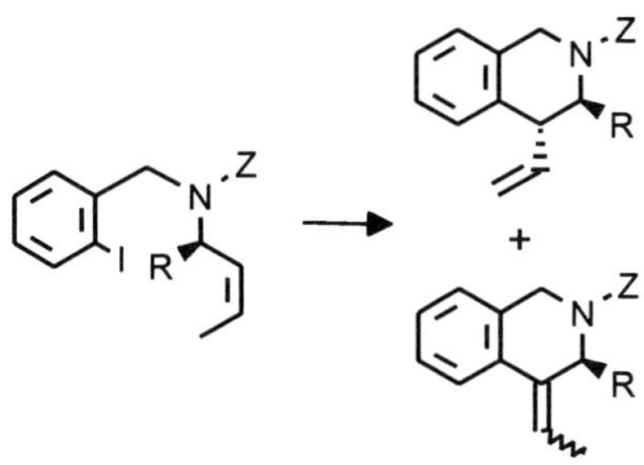	Tietze L.F., *Synthesis*, 1994; 1331
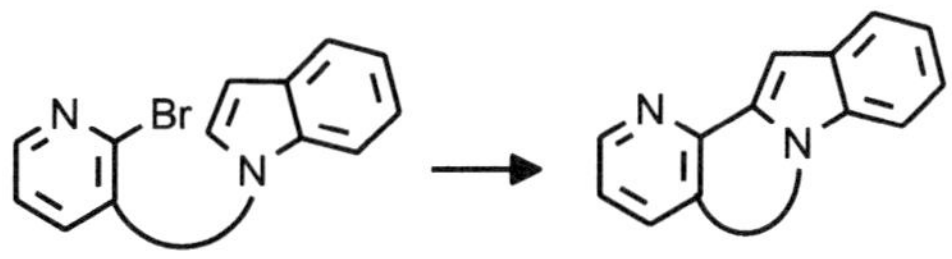	Thal C., *Tetrahedron Lett.*, 1993; 5449 Thal C., *Tetrahedron*, 1995; 1941
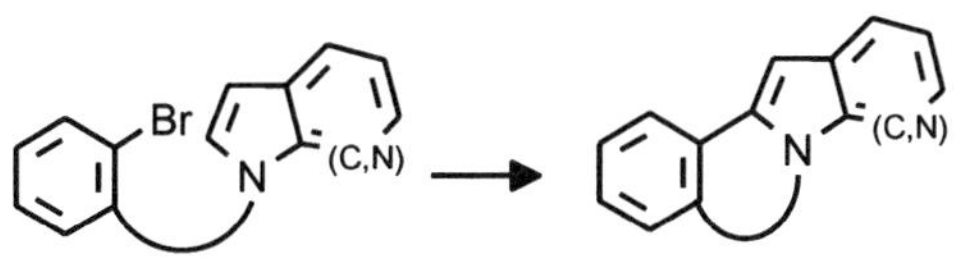	Merour J.Y., *Heterocycles*, 1995; 41, 1987
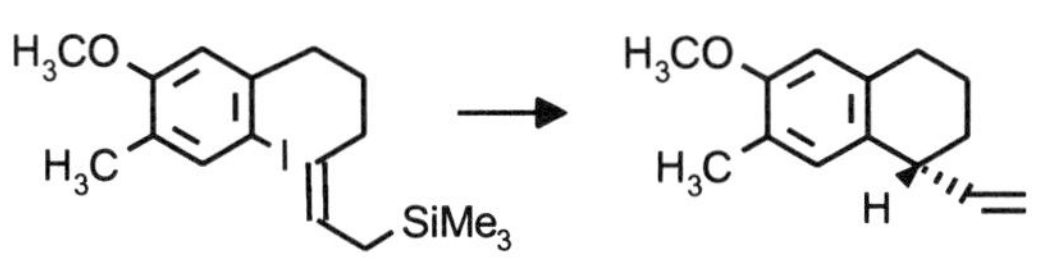	Tietze L.F., *Synlett.*, 1995; 597 (-IPdSi , asym. catalysis)
	Negishi E.I., *J. Org. Chem.*, 1989; 2507

RXN7 Intramolecular Coupling of Di(Vinyl Halides)

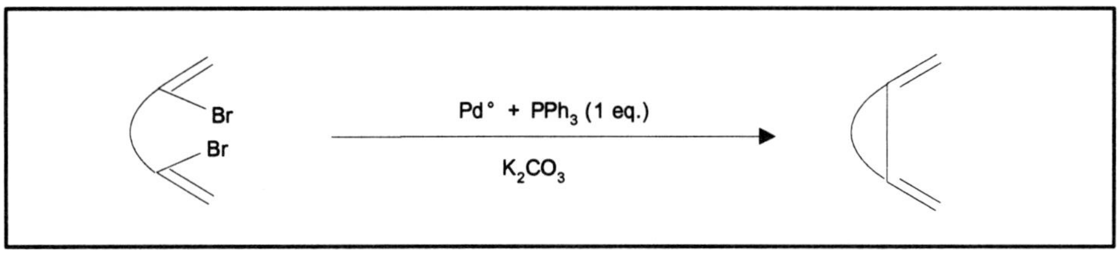

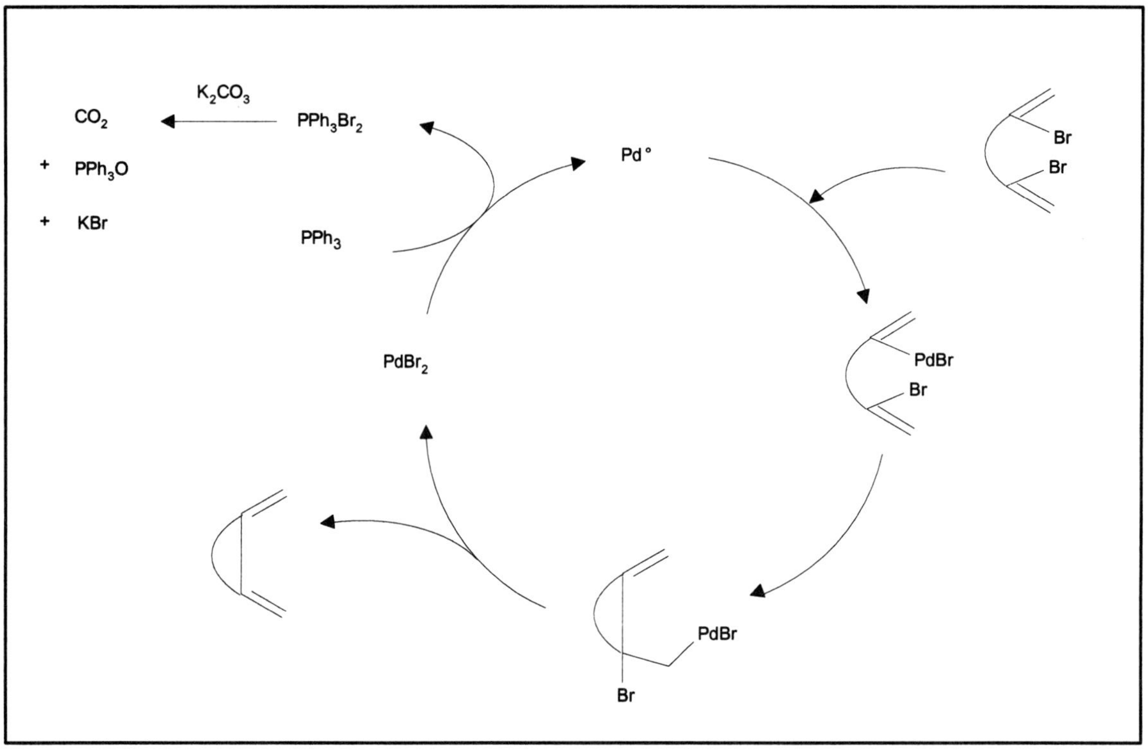

A ---> B	References
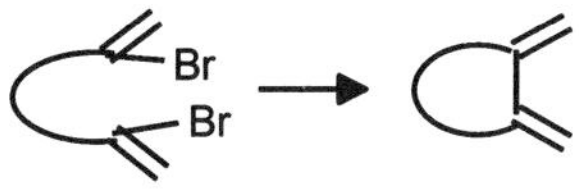	Grigg R., *Tetrahedron*, 1988; 2049

RXN8 Tandem HECK-Anion Capture Process of Alkenes, Alkynes, Allenes and Dienes

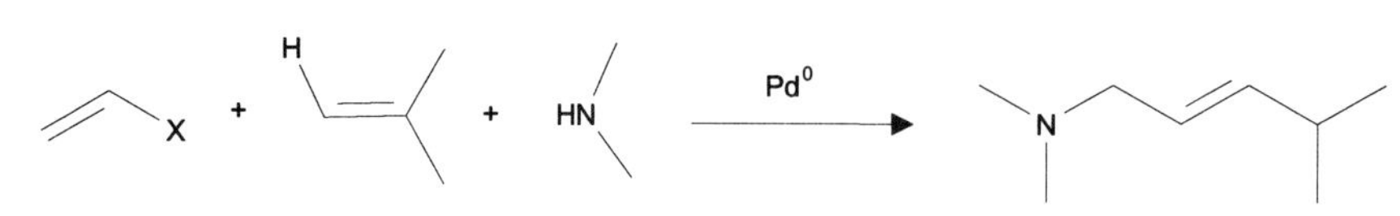

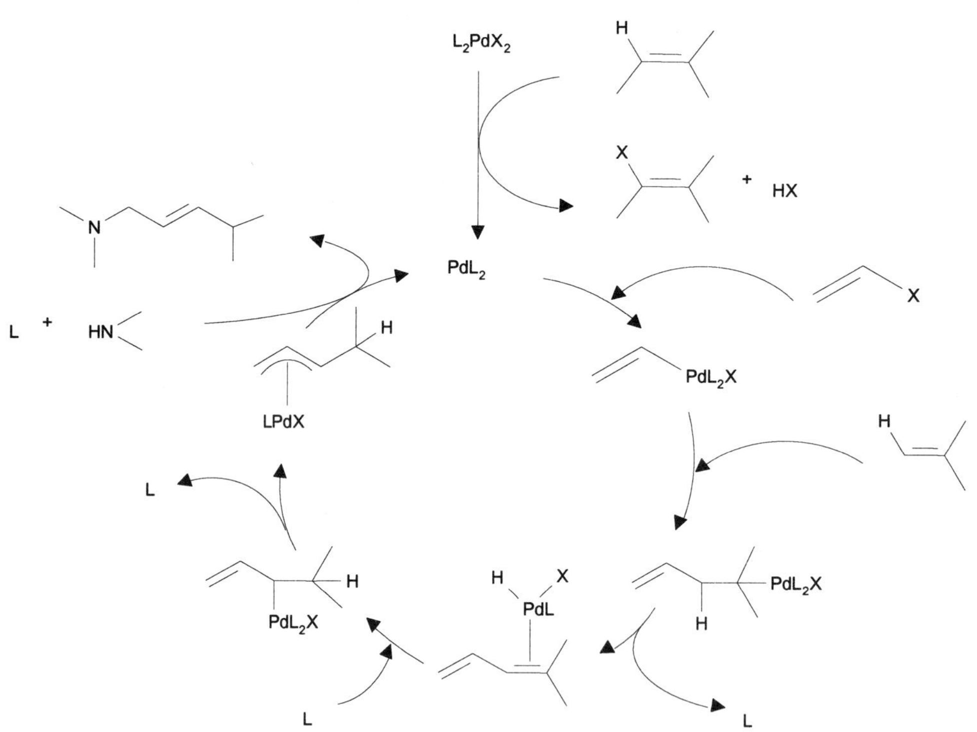

RXN8 Tandem HECK-Anion Capture Process of Alkenes, Alkynes, Allenes and Dienes

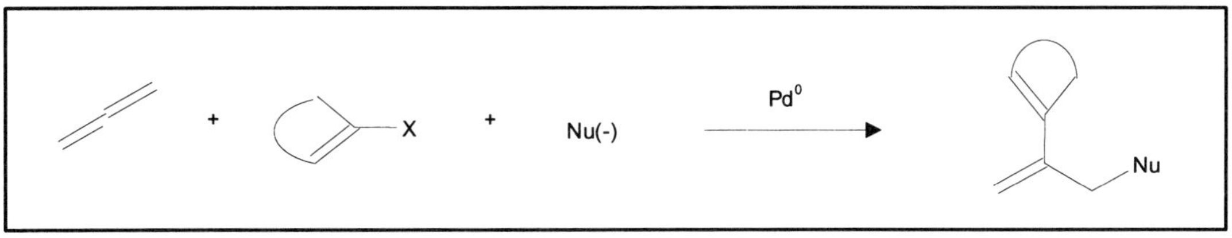

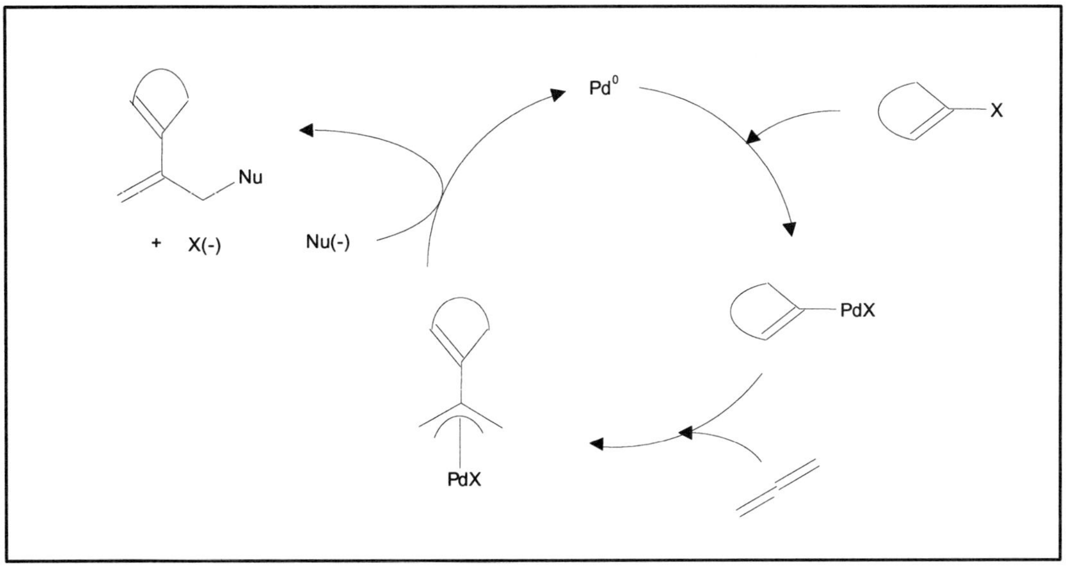

RX	C=C	Nucleophile	References
RX	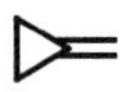	Z-CH-Z'	Gore J., *Tetrahedron*, 1988; 5809
RX	C=C=C	NR_2	Tsuji J., *Chem. Lett.*, 1984; 233
RX	C=C=C	Z-CH-Z'	Gore J., *Tetrahedron Lett.*, 1989; 3963 Cazes B., *Pure & Appl. Chem.*, 1990; 1867
C=C-Br	C=C	NR_2	Heck R.F., *J. Org. Chem.*, 1978; 3898 Heck R.F., *J. Org. Chem.*, 1978; 3898
C=C-Br		NR_2	Larock R.C., *Tetrahedron*, 1995; 6635
C=C-X	C=C-C=C	NR_2	Heck R.F., *J. Org. Chem.*, 1979; 918
C=C-OSO_2CF_3	C=C=C	Z-CH-Z'	Gore J., *Tetrahedron Lett.*, 1988; 4089 Gore J., *Tetrahedron Lett.*, 1995; 5015
PhBr	C=C-C=C	NR_2	Heck R.F., *Org. Reactions*, 1982; 27, 345 Heck R.F., *J. Org. Chem.*, 1980; 3584
RCOCl	C=C-C=C	R'_3Si	Tsuji Y., *J. Am. Chem. Soc.*, 1993; 10414 Tsuji Y., *J. Am. Chem. Soc.*, 1995; 9814
ArI	C=C-C=C	Z-CH-Z'	Takahashi S., *J. Chem. Soc., Chem. Commun.*, 1987; 785 Takahashi S., *J. Chem. Soc., Perkin Trans. 1*, 1990; 647
ArI	C=C-C=C	R_3Si	Tsuji Y., *J. Am. Chem. Soc.*, 1995; 9814
ArI		Z-CH-Z'	Larock R.C., *J. Org. Chem.*, 1991; 4589
ArI	C=C-$(C)_n$-C=C	NR_2	Larock R.C., *J. Org. Chem.*, 1994; 8107
I-Ar-NuH	C=C-C=C	intramolecular	Larock R.C., *J. Org. Chem.*, 1990; 3447 Larock R.C., *Synlett.*, 1995; 465
I-Ar-$(CH_2)_n$-CHZZ'	C=C-C=C	intramolecular	Larock R.C., *J. Am. Chem. Soc.*, 1990; 5882
I-Ar-NuH		intramolecular	Larock R.C., *Synlett.*, 1990; 529
I-Ar-NuH	C=C=C	intramolecular	Larock R.C., *J. Org. Chem.*, 1991; 2615 Larock R.C., *J. Org. Chem.*, 1995; 482 (asym. catalysis)
I-Ar-NuH	C=C-C-C=C	intramolecular	Larock R.C., *J. Org. Chem.*, 1993; 4509
I-Ar-NuH	C≡C	intramolecular	Larock R.C., *J. Org. Chem.*, 1995; 3270 Blart E., *Thesis, Université Paris VI*, 1993 Genôt J.P., *J. Org. Chem.*, 1995; 6829

RX	C=C	Nucleophile	References
I-Ar-OH	C=C-C=C	intramolecular	Overman L.E., *Tetrahedron Lett.*, 1994; 3453
I-Ar-NHR	C≡C	intramolecular	Larock R.C., *J. Am. Chem. Soc.*, 1991; 6689 Gronowitz S., *Tetrahedron Lett.*, 1993; 2823 Gronowitz S., *Tetrahedron Lett.*, 1993; 6471 Chen C-Y., *Tetrahedron Lett.*, 1994; 6981
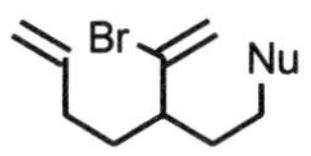	intram.	intramolecular	Weinreb S.M., *Tetrahedron Lett.*, 1994; 4287

RXN9 Tandem HECK-Anion Capture Process of Norbornene and Related Compounds

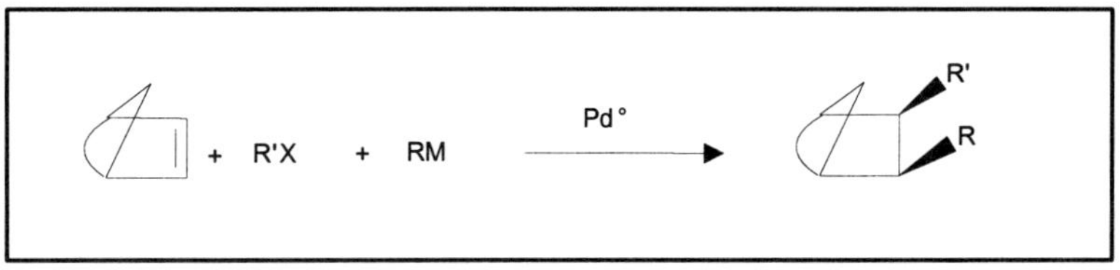

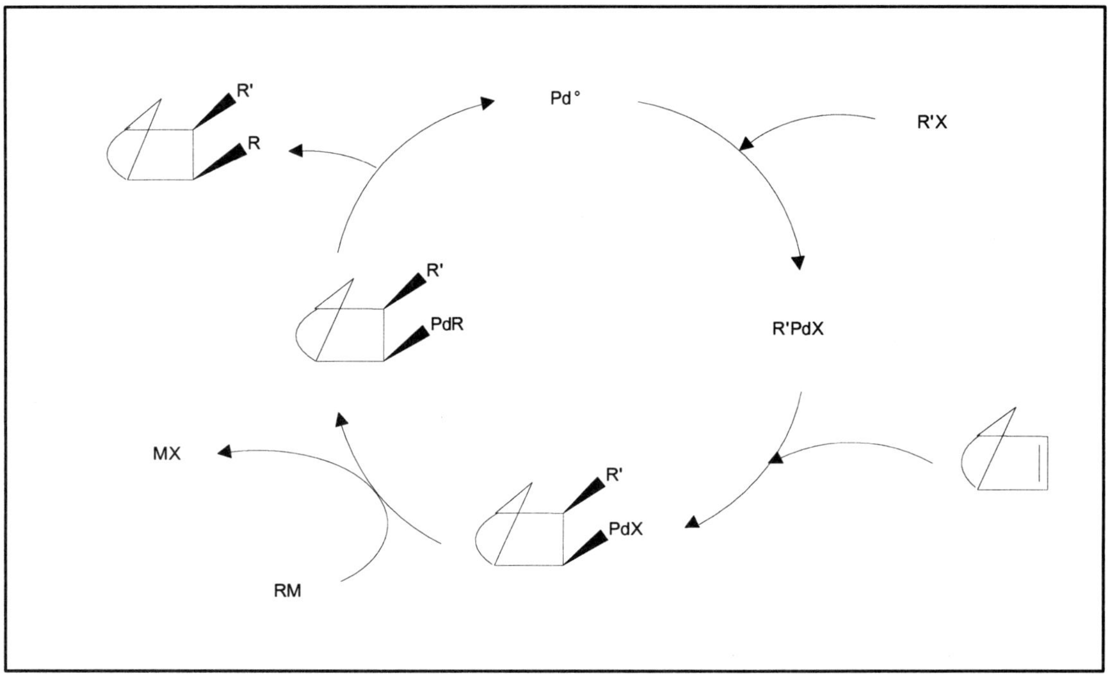

R'-X	References
ArBr	Catellani M., *Tetrahedron*, 1989; 961 Kosugi M., *Tetrahedron*, 1989; 5263 Catellani M., *Pure & Appl. Chem.*, 1990; 623 (review)
ArI	Larock R.C., *J. Chem. Soc., Chem. Commun.*, 1989; 1368
ArX	Kosugi M., *Chem. Lett.*, 1987; 193
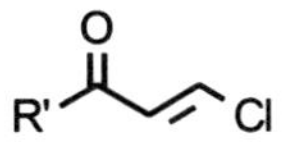	Kosugi M., *Tetrahedron*, 1995; 695
R'-C=C-Br	Catellani M., *Tetrahedron Lett.*, 1982; 4517
Ph-C=C-Br	Catellani M., *J. Organomet. Chem.*, 1984; 275, 129
R'-C=C-I	Torii S., *Tetrahedron Lett.*, 1992; 3499 Torii S., *Tetrahedron Lett.*, 1992; 3503
R'-C=C-X	Blart E., *Thesis, Université Paris VI*, 1993 Kosugi M., *Tetrahedron*, 1995; 695
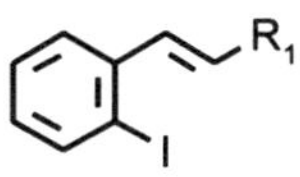	Grigg R., *Tetrahedron Lett.*, 1993; 153 (cyclisation)

R-M	References
R−C≡C−$SnBu_3$	Kosugi M., *Chem. Lett.*, 1987; 193
C=C-$SnBu_3$	Kosugi M., *Chem. Lett.*, 1987; 193 Kosugi M., *Tetrahedron*, 1995; 695
C=C-C-$SnBu_3$	Kosugi M., *Chem. Lett.*, 1987; 193
C=C(OEt)-$SnBu_3$	Kosugi M., *Tetrahedron*, 1989; 961
$NaB(Ph)_4$	Catellani M., *Tetrahedron*, 1989; 5263
KCN	Torii S., *Tetrahedron Lett.*, 1990; 5319
R−C≡CH	Catellani M., *Tetrahedron Lett.*, 1982; 4517 Catellani M., *J. Organomet. Chem.*, 1984; 275, 129 Blart E., *Thesis, Université Paris VI*, 1993
HCO_2NH_4	Brunner H., *Synthesis*, 1991; 1121 (asymmetric catalysis) Larock R.C., *J. Chem. Soc., Chem. Commun.*, 1989; 1368 Regan A.C., *Tetrahedron Lett.*, 1993; 7493
HCO_2H / NEt_3	Achiwa K., *Synlett.*, 1994; 291 Achiwa K., *Chem. Pharm. Bull.*, 1995; 927 (asym. catalysis)

R-M	References
H_2O (H donor)	Cheng C.H., *J. Chem. Soc., Chem. Commun.*, 1990; 1774 Cheng C.H., *Organometallics*, 1993; 3945
CO / AcOK	Catellani M., *Tetrahedron Lett.*, 1983; 813
CO (cyclisation)	Grigg R., *Tetrahedron Lett.*, 1994; 3197

RXN10 Tandem Arylsulfonation-Cyclization Process

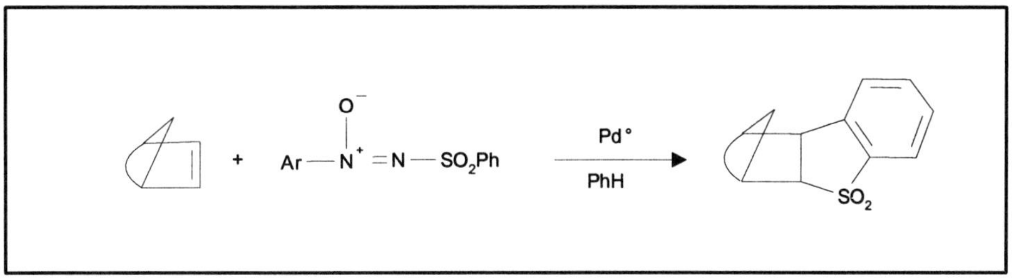

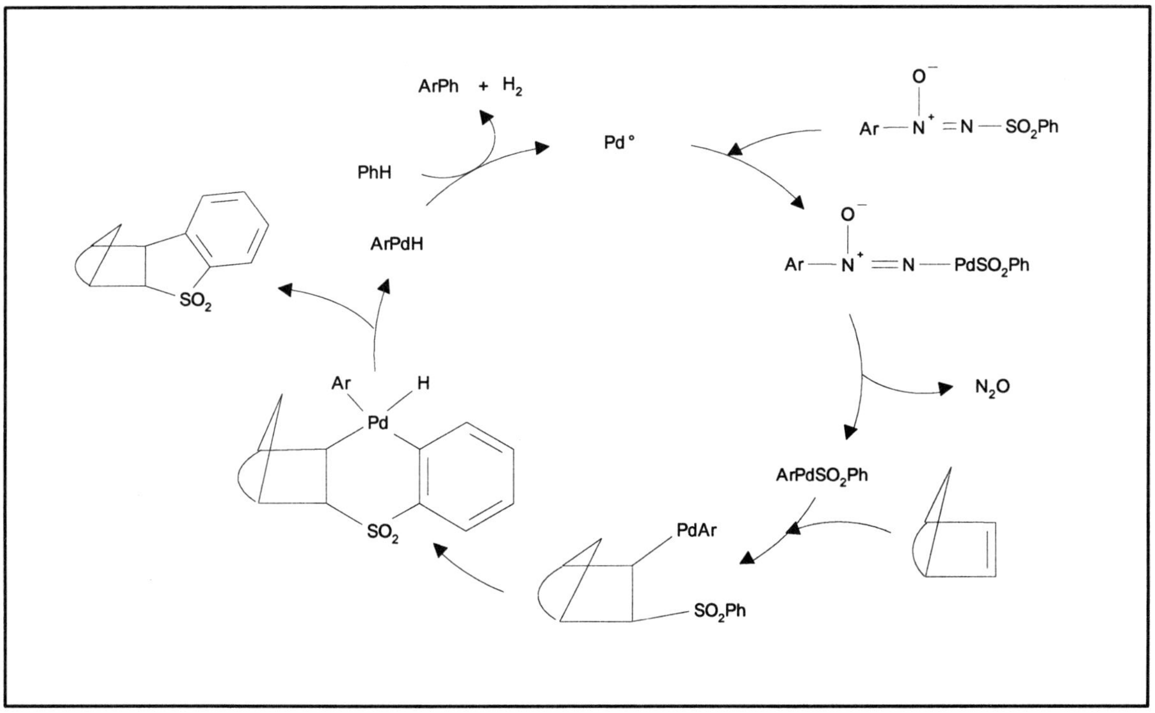

RXN10

A	B	References
	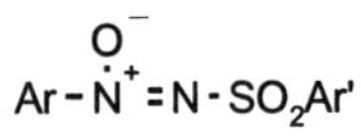	Kamigata N., *J. Organomet. Chem.*, 1991; 401, C26

RXN11 Tandem Cyclization-Anion Capture (-Carbonylation) Process of Alkenes, Allenes and Alkynes

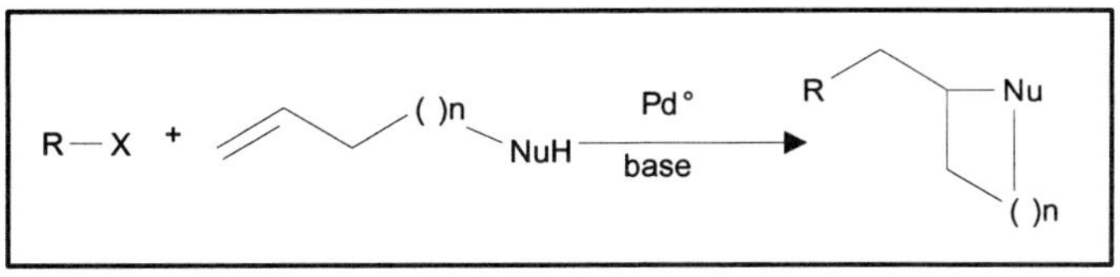

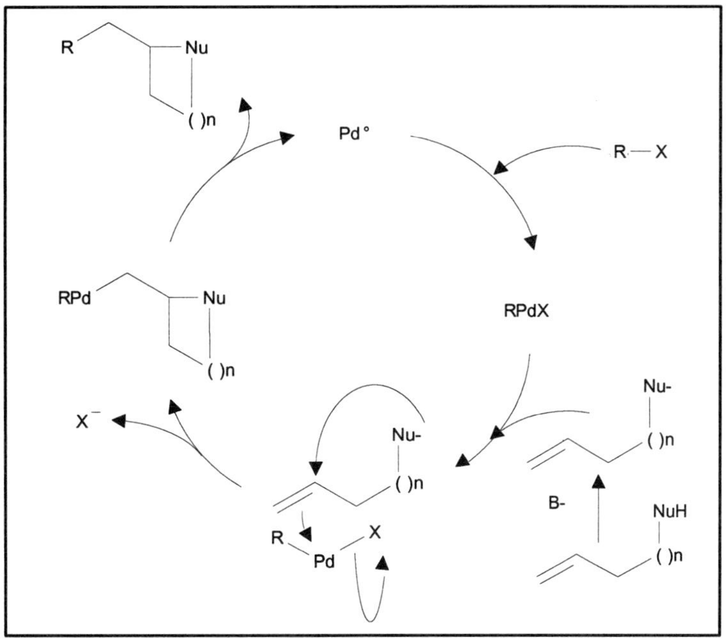

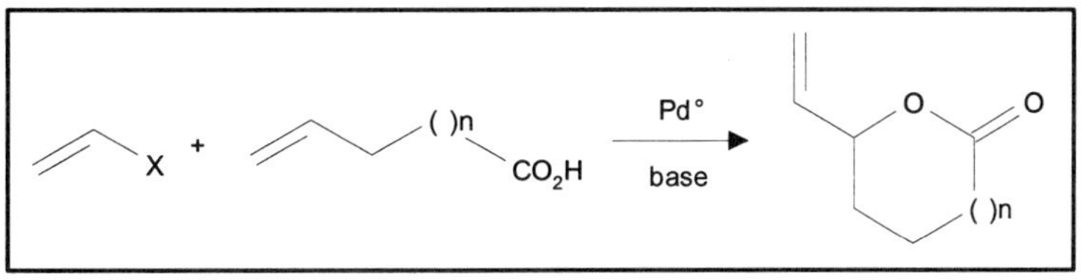

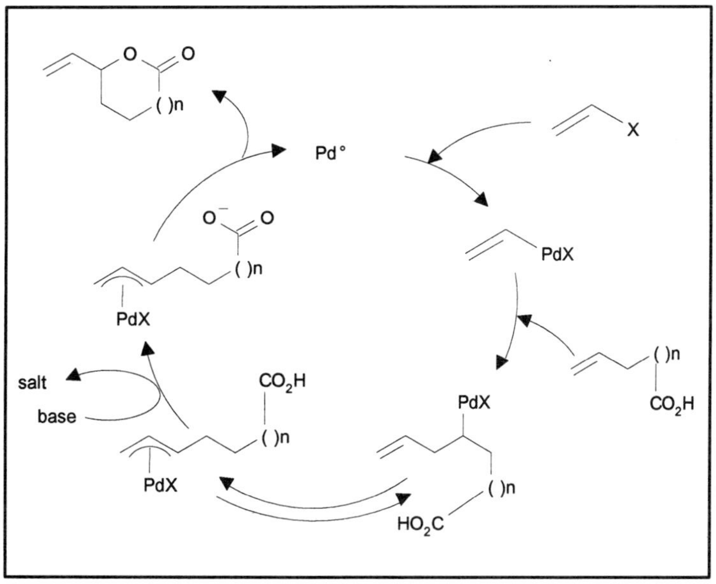

RXN11 Tandem Cyclization-Anion Capture (-Carbonylation) Process of Alkenes, Allenes and Alkynes

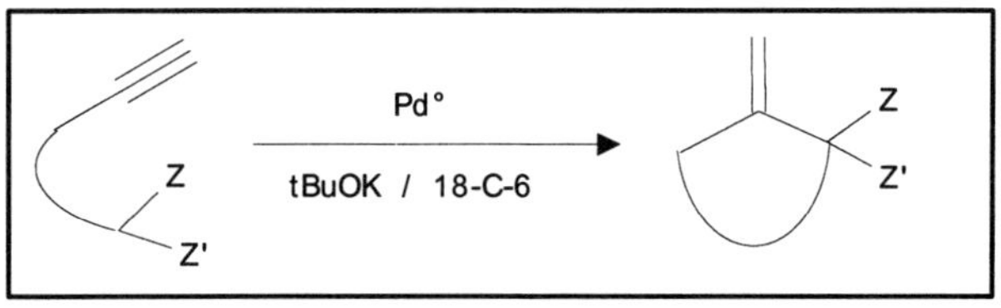

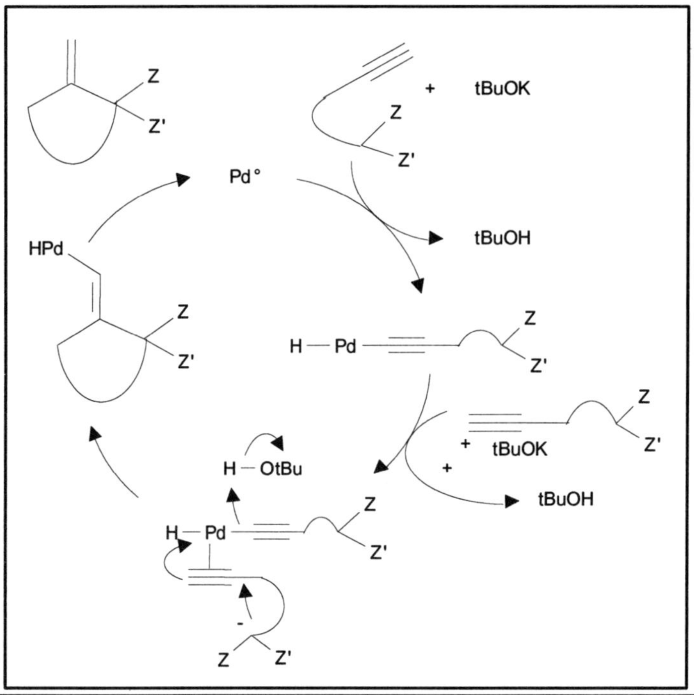

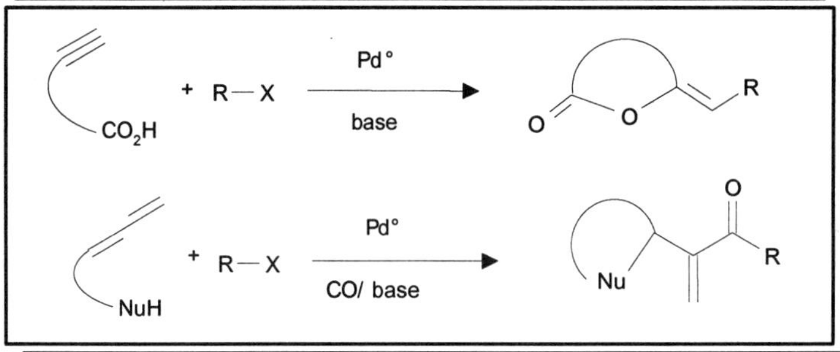

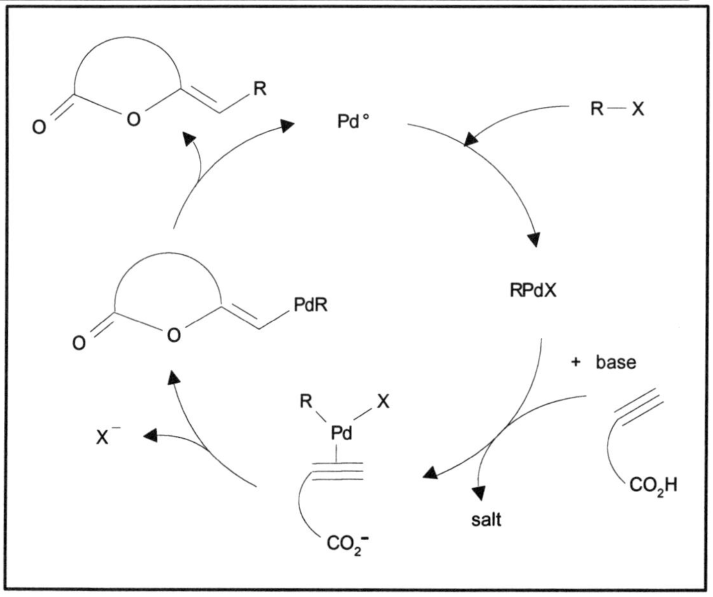

RX	C=C or C≡C	References
RX	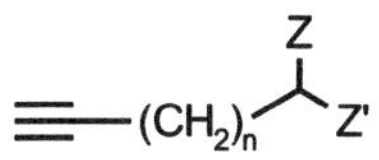	Gore J., *Tetrahedron Lett.*, 1990; 5147 Gore J., *Tetrahedron*, 1991; 6293 Balme G., *Tetrahedron Lett.*, 1995; 8019
RX	C=C=C-C-C(Z)-Z'	Gore J., *Tetrahedron*, 1987; 3453
RX	$C\equiv C-(CH_2)_n-CO_2H$	Cacchi S., *J. Org. Chem.*, 1992; 976 Cacchi S., *Gazz. Chim. Ital.*, 1992; 11 Gore J., *Tetrahedron Lett.*, 1992; 2811
RX	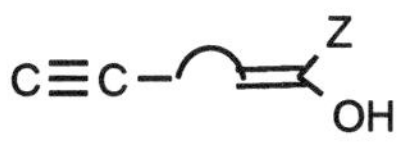	Cacchi S., *Tetrahedron Lett.*, 1993; 2813
RX	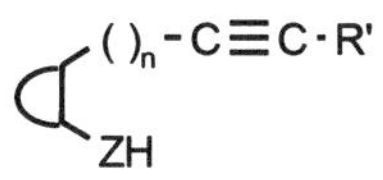	Luo F.T., *J. Org. Chem.*, 1992; 2213 Luo F.T., *Tetrahedron Lett.*, 1992; 6835
RX	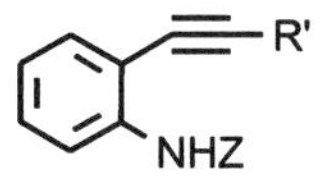	Cacchi S., *Tetrahedron Lett.*, 1992; 3915 Cacchi S., *Tetrahedron*, 1994; 437
RX	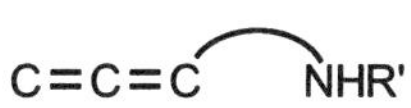	Gallagher T., *Synlett.*, 1993; 85
RX	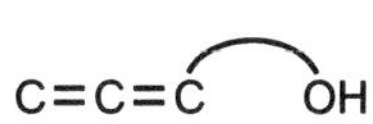	Walkup R.D., *Synlett.*, 1993; 88 Walkup R.D., *Tetrahedron Lett.*, 1995; 3805 (carbonylation)
RX	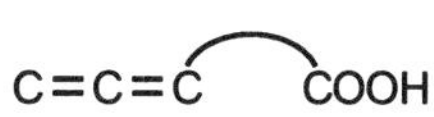	Walkup R.D., *Synlett.*, 1993; 88
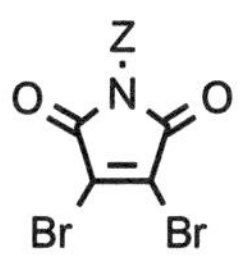	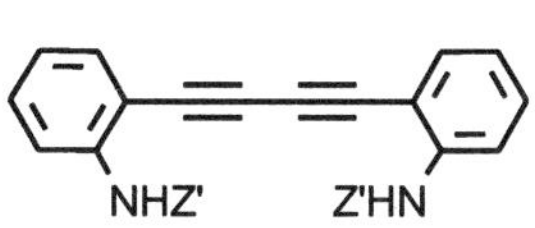	Saulnier M.G., *Tetrahedron Lett.*, 1995; 7841
ArX	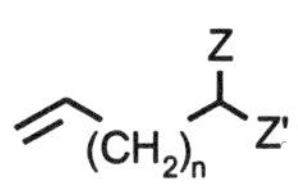	Gore J., *Tetrahedron*, 1990; 7763 Gore J., *Tetrahedron Lett.*, 1991; 1641
ArX	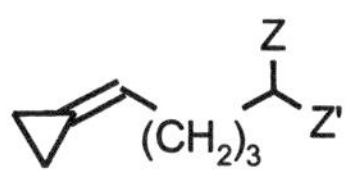	Gore J., *Tetrahedron*, 1988; 5821
C=C-X	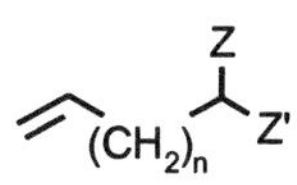	Gore J., *Tetrahedron*, 1992; 3891 Gore J., *Tetrahedron Lett.*, 1989; 69
C=C-X	$C=C-(CH_2)_n-CO_2H$	Larock R.C., *Tetrahedron Lett.*, 1988; 6399
C=C-X	$C=C-(CH_2)_n-NHZ$	Larock R.C., *J. Org. Chem.*, 1994; 4172

RX	C=C or C≡C	References
C=C-X	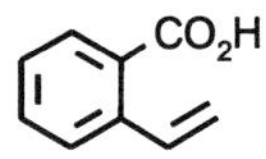	Larock R.C., *Synlett.*, 1994; 748
C=C-C-Br	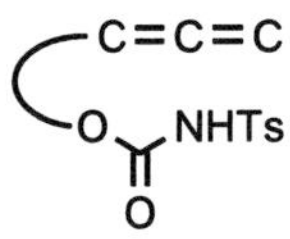	Tamaru Y., *J. Org. Chem.*, 1992; 6377 Tamaru Y., *J. Org. Chem.*, 1995; 3764
C=C-C-OAc	$C{\equiv}C-(CH_2)_n-CO_2H$	Saegusa T., *J. Org. Chem.*, 1988; 2650 Tsuji J., *Synlett.*, 1992; 671
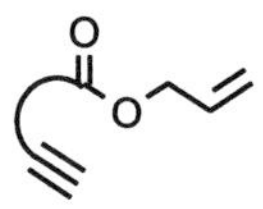	intramolecular	Saegusa T., *J. Org. Chem.*, 1988; 2650
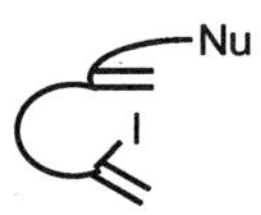	intramolecular	Balme G., *Tetrahedron*, 1994; 403
C≡C−X	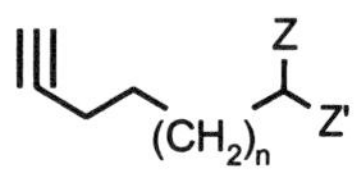	Gore J., *Tetrahedron Lett.*, 1991; 6541
HX	$C{\equiv}C-C{=}C-CO_2H$	Lu X., *Tetrahedron Lett.*, 1993; 5963

RXN12 Tandem Cyclization-Anion-Capture Process of Ene-Vinyl, Ene-Aryl and Ene-Alkyl Halides

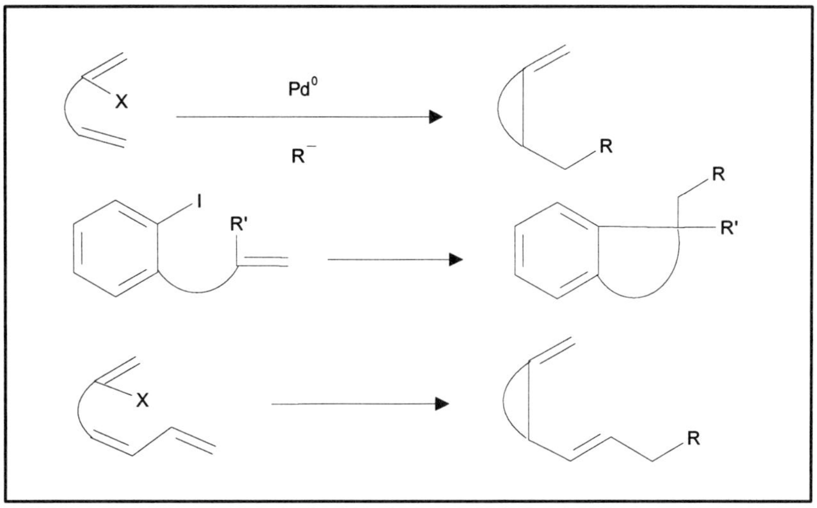

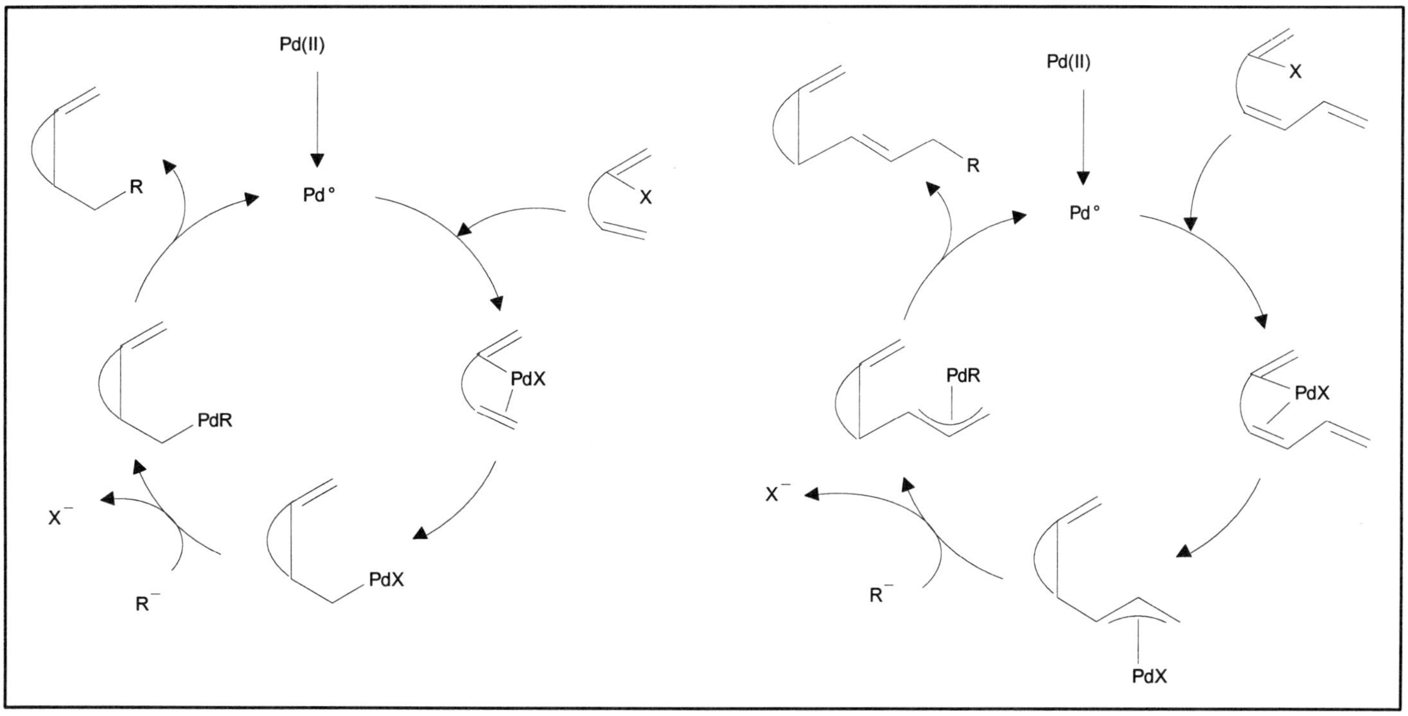

RXN12 Tandem Cyclization-Anion-Capture Process of Ene-Vinyl, Ene-Aryl and Ene-Alkyl Halides

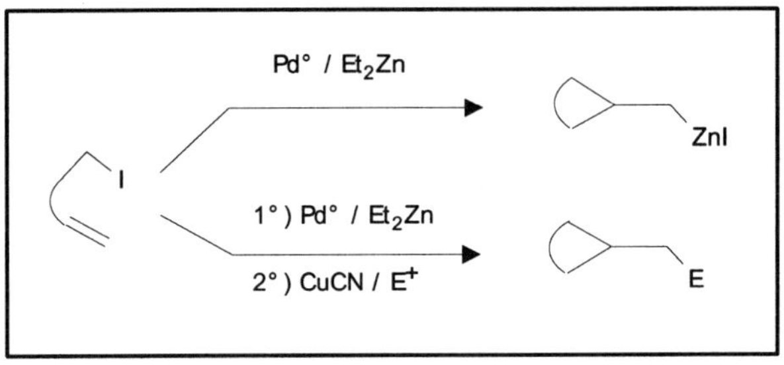

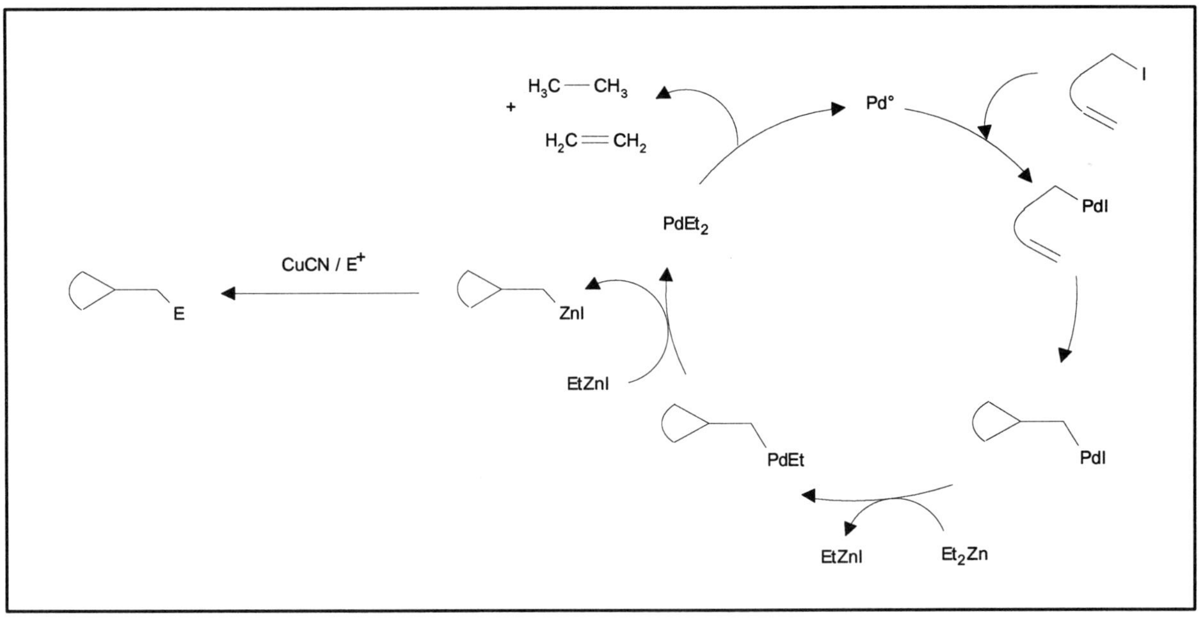

 J.-L. Malleron et al

A ---> B	References
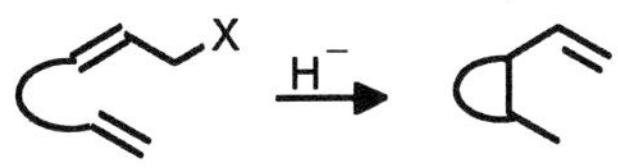	Grigg R., *Tetrahedron Lett.*, 1991; 3855 Grigg R., *Tetrahedron*, 1992; 7297
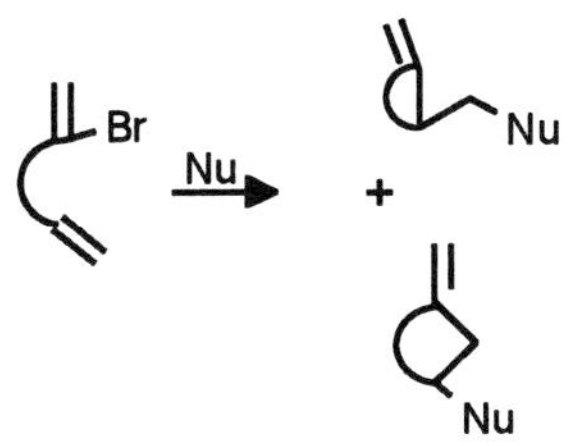	Weinreb S.M., *Tetrahedron*, 1995; 9301
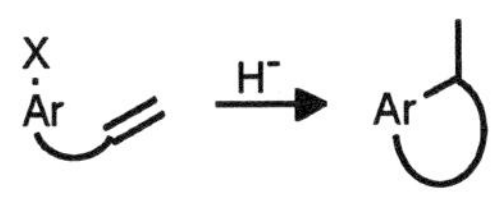	Grigg R., *Tetrahedron*, 1992; 7297 Grigg R., *Tetrahedron Lett.*, 1988; 4329 Martin H., *Tetrahedron*, 1989; 6113
	Weinreb S.M., *J. Org. Chem.*, 1993; 5452
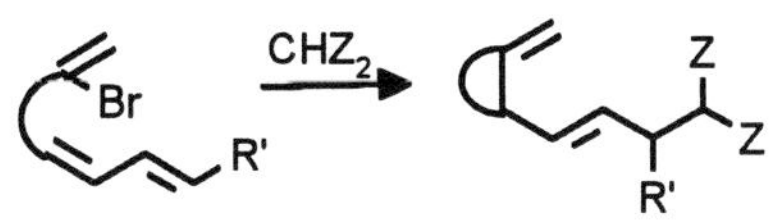	Grigg R., *Tetrahedron Lett.*, 1989; 1139
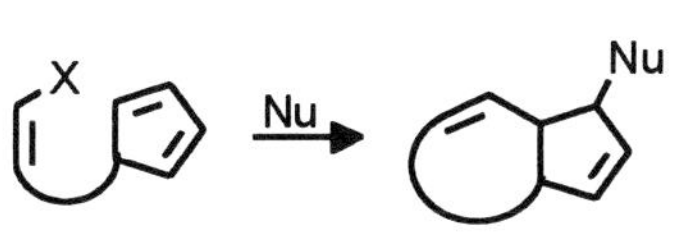	Shibasaki M., *J. Org. Chem.*, 1991; 4093 (asymmetric catalysis) Shibasaki M., *Tetrahedron*, 1993; 1773 (asymmetric catalysis)
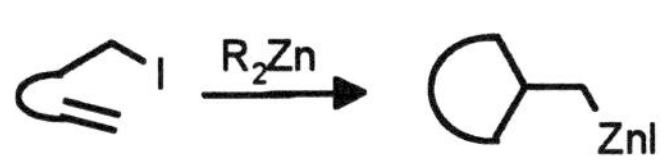	Knochel P., *J. Am. Chem. Soc.*, 1993; 7027
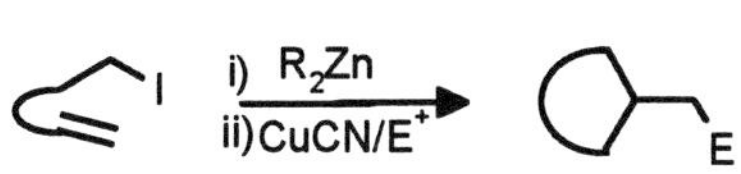	Knochel P., *Tetrahedron Lett.*, 1993; 7911

A ---> B	References
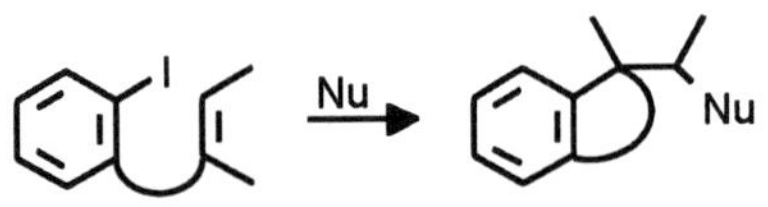	Grigg R., *Tetrahedron Lett.*, 1989; 1139 Grigg R., *Tetrahedron Lett.*, 1989; 1135 Weinreb S.M., *J. Org. Chem.*, 1992; 2528 Grigg R., *J. Heterocycl. Chem.*, 1994; 631
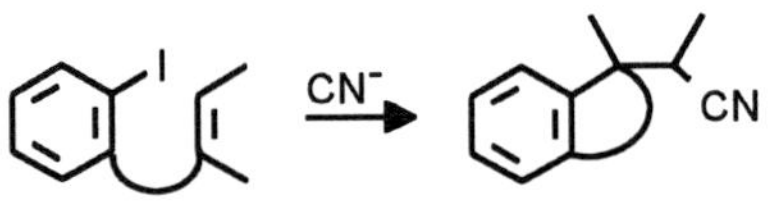	Grigg R., *Tetrahedron Lett.*, 1993; 3163
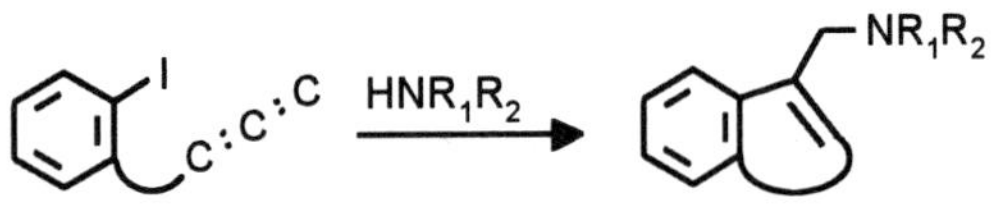	Grigg R., *J. Hetrocycl. Chem.*, 1994; 631 Grigg R., *J. Chem. Soc., Chem. Commun.*, 1995; 1903
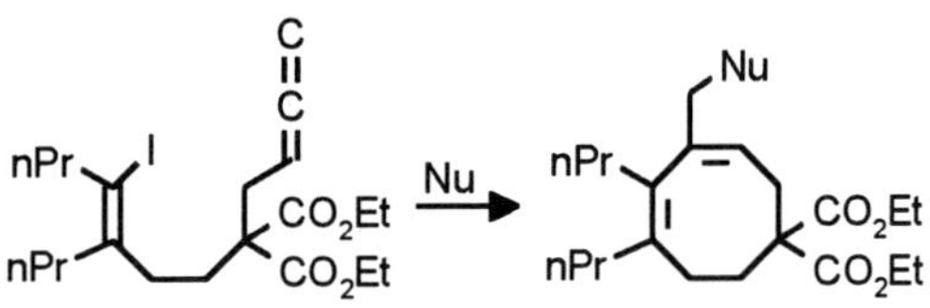	Negishi E-I., *J. Am. Chem. Soc.*, 1995; 6345
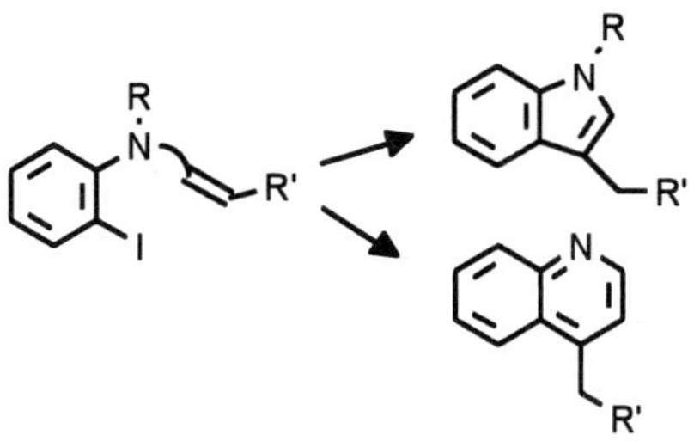	Larock R.C., *Tetrahedron Lett.*, 1987; 5291
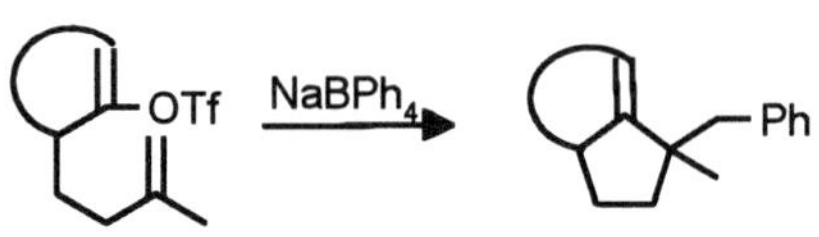	Grigg R., *Tetrahedron Lett.*, 1994; 2753

RXN13 Tandem Cyclization-Anion Capture (-Carbonylation) Process of Yne-Vinyl and Yne-Aryl Halides

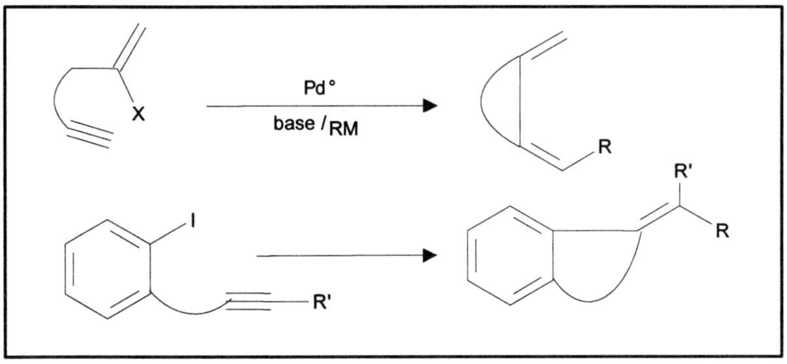

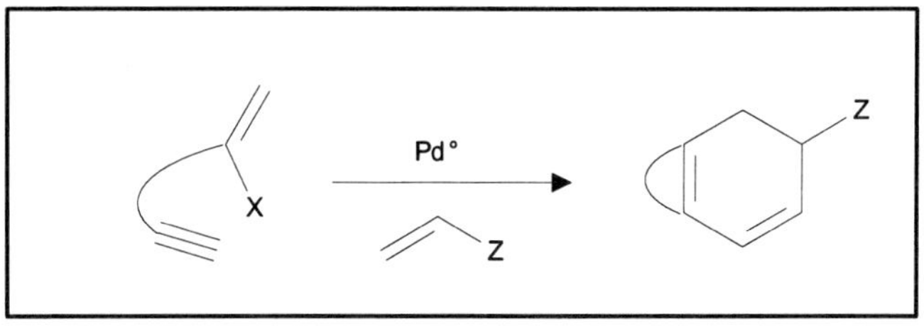

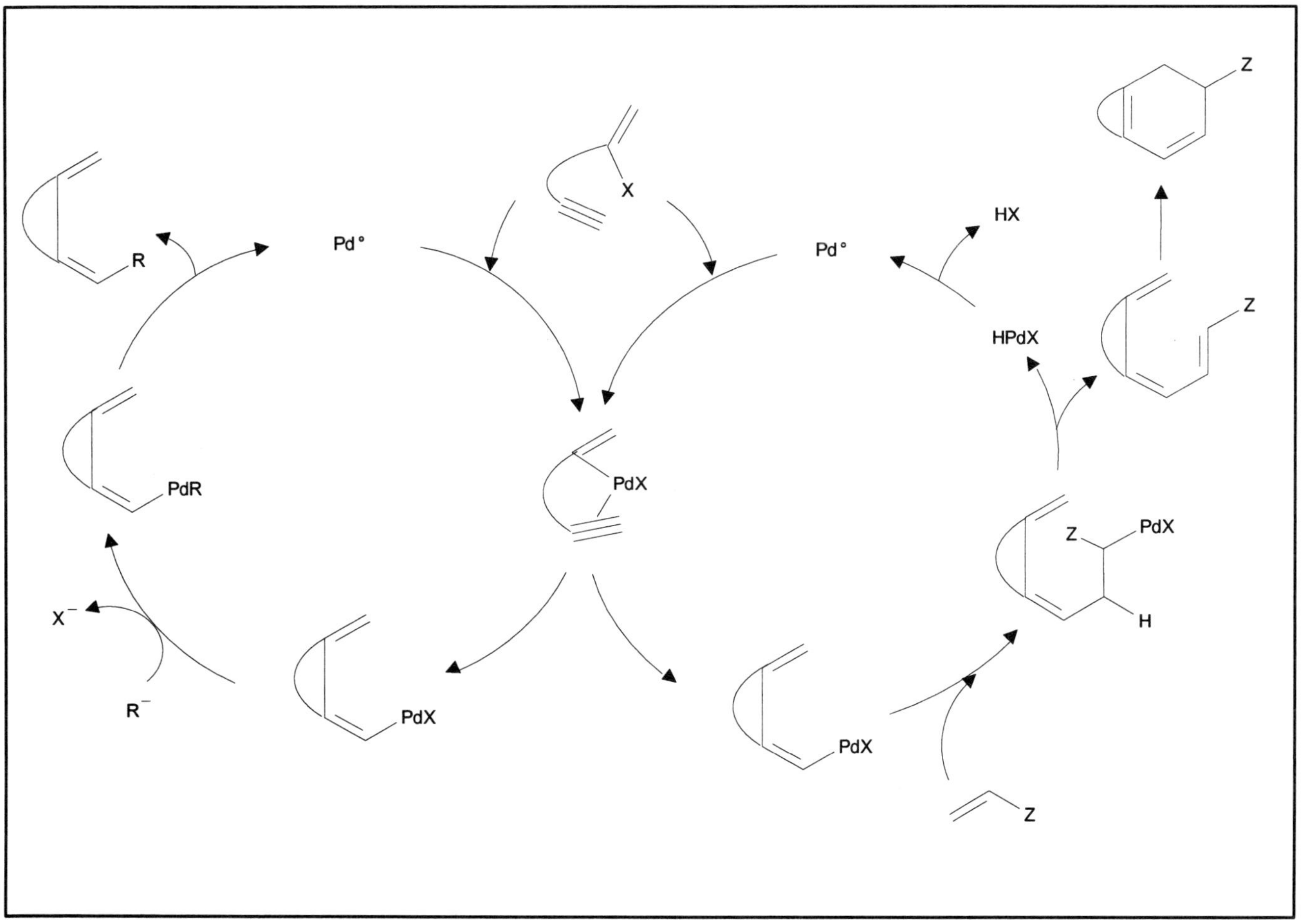

A ----> B	References
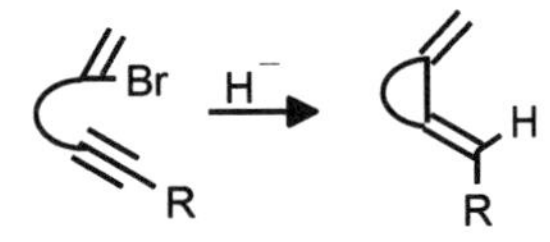	Grigg R., *Tetrahedron Lett.*, 1988; 4325
	Ishikura M., *J. Chem. Soc., Chem. Commun.*, 1995; 409
	Negishi E.I., *Tetrahedron Lett.*, 1990; 4393
	Torii S., *Tetrahedron Lett.*, 1993; 2139
	Falck J.R., *Tetrahedron Lett.*, 1992; 4885
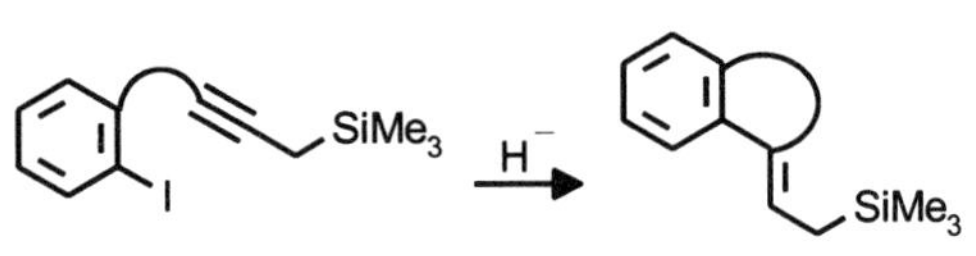	Tietze L. F., *Chem. Ber.*, 1994; 2235
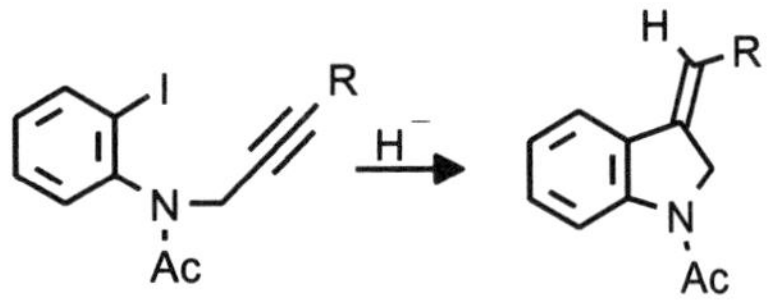	Grigg R., *Tetrahedron Lett.*, 1988; 4325
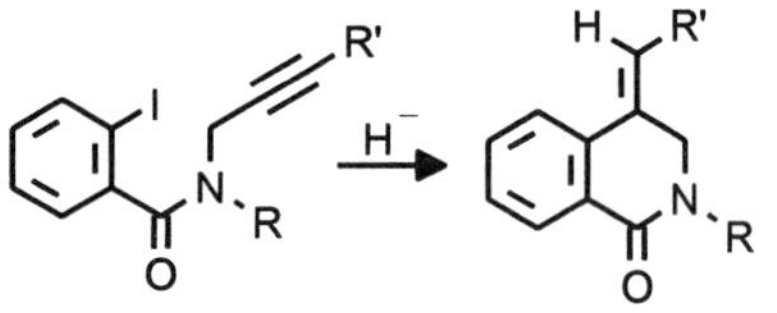	Grigg R., *Tetrahedron Lett.*, 1988; 4325
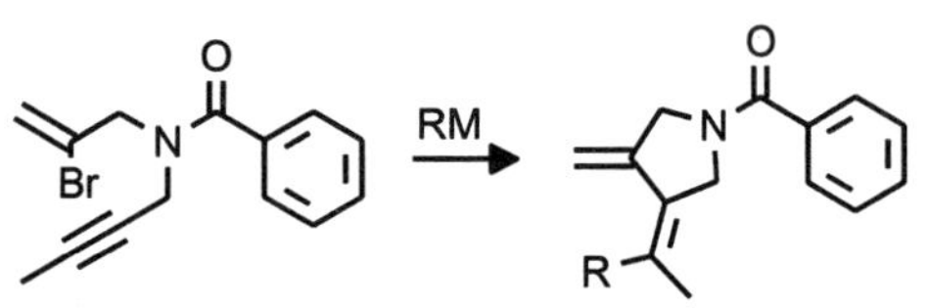	Ishikura M., *J. Chem. Soc., Chem. Commun.*, 1995; 409

A ----> B	References
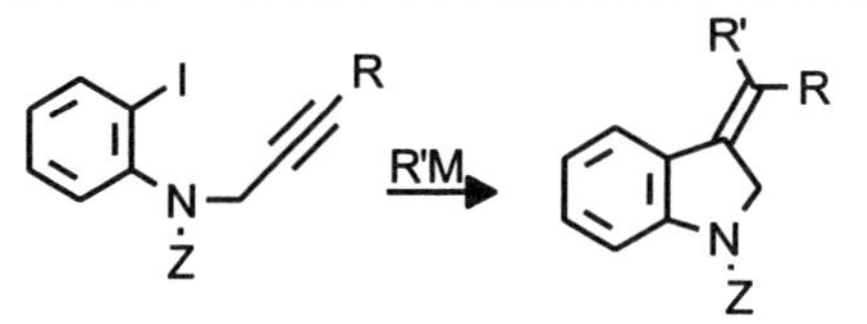	Grigg R., *Tetrahedron Lett.*, 1988; 5565 Grigg R., *Tetrahedron Lett.*, 1989; 1135 Luo F.T., *Heterocycles*, 1991; 32, 2365
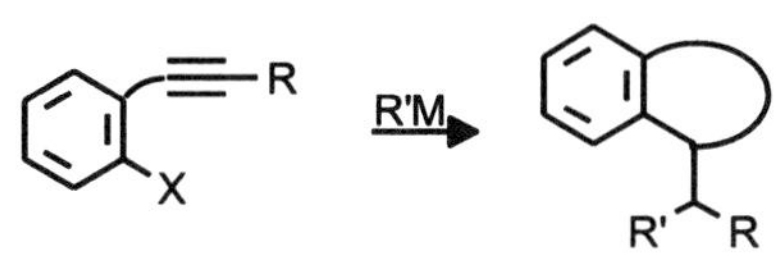	Luo F.T., *J. Org. Chem.*, 1990; 4846 Negishi E.I., *Tetrahedron Lett.*, 1990; 4393 Luo F.T., *Heterocycles*, 1990; 31, 2181 Finch H., *Tetrahedron Lett.*, 1993; 8353
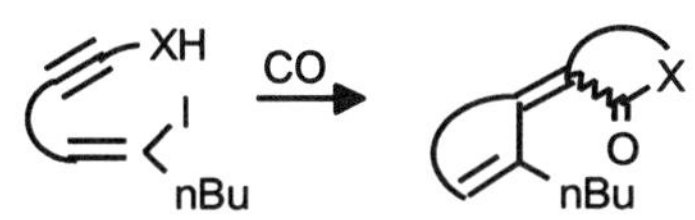	Negishi E.I., *Tetrahedron Lett.*, 1995; 1771
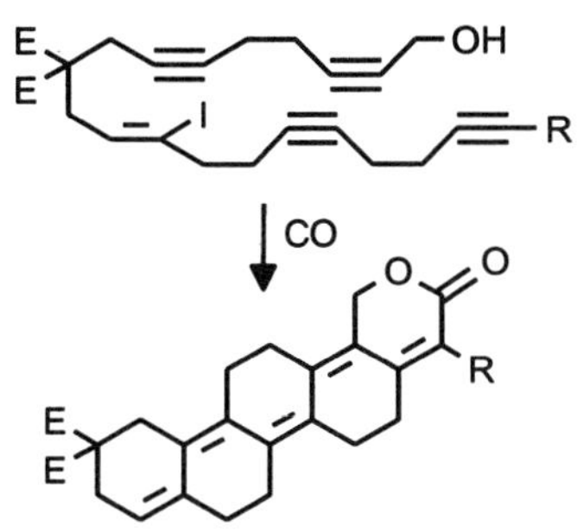	Negishi E.I., *J. Am. Chem. Soc.*, 1994; 7923
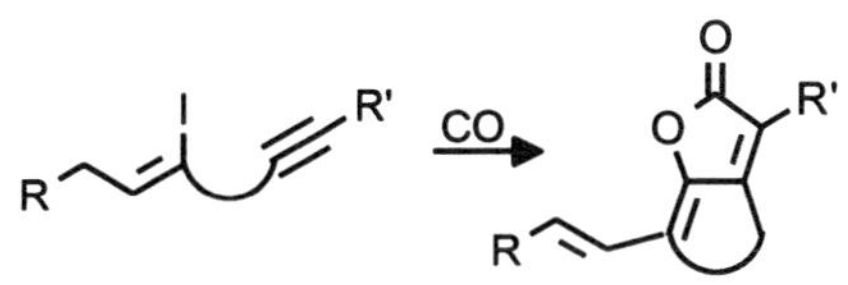	Negishi E.I., *J. Am. Chem. Soc.*, 1994; 7923
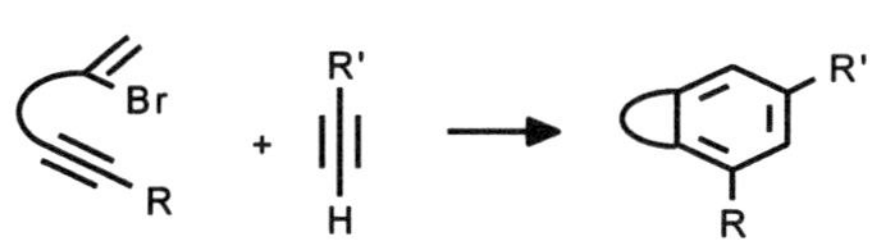	Torii S., *Tetrahedron Lett.*, 1991; 4167 Negishi E.I., *Tetrahedron*, 1993; 5471
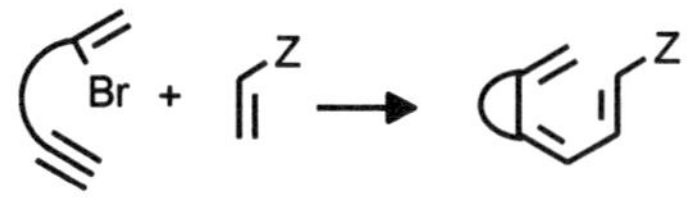	Parsons P.J., *Synlett.*, 1992; 864 (then aromatisation)
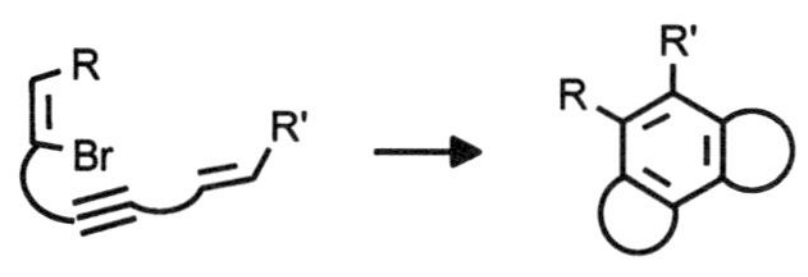	De Meijere A., *Tetrahedron Lett.*, 1992; 8039 (intramolecular reaction then cyclisation)

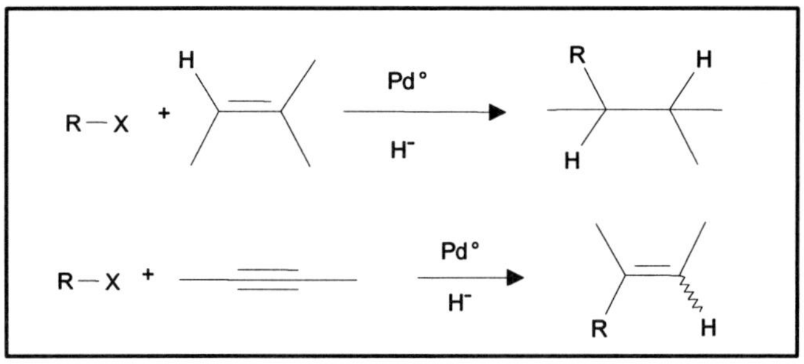
R—X
+
H
Pd°
H⁻
R
H
H
R—X
+
Pd°
H⁻
R
H

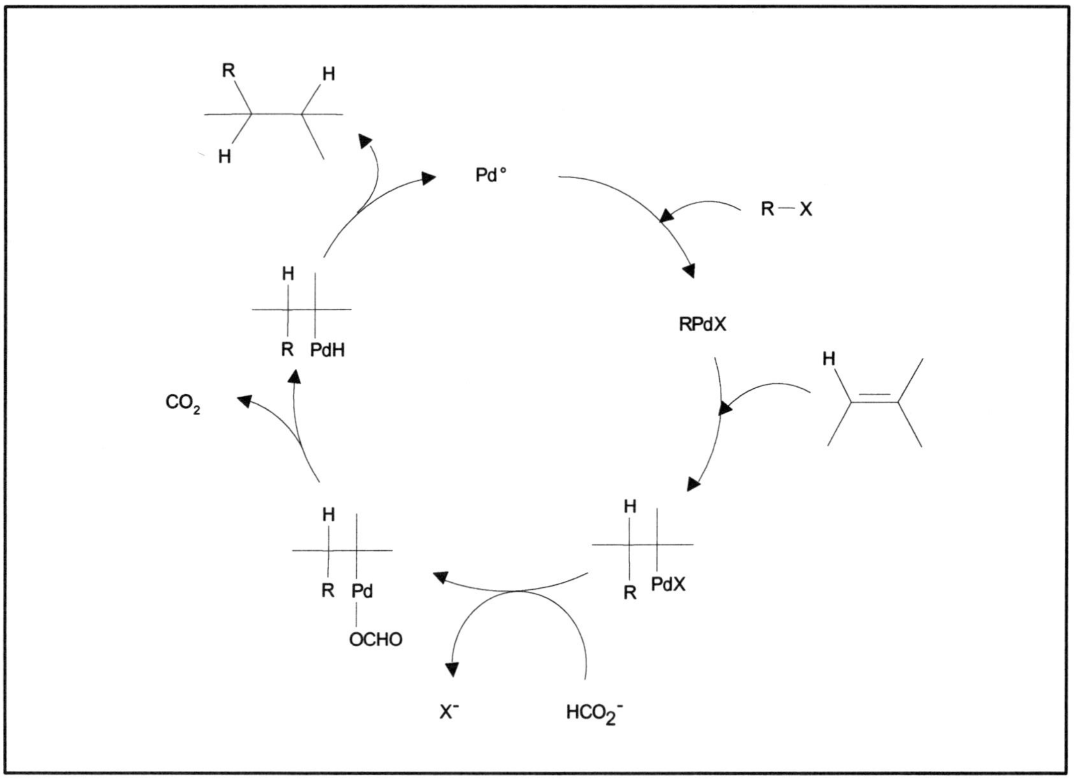
R
H
H
Pd°
R—X
RPdX
H
H
R PdH
CO2
H
R Pd
OCHO
H
R PdX
X⁻
HCO2⁻

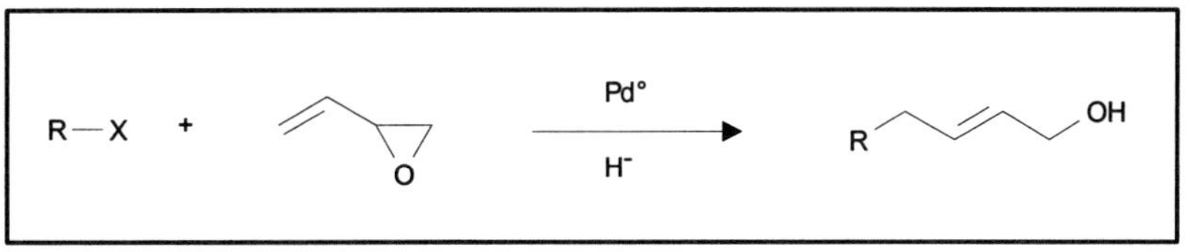
R—X
+
Pd°
H⁻
R
OH
O

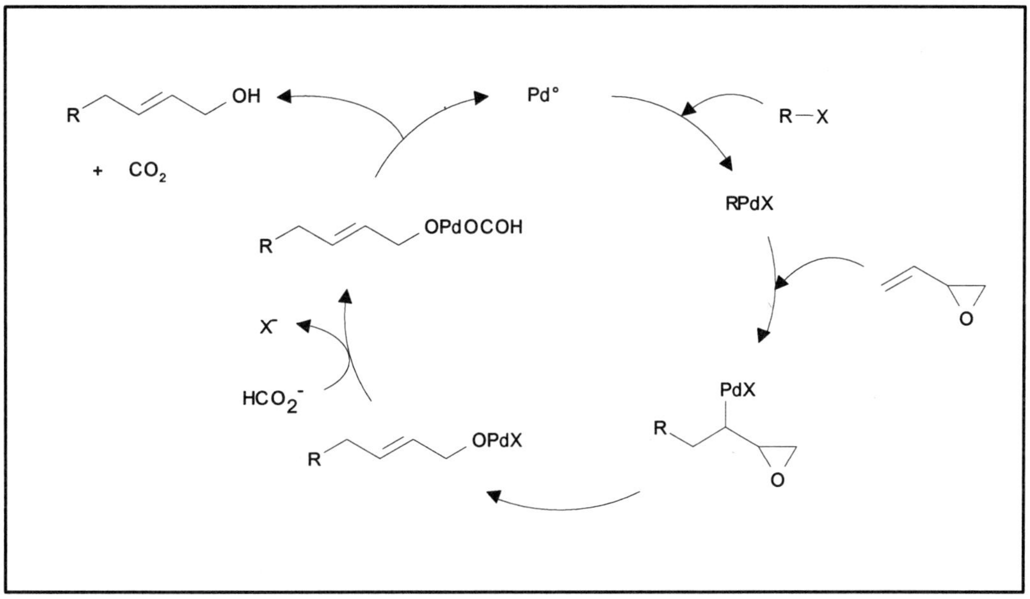
R
OH
+ CO_2
Pd°
R—X
RPdX
OPdOCOH
X⁻
HCO_2^-
PdX
R
O
OPdX

Hydrovinylation	References
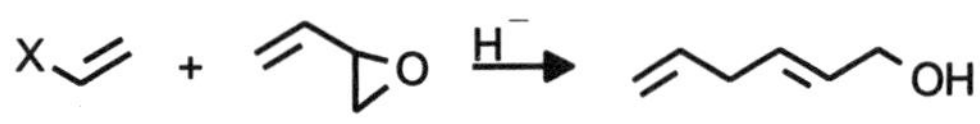	Larock R.C., *J. Org. Chem.*, 1993; 804
	Larock R.C., *Synlett.*, 1993; 145
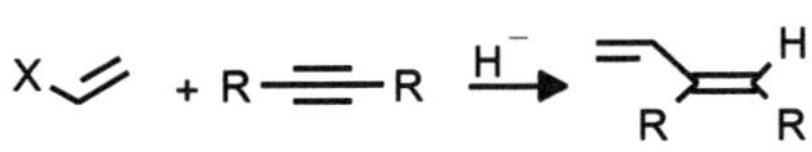	Cacchi S., *Tetrahedron Lett.*, 1989; 3465

Hydroarylation	References
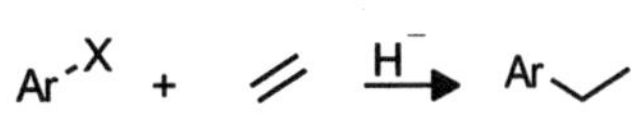	Cacchi S., *Synthesis*, 1984; 575 Cacchi S., *Tetrahedron*, 1989; 813
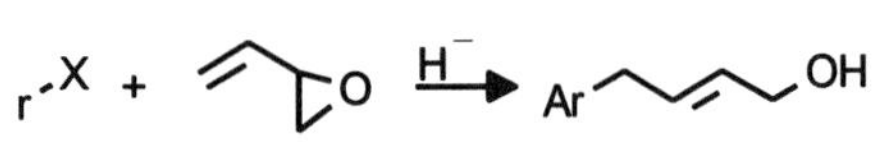	Larock R.C., *J. Org. Chem.*, 1993; 804 Larock R.C., *J. Org. Chem.*, 1990; 6244
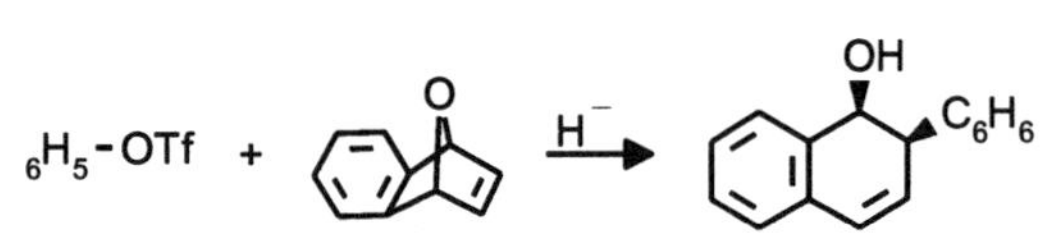	Fiaud J.C., *Tetrahedron Lett.*, 1995; 2051 (asymmetric catalysis)
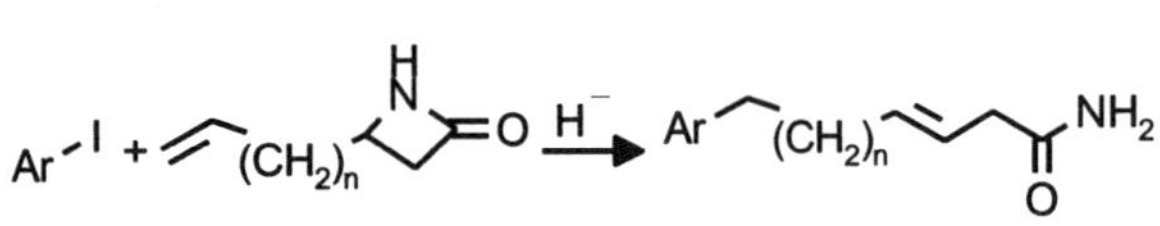	Larock R.C., *Tetrahedron Lett.*, 1993; 979
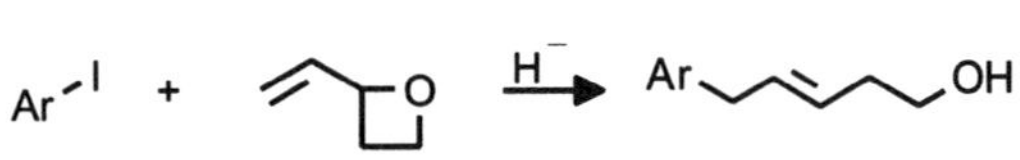	Larock R.C., *Synlett.*, 1993; 145
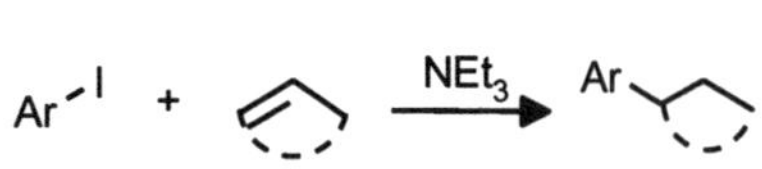	Stokker G.E., *Tetrahedron Lett.*, 1987; 3179 (hydride from NEt_3)
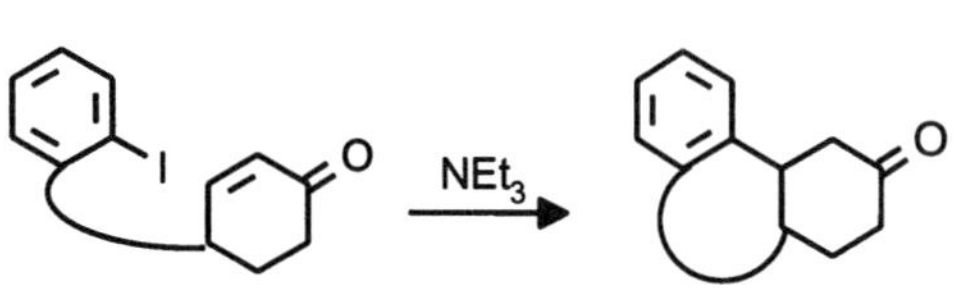	Overman L.E., *Tetrahedron Lett.*, 1994; 3453 (hydride from NEt_3)

Hydroarylation	References
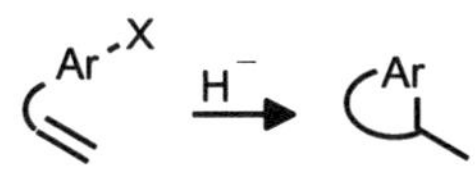	Grigg R., *Tetrahedron Lett.*, 1988; 4329 Grigg R., *Tetrahedron*, 1992; 7297 Ghatak U.R., *Synth. Commun.*, 1995; 1641
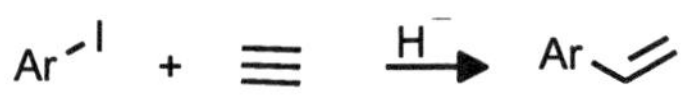	Cacchi S., *Tetrahedron Lett.*, 1984; 3137
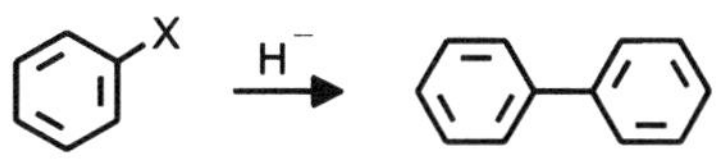	Bamfield P., *Synthesis*, 1978; 537

RXN15 Reduction of Alkenes

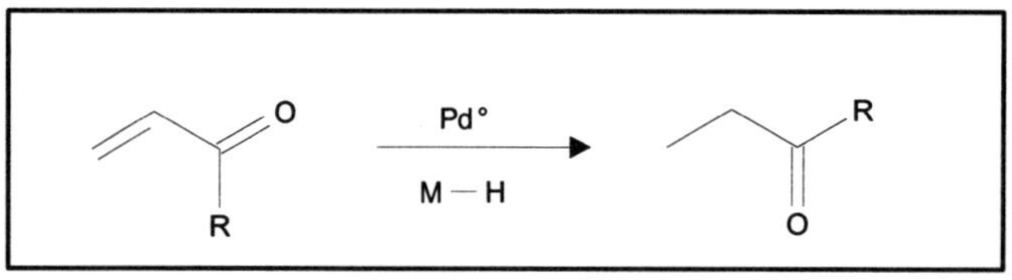

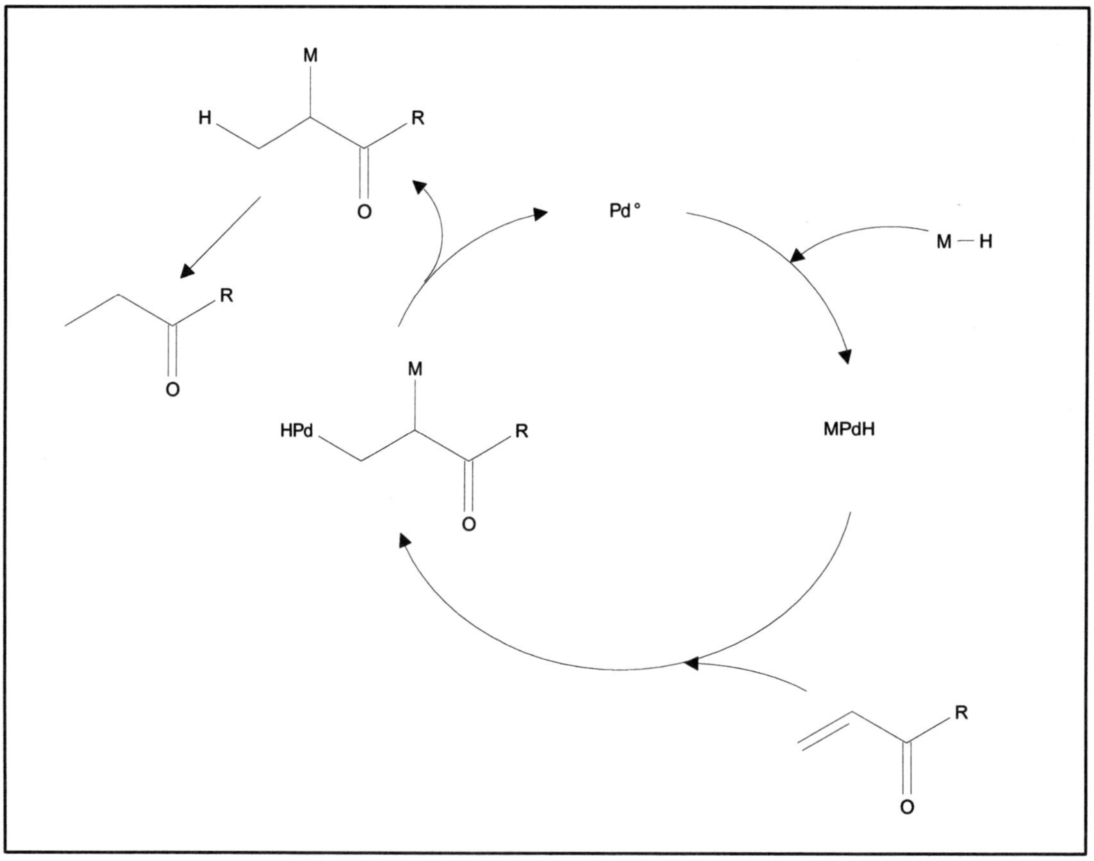

Reduction	References
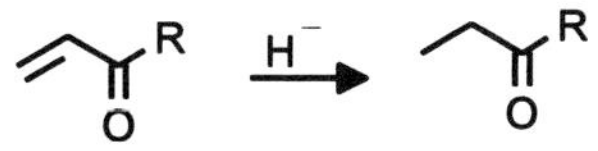	Keinan E., *Tetrahedron Lett.*, 1982; 477 Keinan E., *Tetrahedron Lett.*, 1985; 1353 Keinan E., *J. Org. Chem.*, 1983; 3545 Tour J.M., *J. Org. Chem.*, 1990; 3452 Cacchi S., *Synlett.*, 1991; 27

RXN16 Semihydrogenation of Alkynes and 1,3-Dienes

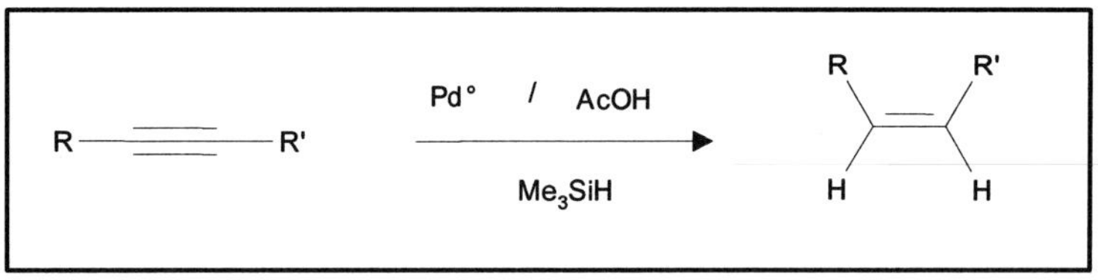

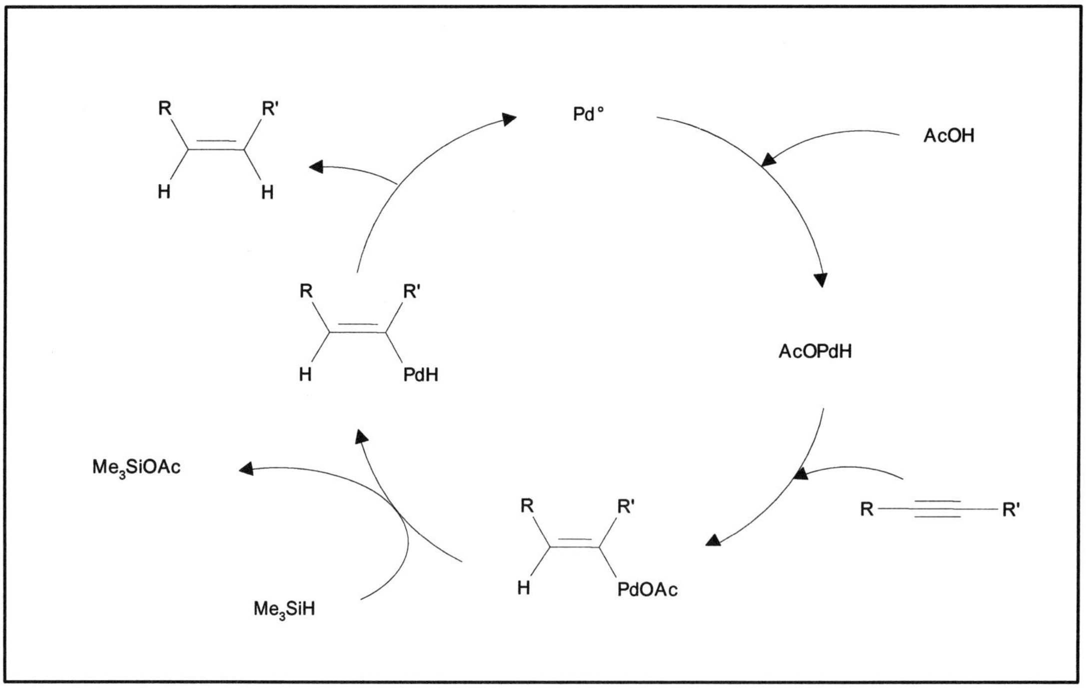

Semihydrogenation	References
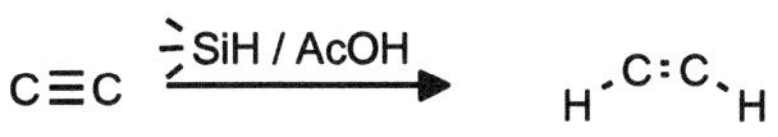	Trost B.M., *Tetrahedron Lett.*, 1989; 4657 Fry J.L., *Tetrahedron Lett.*, 1994; 1507
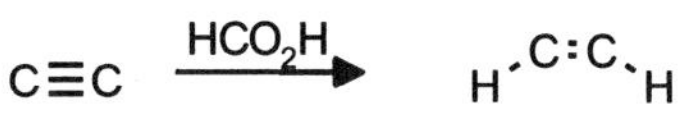	Sato F., *J. Chem. Soc., Chem. Commun.*, 1993; 386
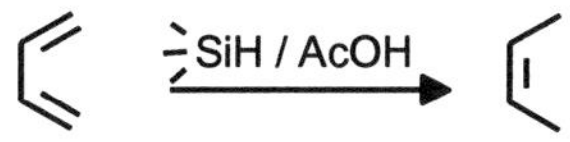	Trost B.M., *Tetrahedron Lett.*, 1989; 4657

RXN17 Hydroboration, Hydrogermylation, Hydrosilylation and Hydrostannation of Alkynes, Allenes, Dienes and Enynes

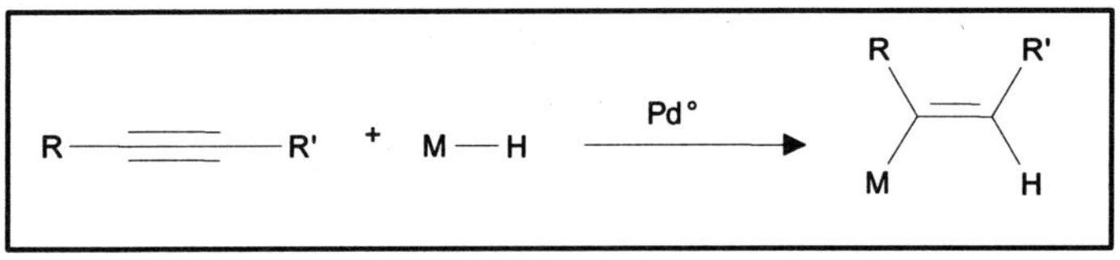

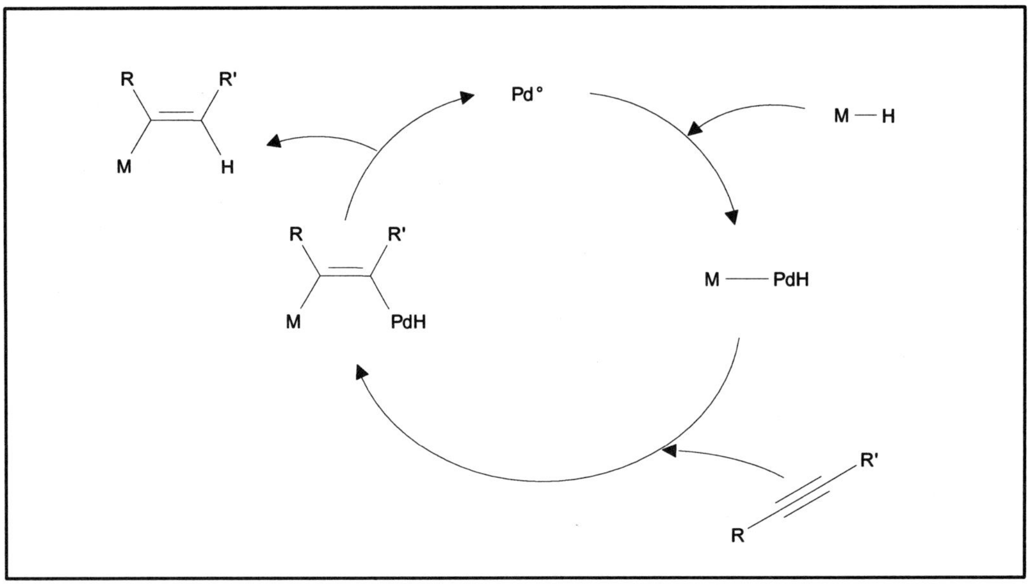

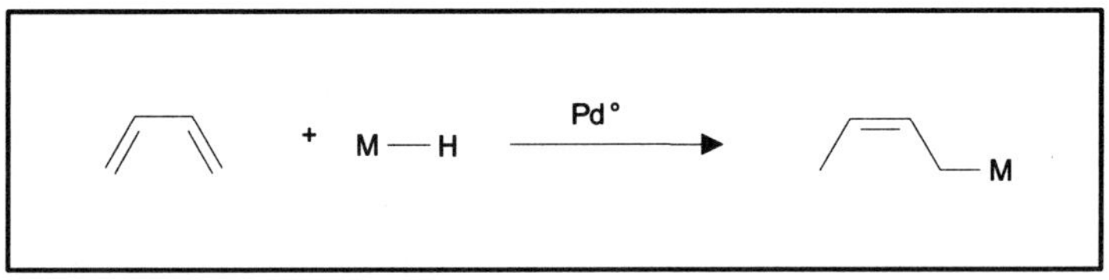

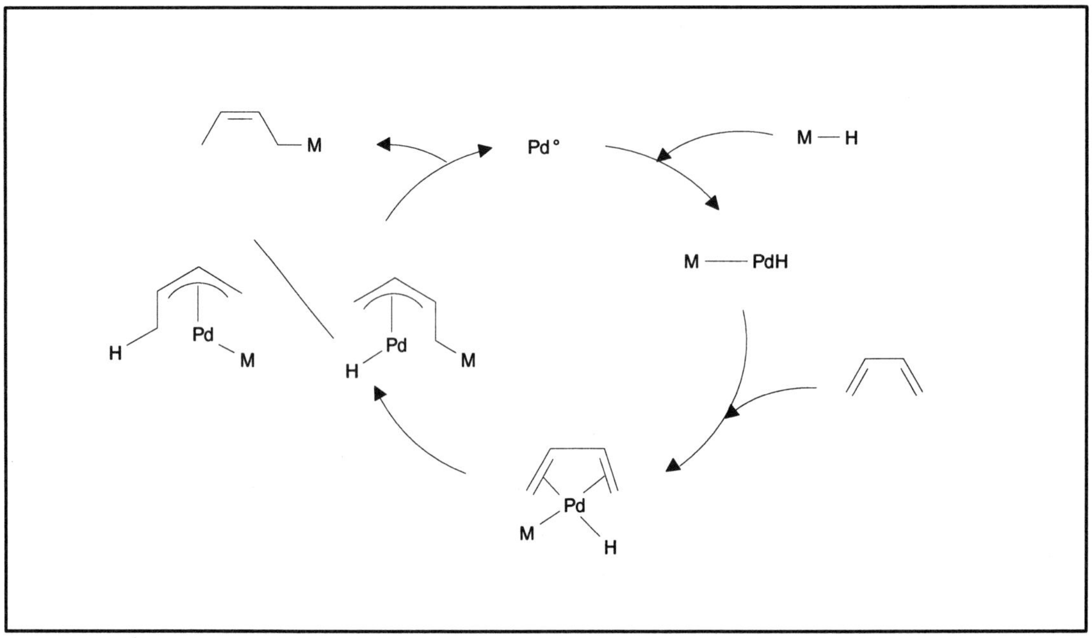

Hydroboration	References
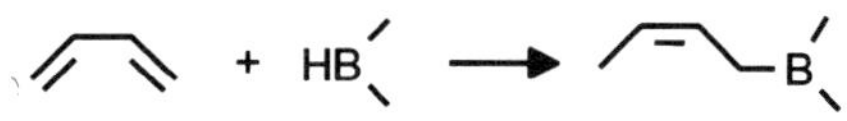	Suzuki A., *Tetrahedron Lett.*, 1989; 3789

Hydrosilylation	References
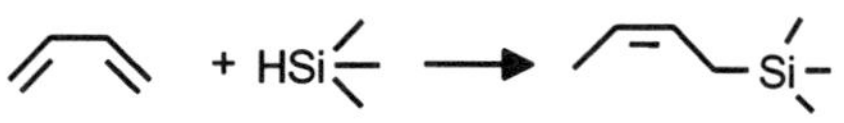	Ojima I., *J. Organomet. Chem.*, 1978; 157, 359 Hayashi T., *Chem. Lett.*, 1990; 1377 Kobayashi S., *Synthesis*, 1994; 457 Hatanaka Y., *Tetrahedron Lett.*, 1994; 7981 (asymmetric catalysis) Achiwa K., *Chem. Pharm. Bull.*, 1995; 927 (asymmetric catalysis)
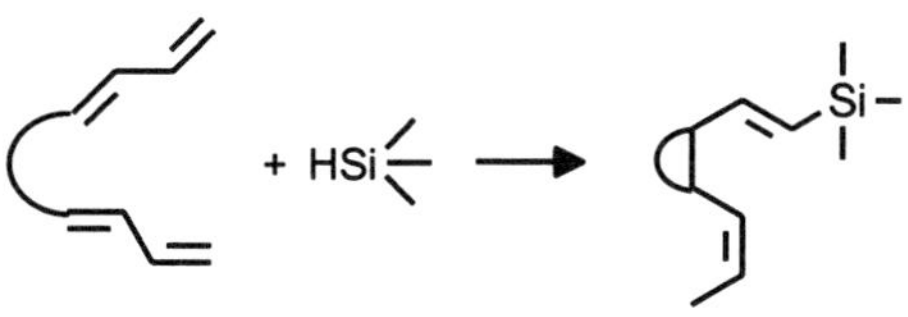	Takacs J.M., *Organometallics*, 1990; 2877
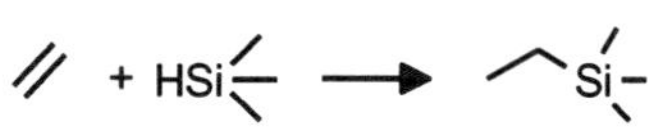	Achiwa K., *Chem. Pharm. Bull.*, 1995; 927 (asym. catalysis) Hayashi T., *J. Chem. Soc., Chem. Commun.*, 1995; 1533 (asym. catal.)

Hydrostannation	References
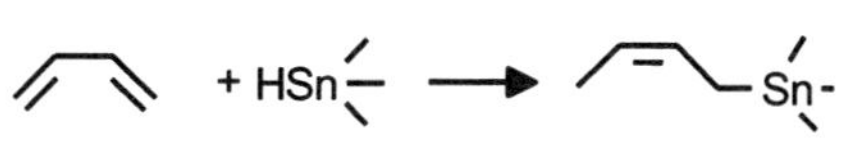	Miyake H., *Chem. Lett.*, 1992; 507 Miyake H., *Chem. Lett.*, 1992; 1099
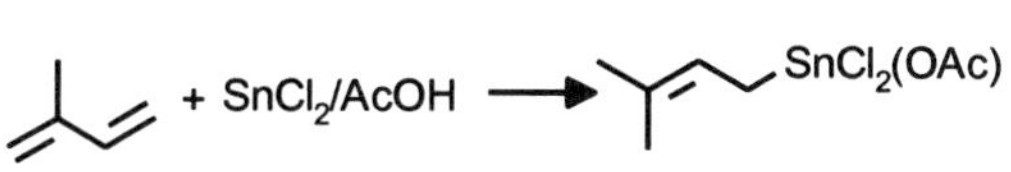	Masuyama Y., *J. Chem. Soc., Chem. Commun.*, 1994; 1451
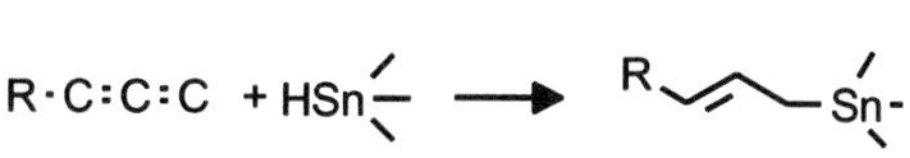	Utimoto K., *Bull. Chem. Soc. Jpn.*, 1988; 2693 Gore J., *Tetrahedron Lett.*, 1991; 1187
	Miyake H., *Chem. Lett.*, 1989; 981 Guibé F., *J. Org. Chem.*, 1990; 1857 Cochran J.C., *Tetrahedron Lett.*, 1990; 6621 Carpita A., *Tetrahedron*, 1994; 4853 Cossi P., *Tetrahedron*, 1992; 8801 Finch H., *Tetrahedron Lett.*, 1993; 8353

Hydrostannation	References
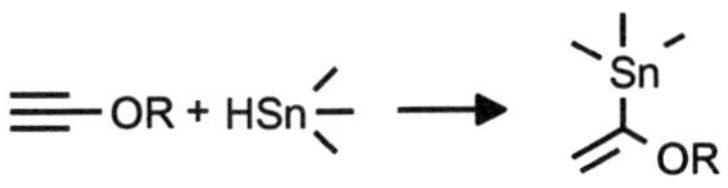	Kocienski P., *J. Chem. Soc., Perkin Trans. 1*, 1994; 1187
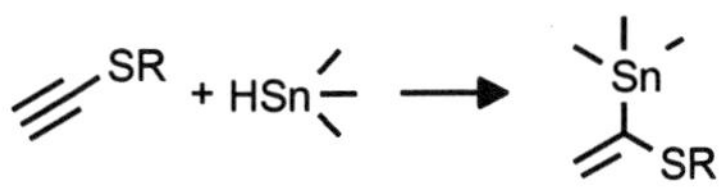	Magriotis P.A., *Tetrahedron Lett.*, 1991; 5047
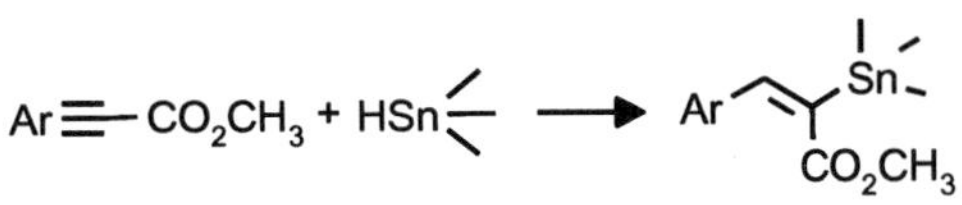	Horikawa H., *Synthesis*, 1995; 582
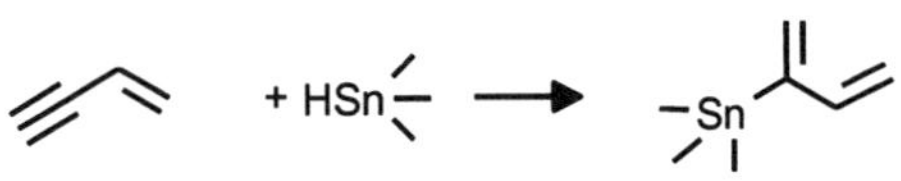	Trost B.M., *Synthesis*, 1994; 1267

Hydrogermylation	References
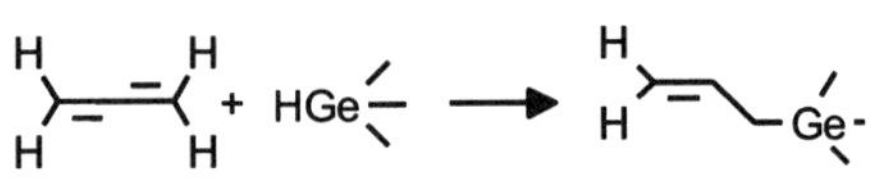	Utimoto K., *Bull. Chem., Soc., Jpn.*, 1988; 2693
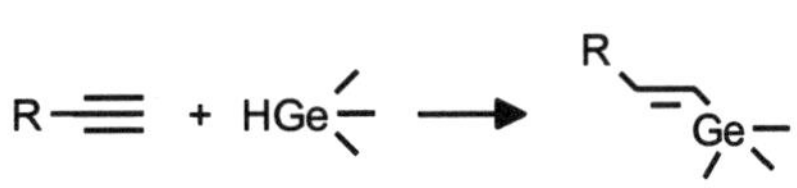	Utimoto K., *Bull. Chem. Soc. Jpn.*, 1987; 3468

RXN18 Hydroselenation of Alkynes

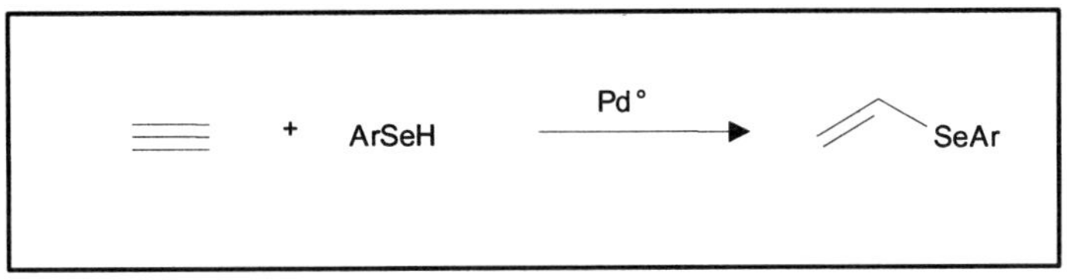

No mechanism is described but formally the results are the same as for RXN17

RXN18

Hydroselenation	References
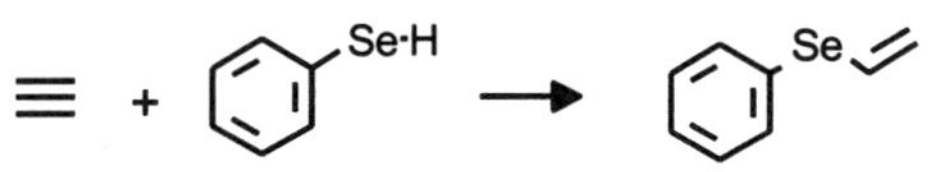	Ogawa A., *Tetrahedron Lett.*, 1992; 5525

RXN19 1,4- Disilylation of Conjugated Enones

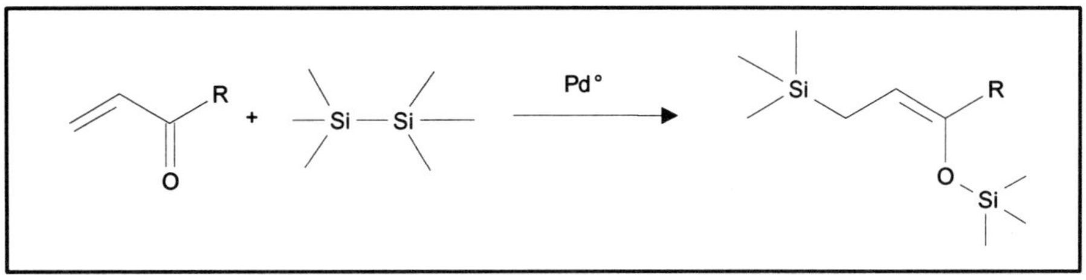

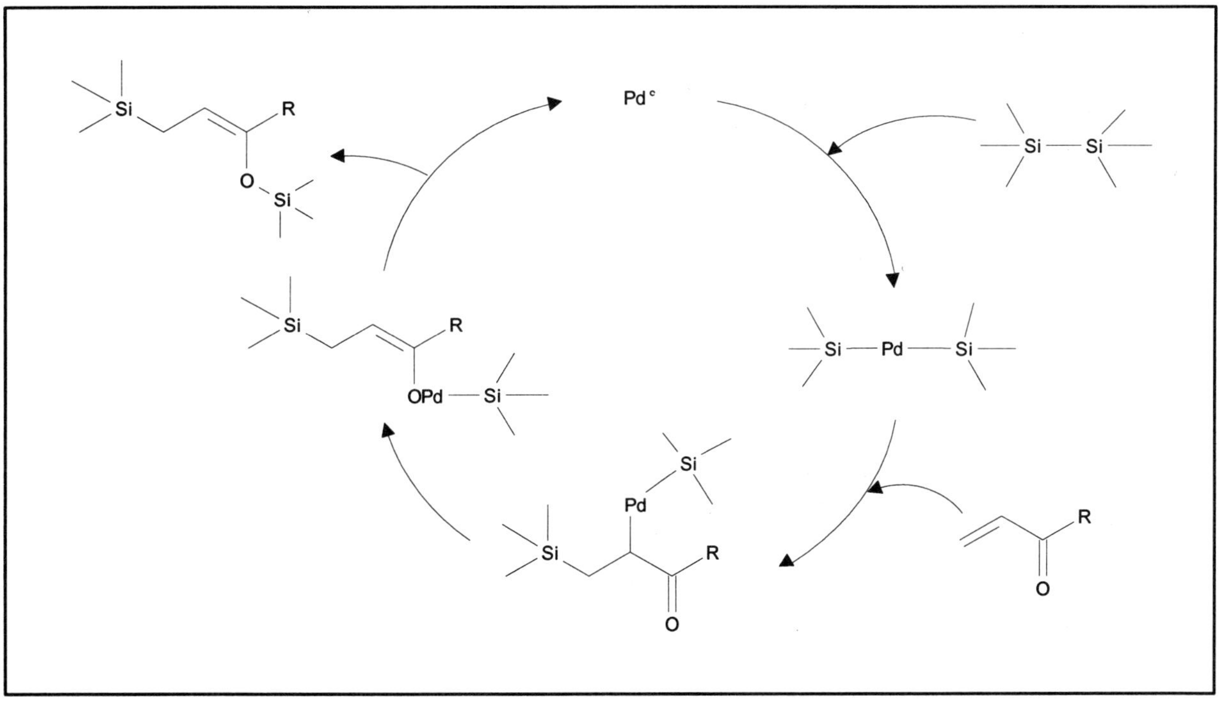

RXN19

	References
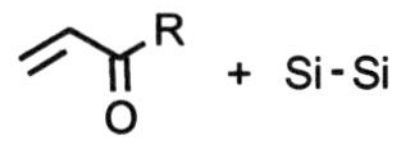	Hayashi T., *Tetrahedron Lett.* , 1988; 4147

RXN20 Hydrocarboxylation, Hydrocarboalkoxylation and Hydrocarboamination of Alkenes and Alkynes

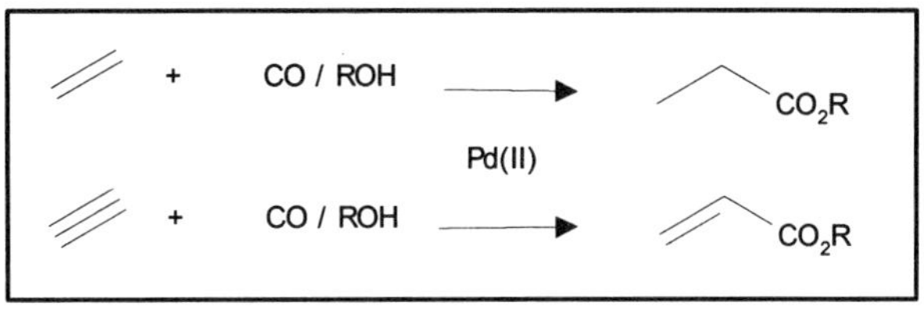

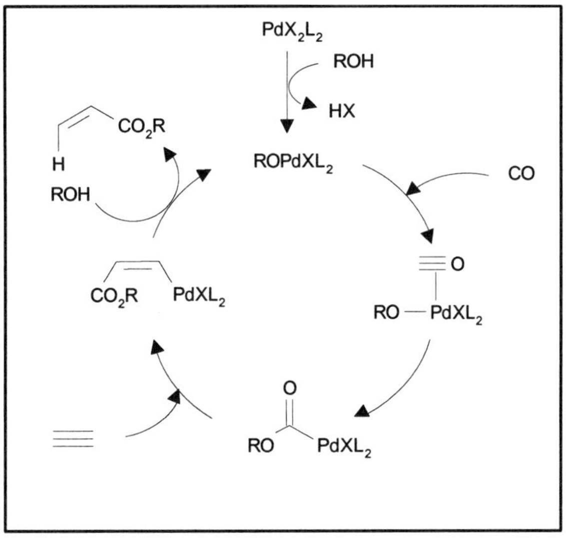

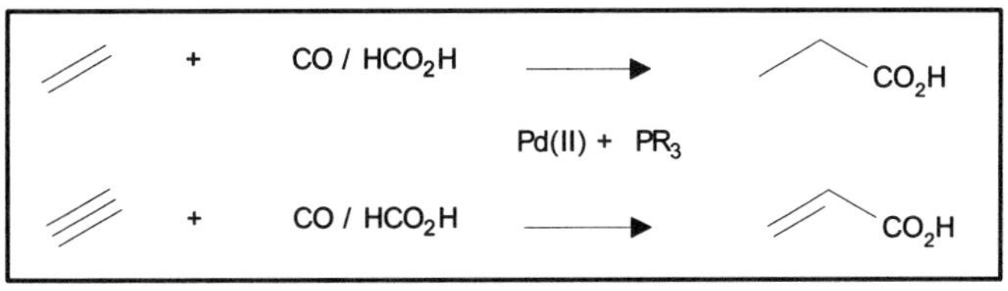

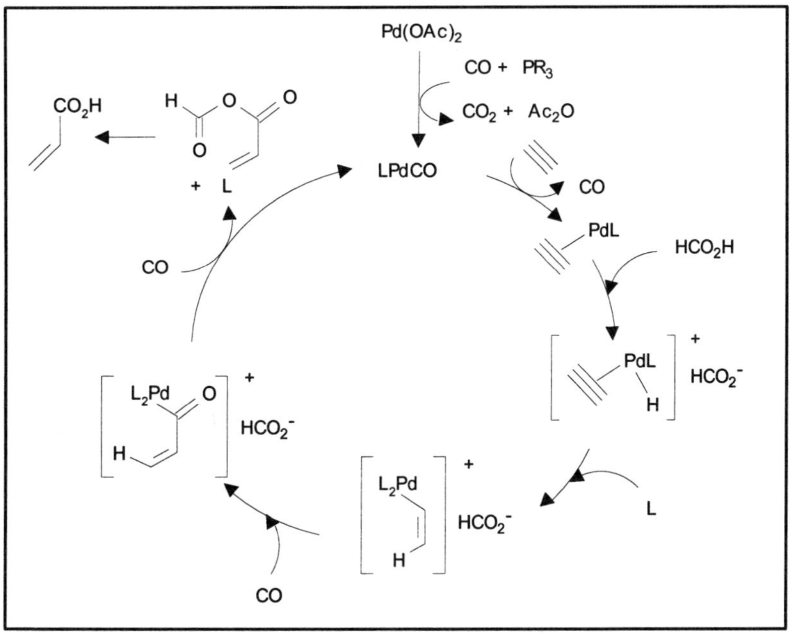

RXN20 Hydrocarboxylation, Hydrocarboalkoxylation and Hydrocarboamination of Alkenes and Alkynes

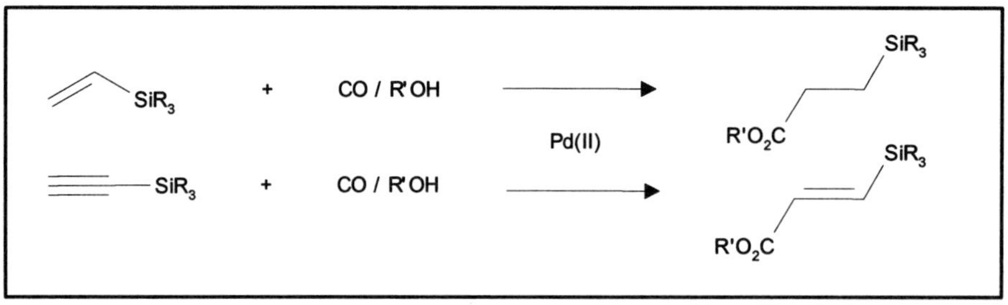

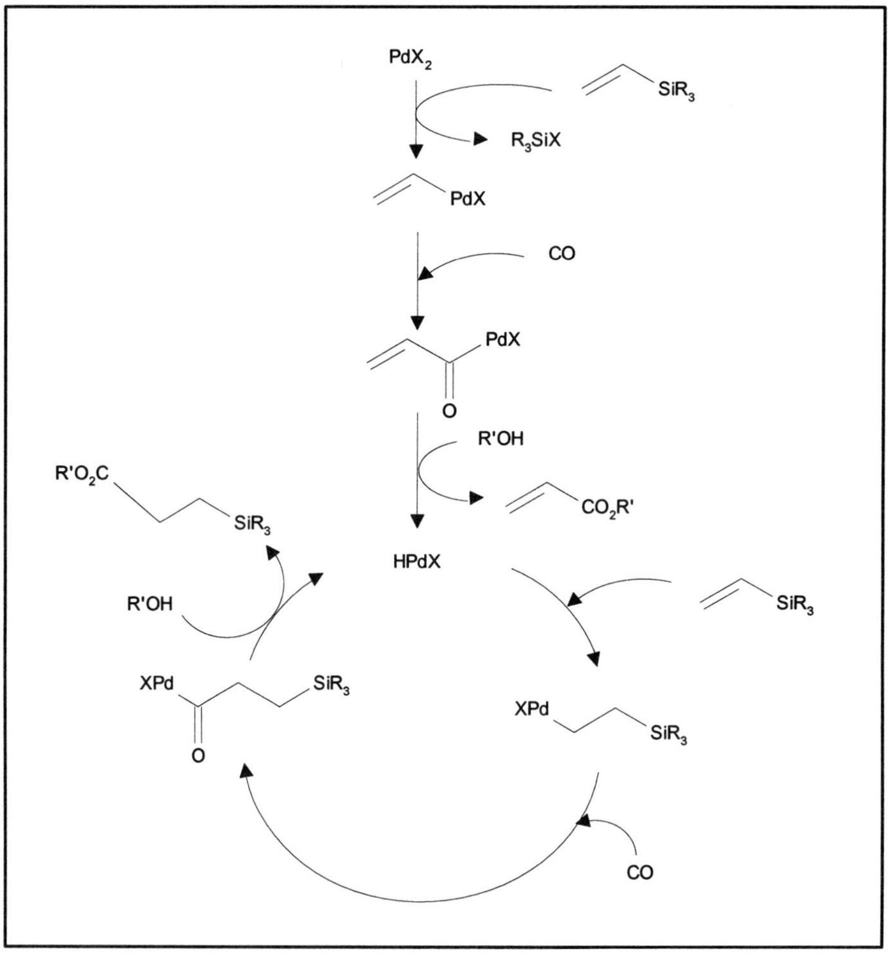

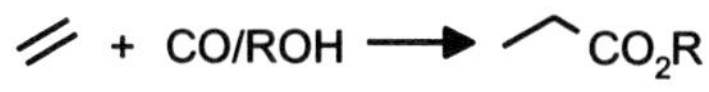	References
	Törös S., *Tetrahedron Lett.*, 1992; 3667 Kalck Ph., *J. Organomet. Chem.*, 1994; 480, 91
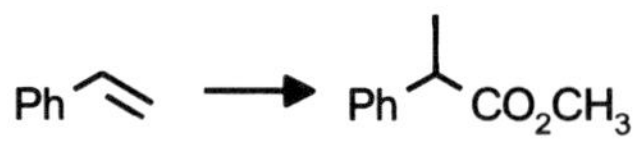	Alper H., *J. Org. Chem.*, 1993; 3595
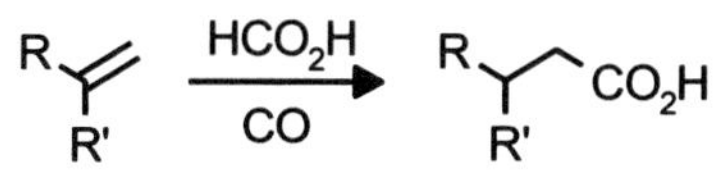	Alper H., *J. Mol. Catal.*, 1992; 77, 7 Alper H., *J. Org. Chem.*, 1993; 3595
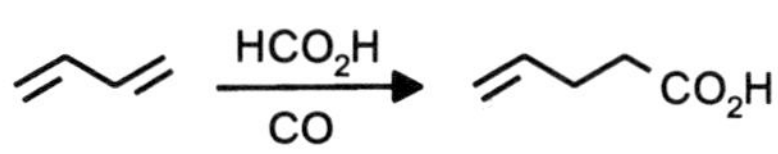	Alper H., *Tetrahedron Lett.*, 1994; 6203
	Alper H., *J. Mol. Catal.*, 1991; 67, 29 Miura M., *Tetrahedron Lett.*, 1992; 5369
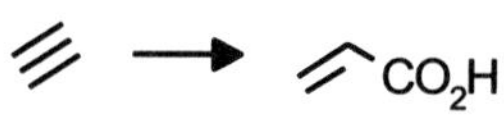	Alper H., *Organometallics*, 1993; 712
	Alper H., *J. Mol. Catal.*, 1995; 96, 197
	Torii S., *Chem. Lett.*, 1991; 1673

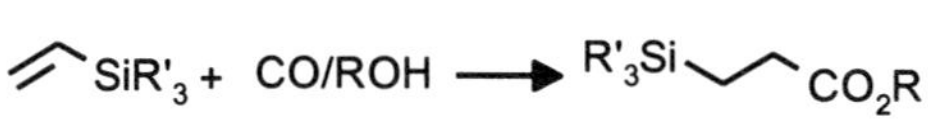	References
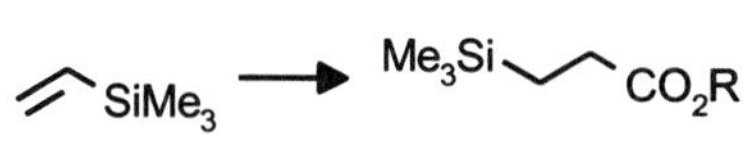	Takeuchi R., *J. Org. Chem.*, 1992; 4189
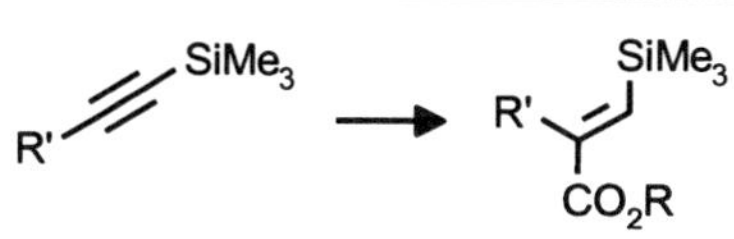	Takeuchi R., *J. Chem. Soc., Chem. Commun.*, 1992; 1358 Takeuchi R., *J. Chem. Soc., Perkin Trans. 1*, 1993; 1031
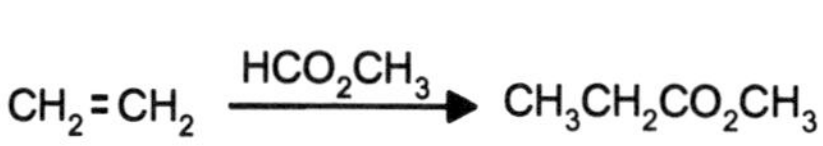	Kalck Ph., *J. Organomet. Chem.*, 1994; 476, C23

RXN21 Tandem Carbonylation-Arylation with Alkynes

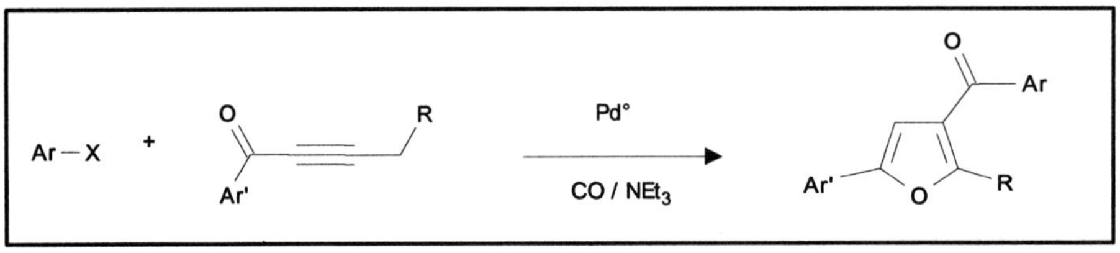

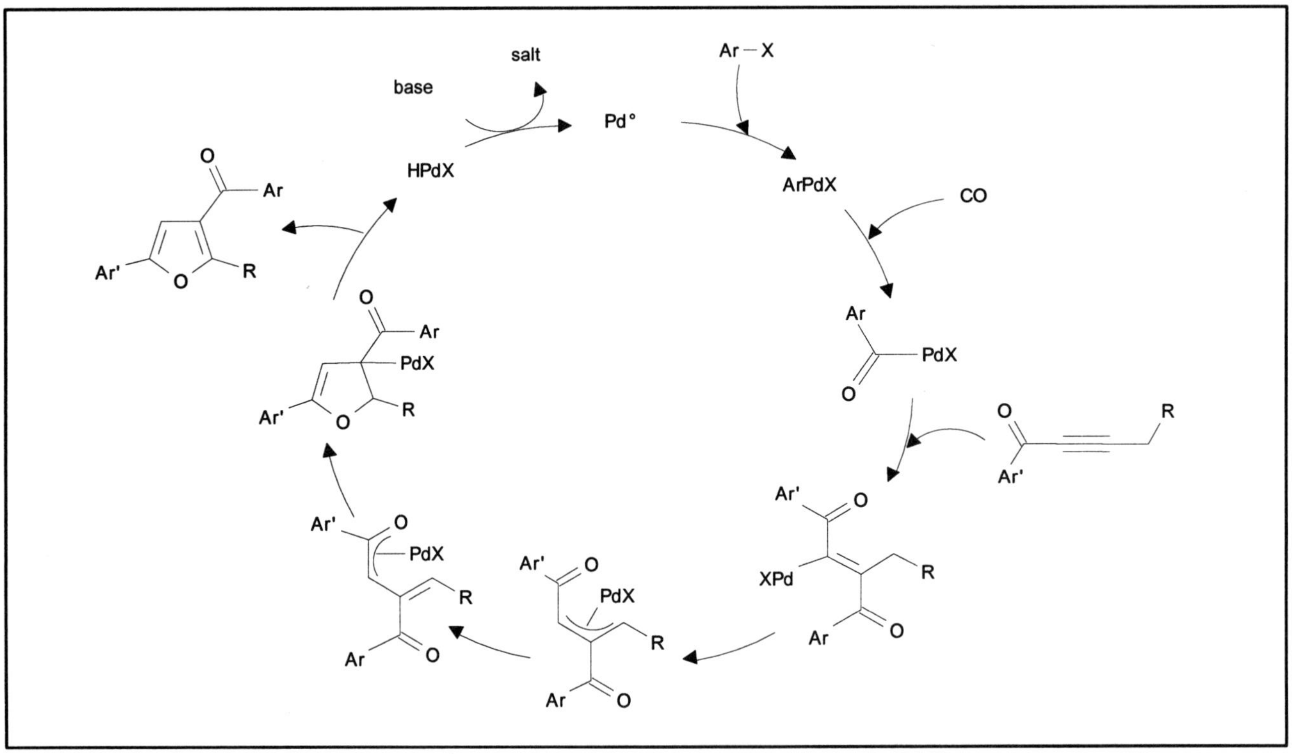

 J.-L. Malleron et al

RXN21 Tandem Carbonylation-Arylation with Alkynes

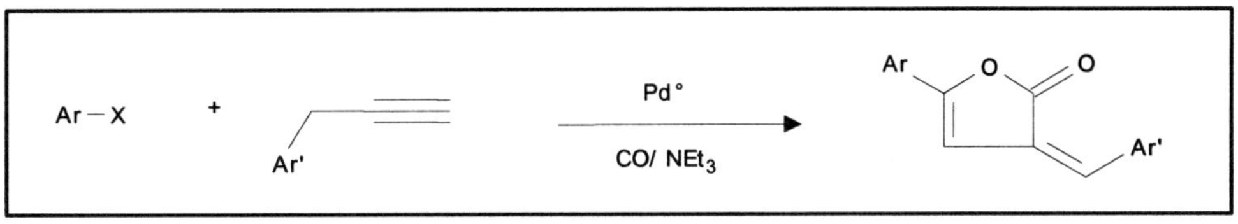

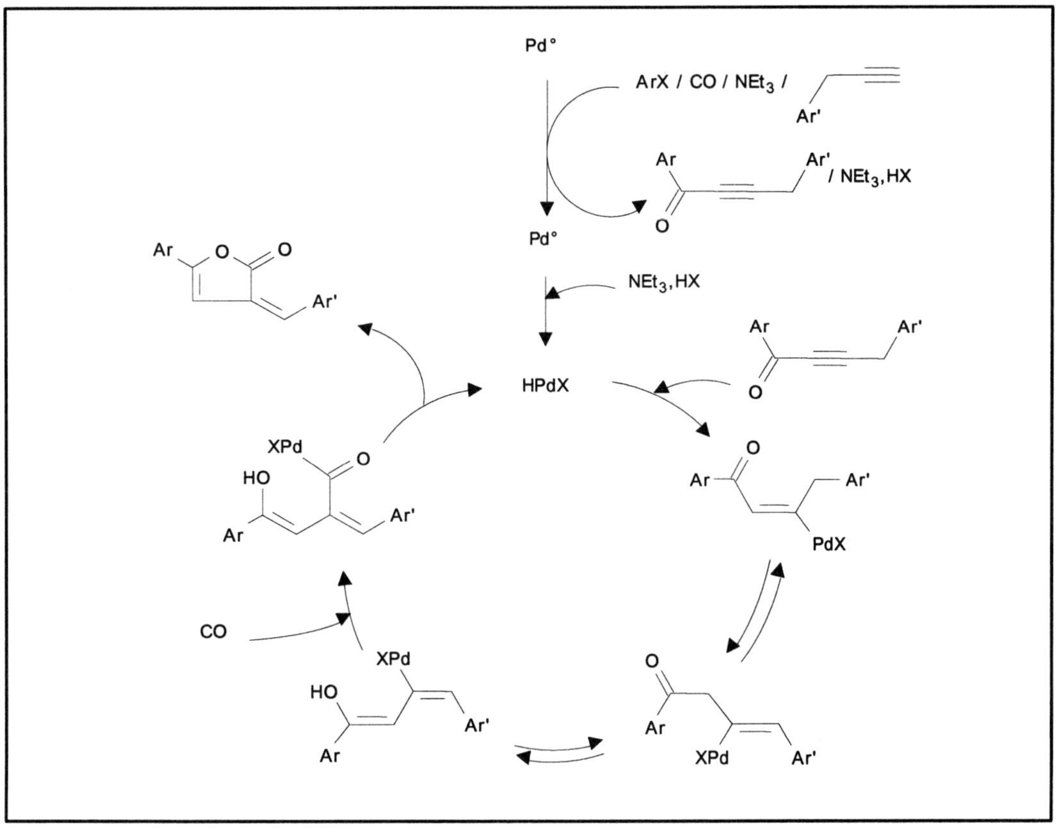

A ---> B	References
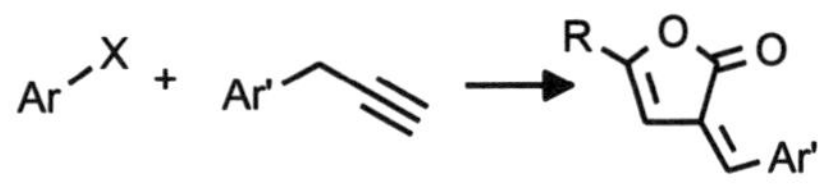	Alper H., *J. Org. Chem.*, 1991; 4534
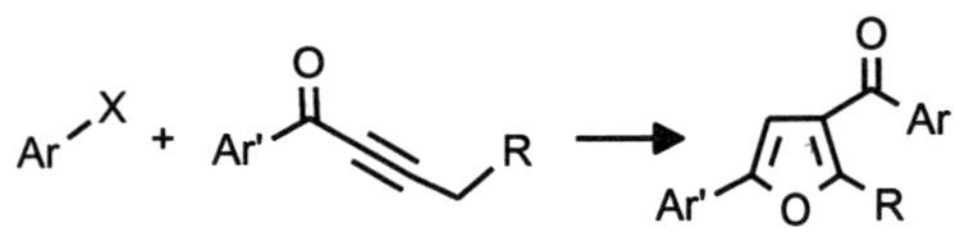	Miura M., *J. Org. Chem.*, 1992; 4754

RXN22 1,2-Dimetallation of Alkynes and Alkenes and Related Reactions

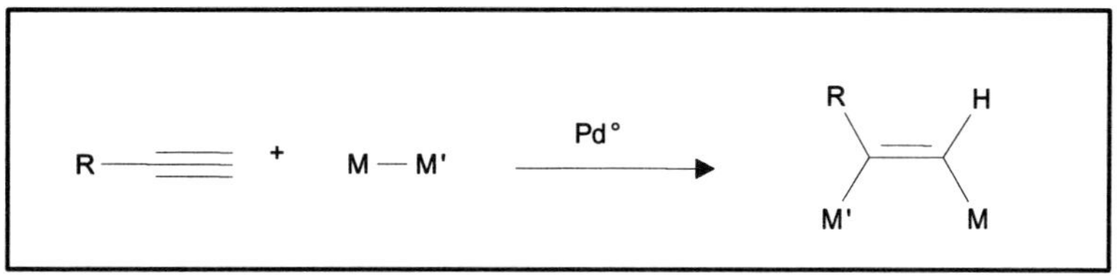

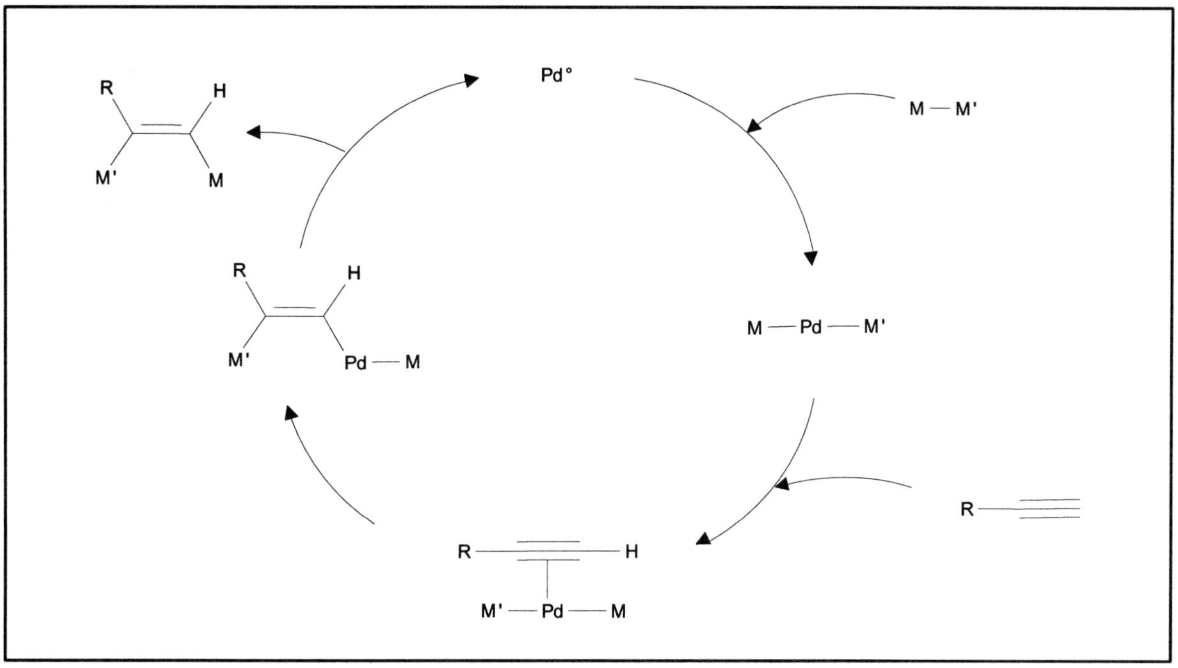

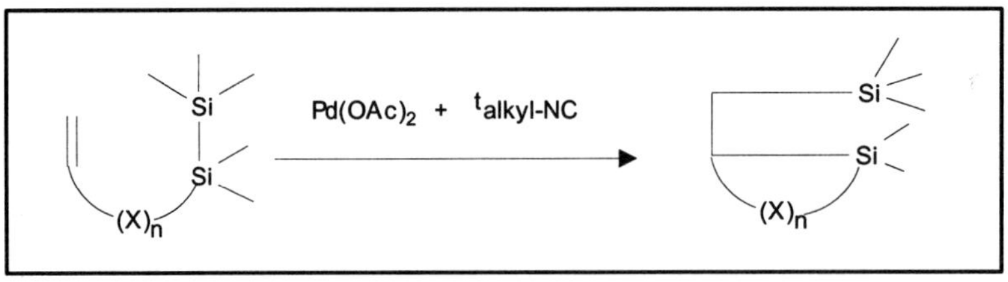
Si
Si
(X)n
Pd(OAc)2 + talkyl-NC
Si
Si
(X)n

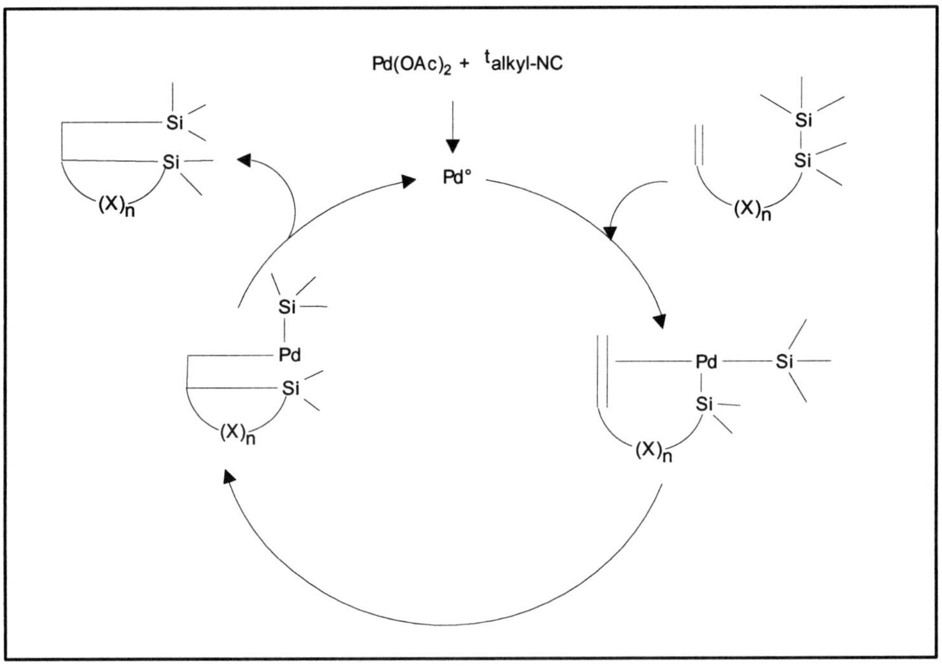
Pd(OAc)2 + talkyl-NC
Pd°
Si
Si
(X)n
Si
Si
(X)n
Pd
Si
Si
(X)n
Si
Pd
Si
(X)n

	References
Si-Si	Tamao K., *J. Organomet. Chem.*, 1976; 114, C19 Watanabe H., *J. Organomet. Chem.*, 1980; 186, 51 Watanabe H., *J. Organomet. Chem.*, 1981; 216, 149 Hayashi T., *Organometallics*, 1994; 3237 Kocienski P., *Synthesis*, 1994; 1301
Si-Sn	Chenard B.L., *J. Org. Chem.*, 1986; 3561 Mitchell T.N., *J. Org. Chem.*, 1987; 4868 Kocienski P., *Synthesis*, 1994; 1301 Mori M., *Chem. Lett.*, 1991; 1615
Sn-Sn	Piers E., *J. Chem. Soc., Chem. Commun.*, 1986; 626 Piers E., *J. Chem. Soc., Perkin Trans. 1*, 1989; 2124 Kocienski P., *Synthesis*, 1994; 1301 Piers E., *Can. J. Chem.*, 1994; 2468 Mitchell T.N., *J. Organomet. Chem.*, 1983; 241, C45

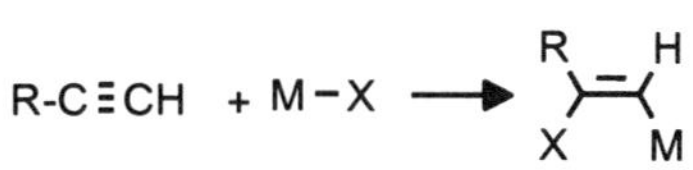	References
Si-CN	Chatani N., *J. Org. Chem.*, 1988; 3539 Ito Y., *Tetrahedron Lett.*, 1994; 8635 (intramolecular)
Ge-CN	Chatani N., *J. Org. Chem.*, 1990; 3393 Chatani N., *J. Organomet. Chem.*, 1994; 473, 335
RS-B(R')$_2$	Suzuki A., *J. Am. Chem. Soc.*, 1993; 7219 (then protonolysis)

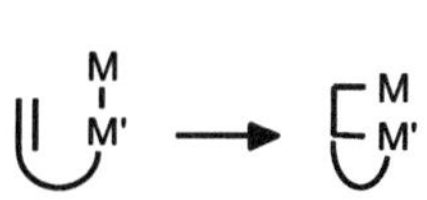	References
Si-Si	Ito Y., *J. Am. Chem. Soc.*, 1993; 6487

RXN23 1,2-Dimetallation of Isonitriles

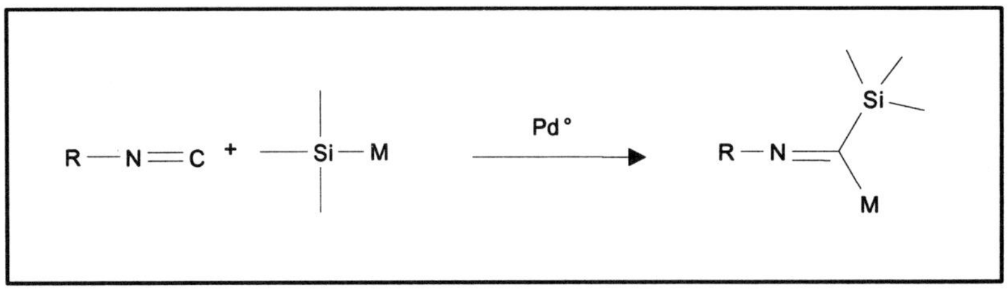

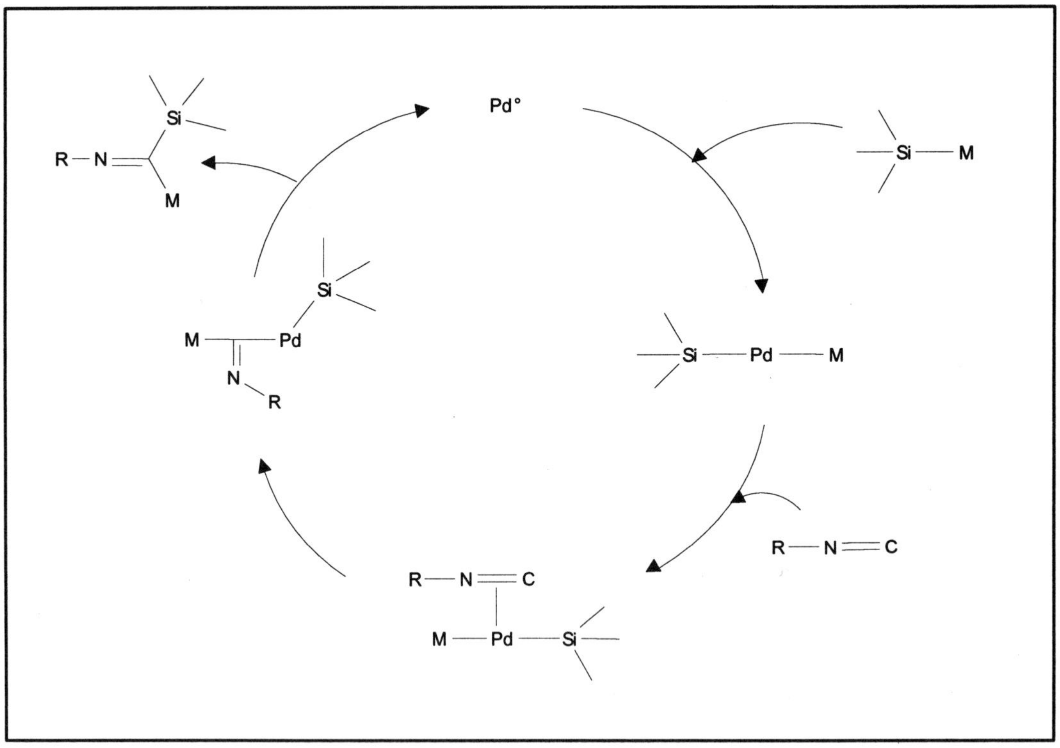

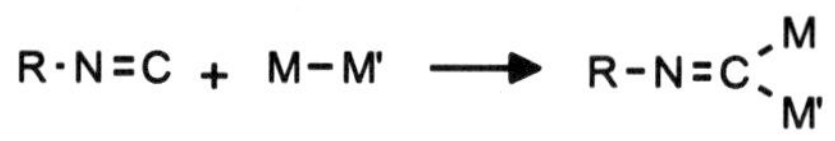	References
Si-Si	Ito Y., *J. Am. Chem. Soc.*, 1991; 8899
Si-Sn	Ito Y., *J. Chem. Soc., Chem. Commun.*, 1986; 980

RXN24 1,2-Dimetallation of Allenes or 1,3-Dienes

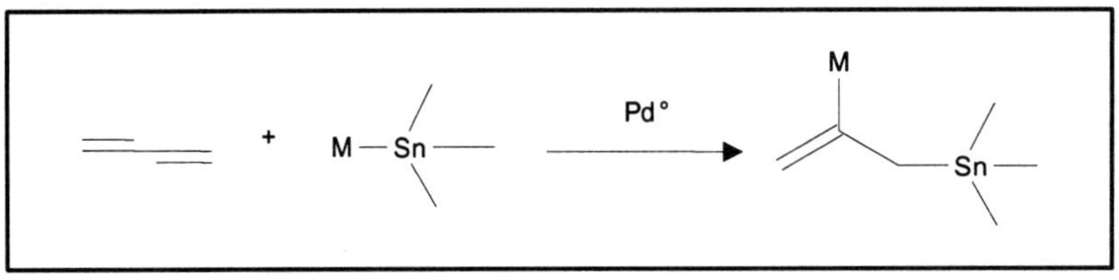

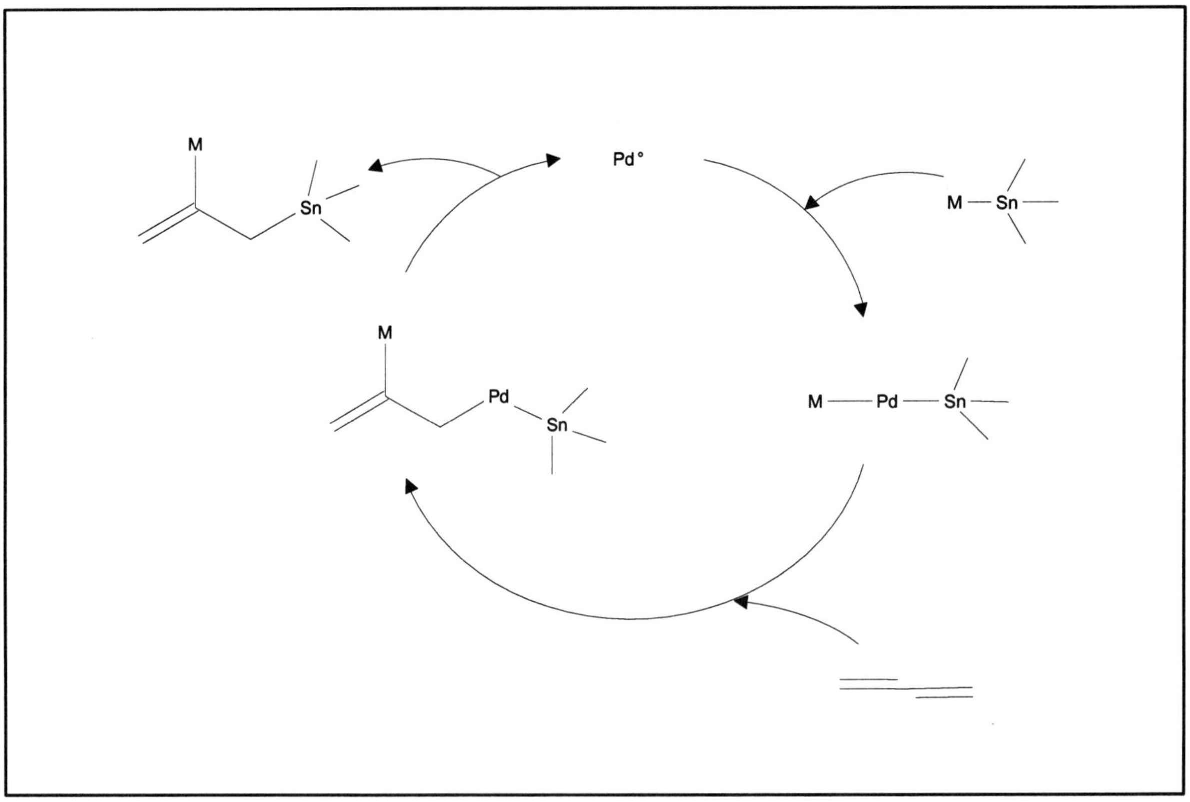

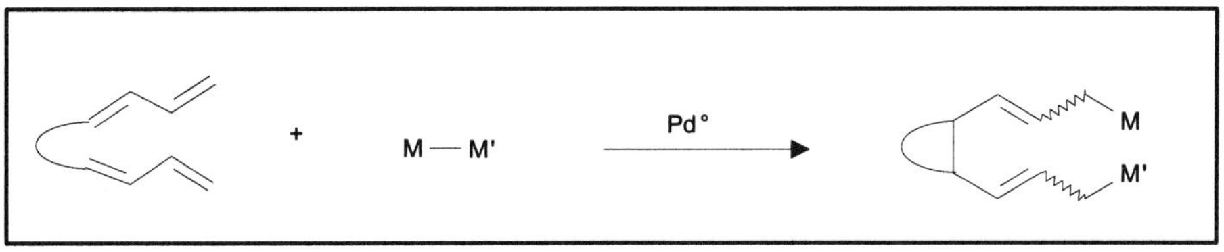
+
M — M'
Pd°
M
M'

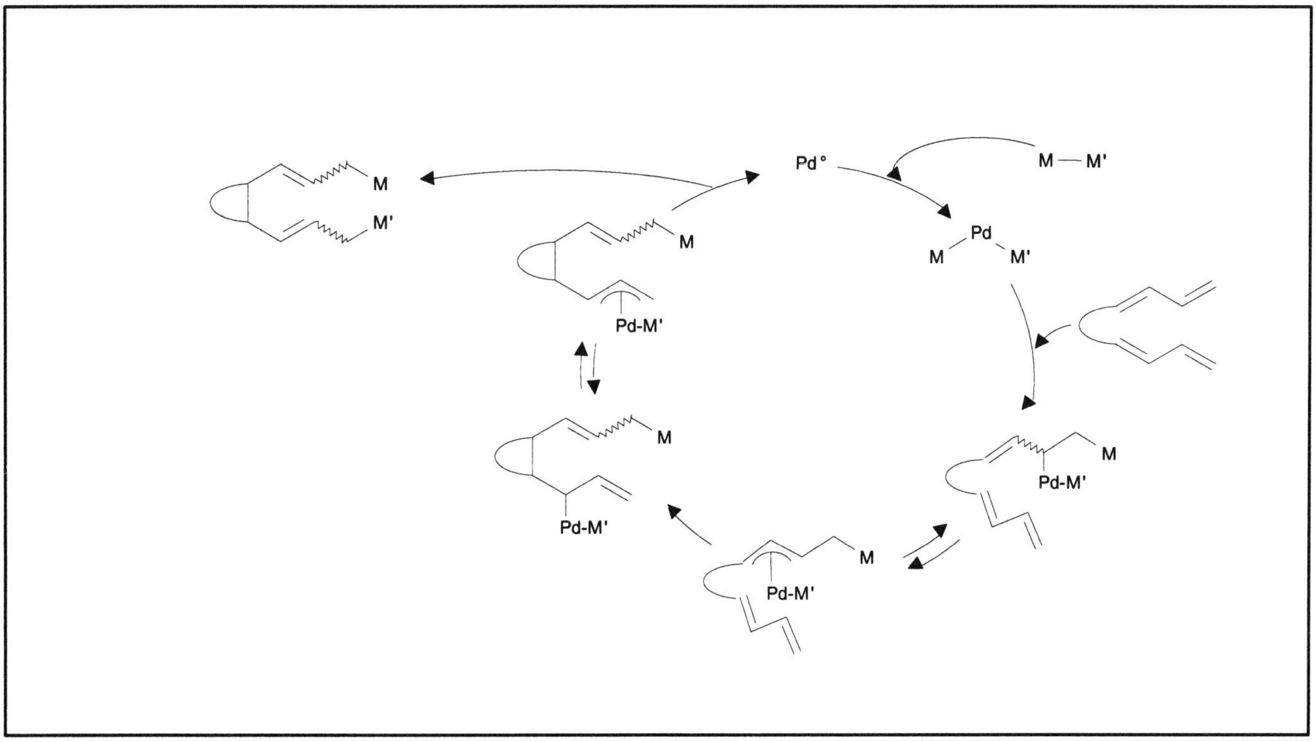
M
M'
Pd°
M — M'
Pd
M
M'
M
Pd-M'
M
Pd-M'
M
Pd-M'
M
Pd-M'

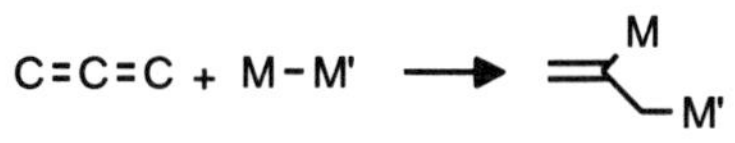	References
M-Sn	Mitchell T.N., *J. Organomet. Chem.*, 1991; 407, 319
Si-Si	Watanabe H., *J. Chem. Soc., Chem. Commun.*, 1981; 617

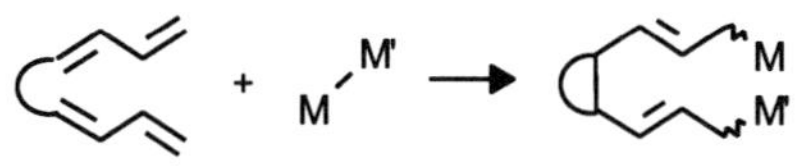	References
Sn-Sn	Tsuji Y., *J. Chem. Soc., Perkin Trans. 1* , 1995; 599
Sn-Si	Tsuji Y., *J. Chem. Soc., Perkin Trans. 1* , 1995; 599
Si-Si	Tsuji Y., *J. Chem. Soc., Perkin Trans. 1* , 1995; 599

RXN25 Coupling of Aryl Derivatives with Alkenes Involving a Pd(II) Catalyst

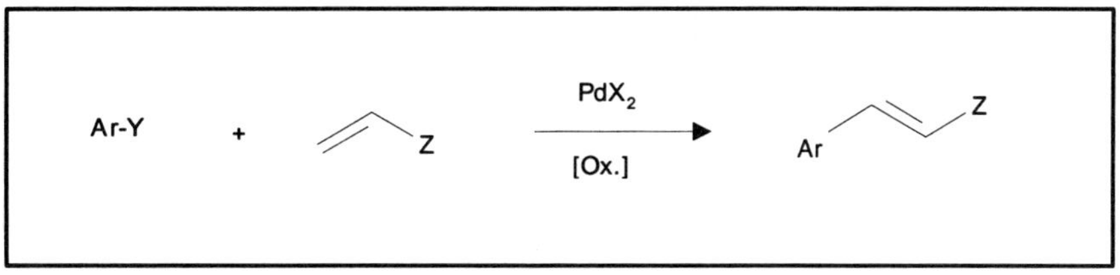

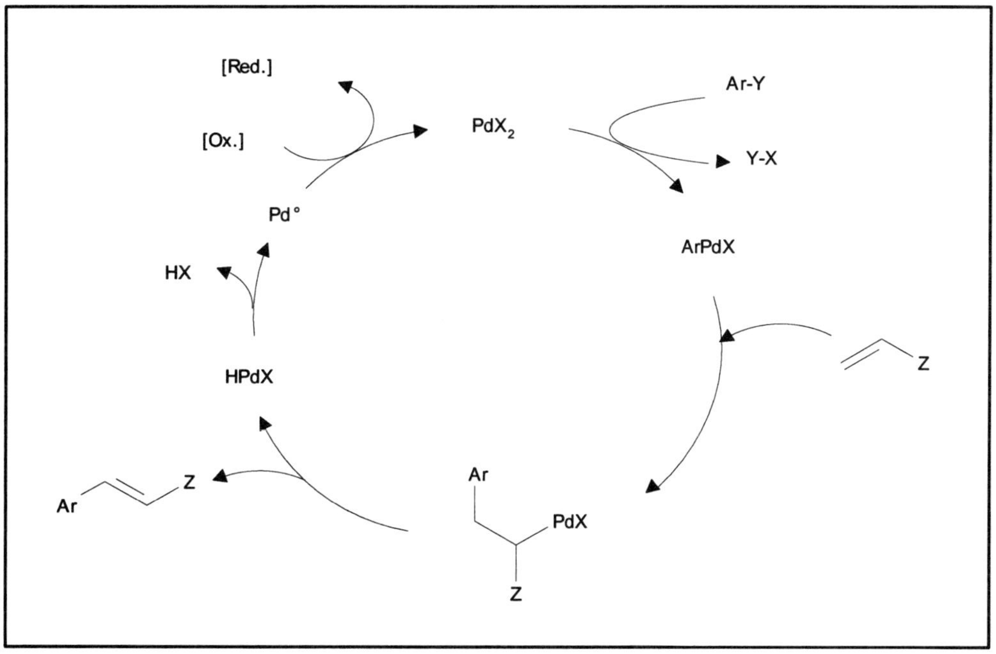

Y	References
H	Itahara T., *J. Org. Chem.*, 1985; 5546 Itahara T., *Synthesis*, 1984; 236 Pindur U., *Helv. Chim. Acta.*, 1990; 827 Tsuji J., *Tetrahedron*, 1984; 2699 Fujiwara Y., *J. Org. Chem.*, 1981; 851 Akermark B., *Tetrahedron Lett.*, 1995; 1325 Thal C., *Tetrahedron*, 1995; 1941
HgCl	Larock R.C., *Tetrahedron Lett.*, 1986; 2211 Larock R.C., *Tetrahedron Lett.*, 1988; 5069 Larock R.C., *J. Org. Chem.*, 1993; 2081
$NaB(Ph)_3$	Uemura S., *J. Organomet. Chem.*, 1993; 443, 253
Ar_2Sb	Uemura S., *Tetrahedron Lett.*, 1994; 1275

RXN26 Homocoupling of Aryl and Vinyl Derivatives

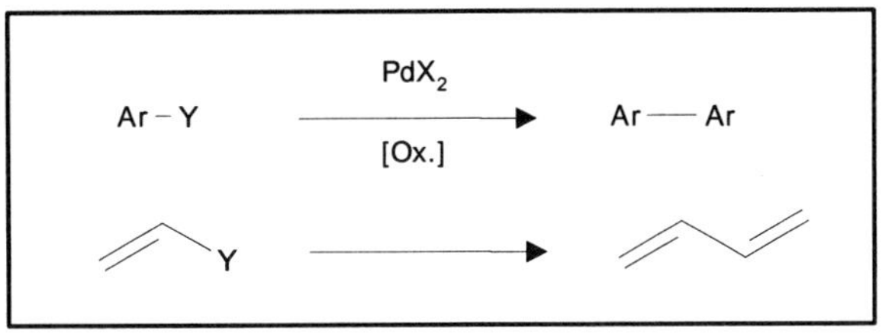

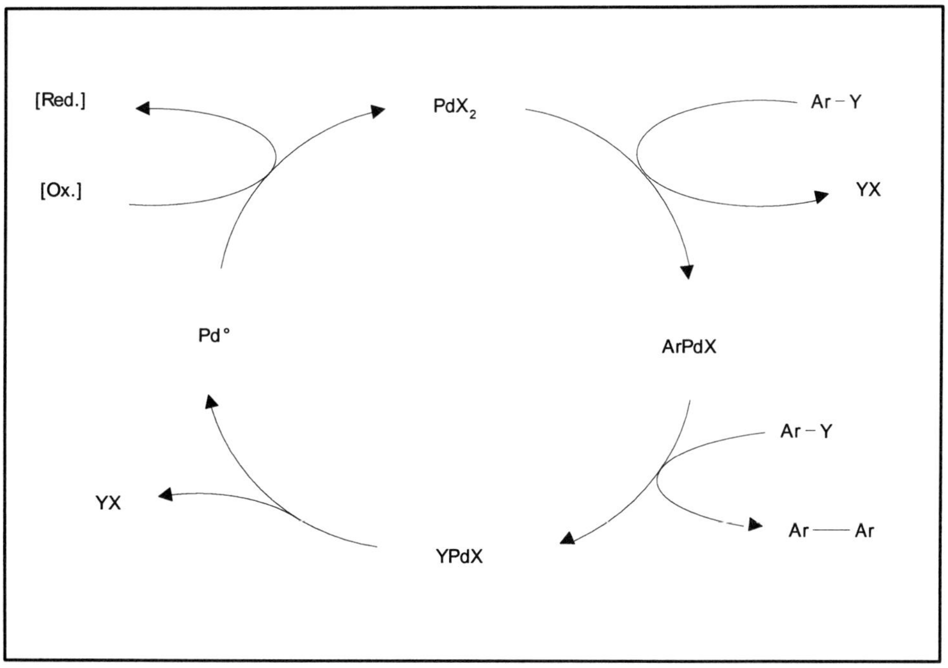

A ---> B	References
Ar−X ⟶ Ar−Ar	Heitz W., *Synlett.*, 1991; 809 Jutand A., *Synlett.*, 1993; 568
Ar−SO_2Cl ⟶ Ar−Ar	Miura M., *Chem. Lett.*, 1990; 459
Ar−HgCl ⟶ Ar−Ar	Heck R.F., *US Patent n°* 3, 539, 622 (1970)
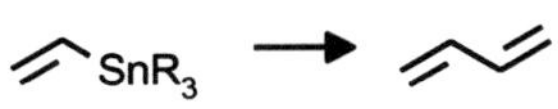	Danilova N.A., *Synthesis*, 1989; 633 Oshima K., *Chem. Lett.*, 1987; 5

RXN27 Codimerization of Alkynes

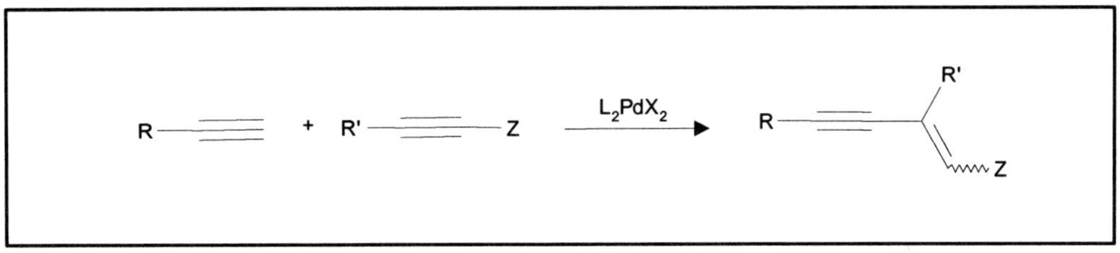

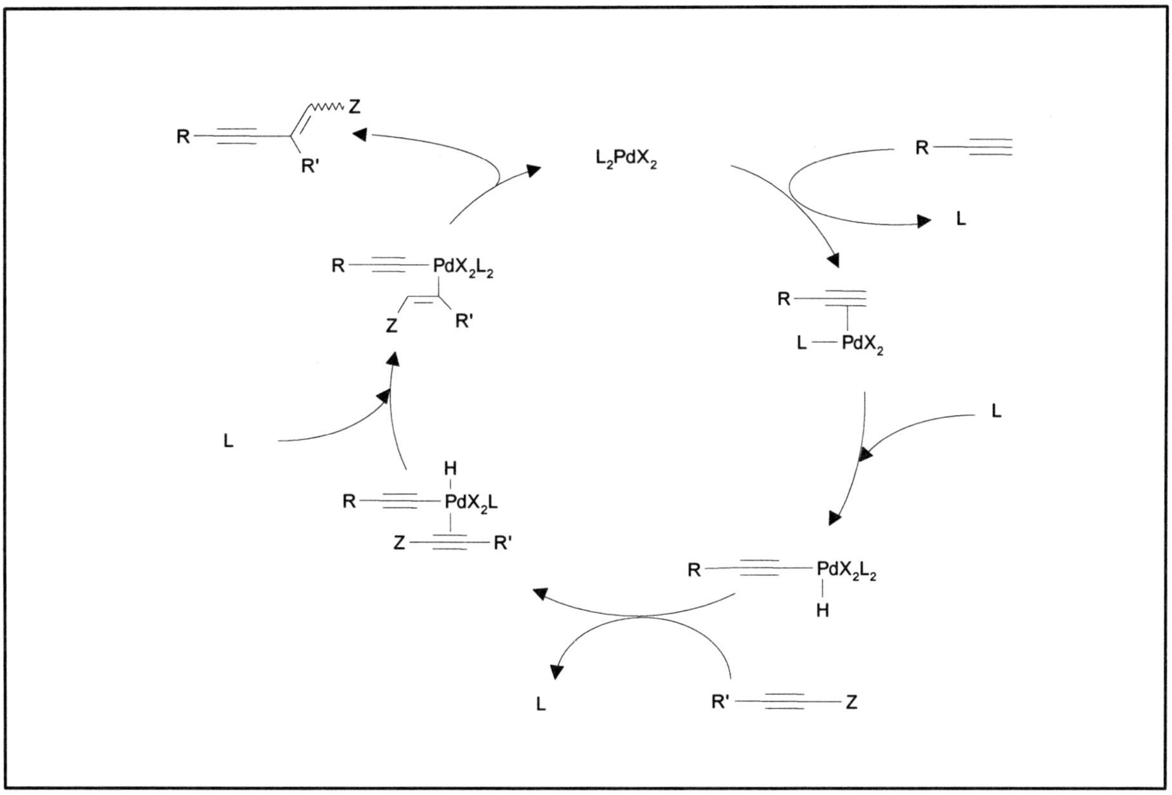

RXN27 Codimerization of Alkynes

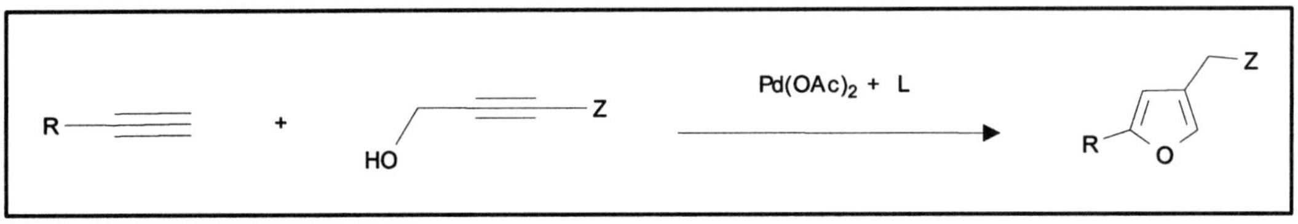

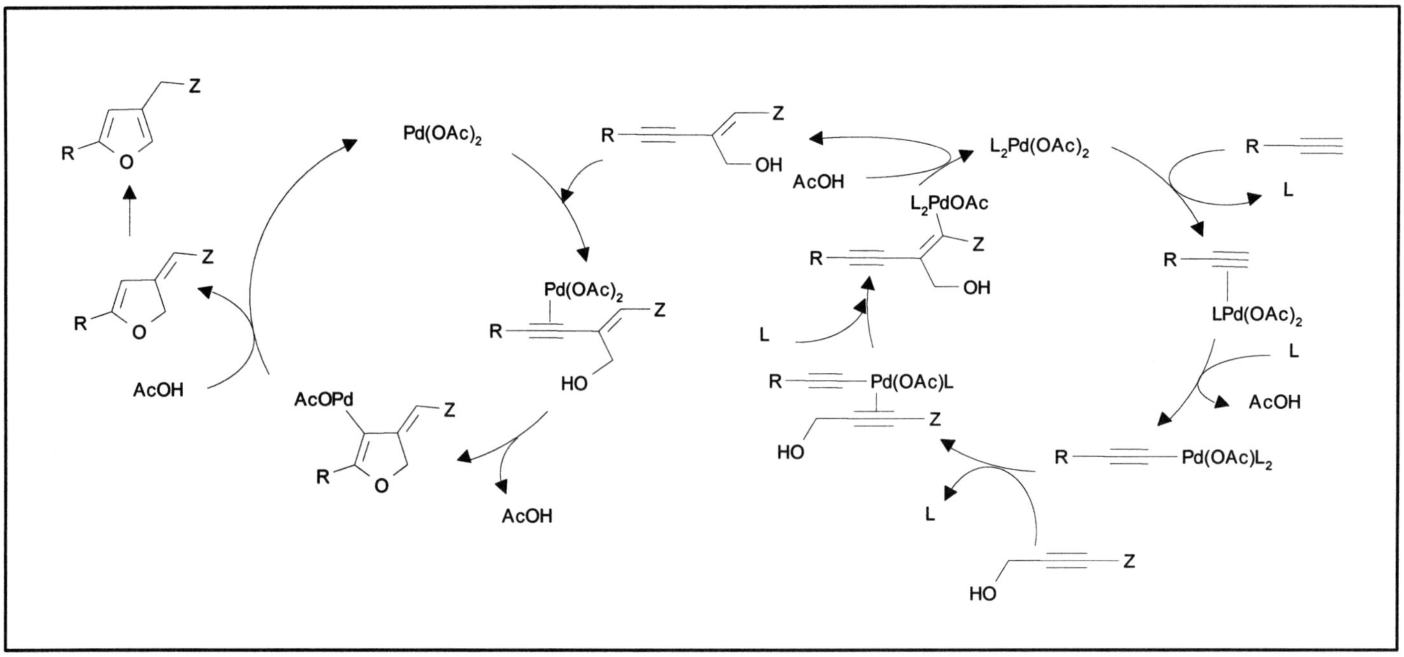

A ----> B	References
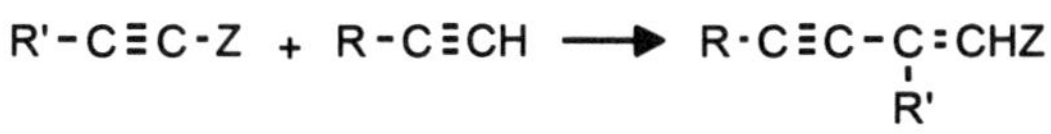	Trost B.M., *J. Am. Chem. Soc.*, 1987; 3486 Trost B.M., *Synthesis*, 1994; 1267
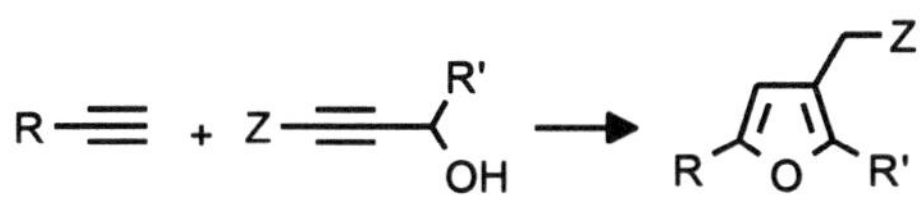	Trost B.M., *J. Am. Chem. Soc.*, 1995; 7255 (tandem reaction)

RXN28 Codimerization of Terminal Alkynes with Allenes

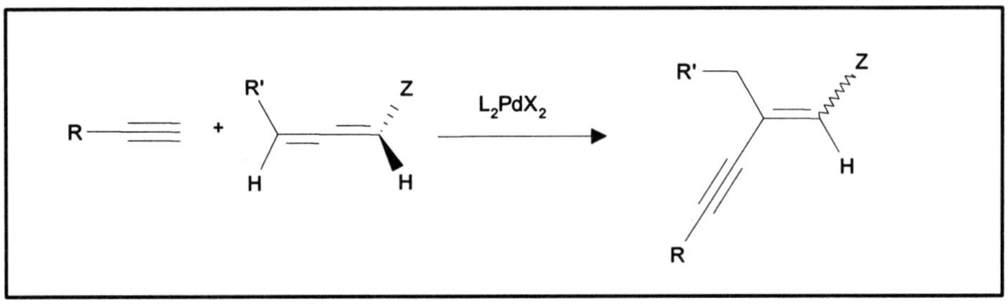

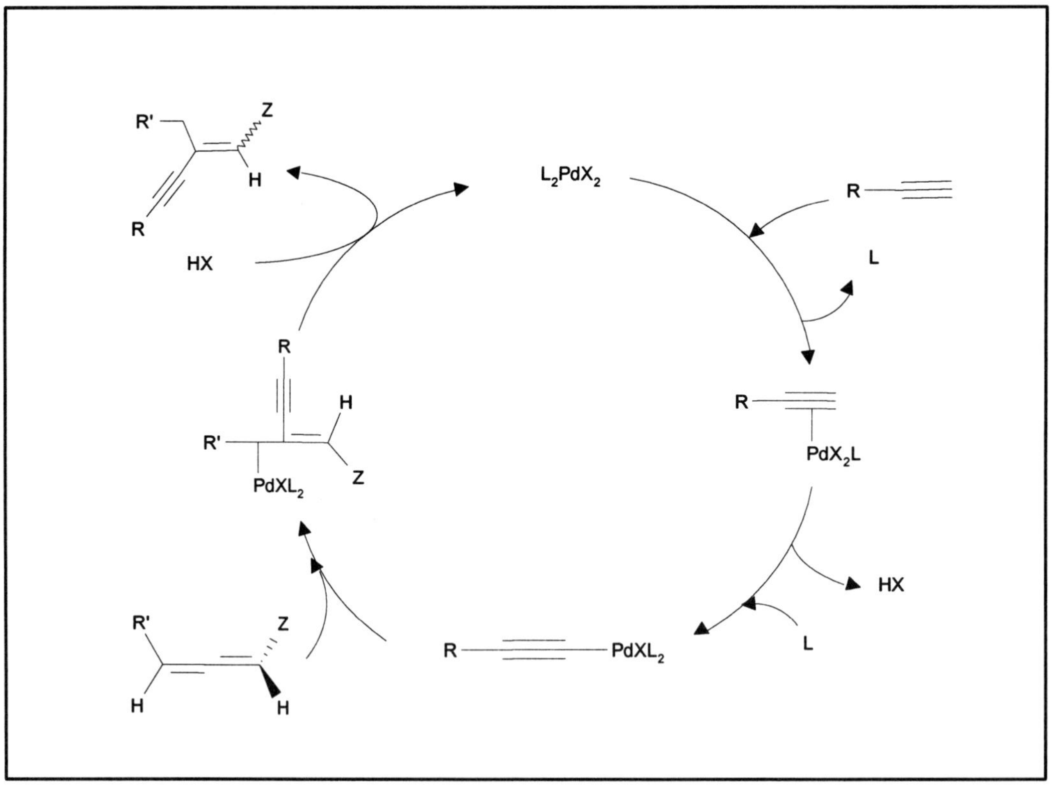

 J.-L. Malleron et al

A ----> B	References
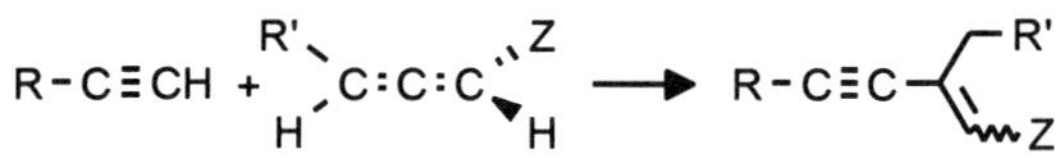	Trost B.M., *J. Am. Chem. Soc.*, 1990; 2816

RXN28

RXN29 Codimerization of Alkynes and Allyl Halides

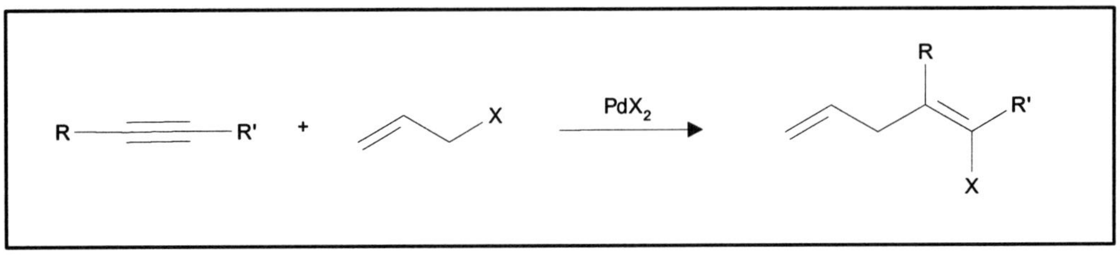

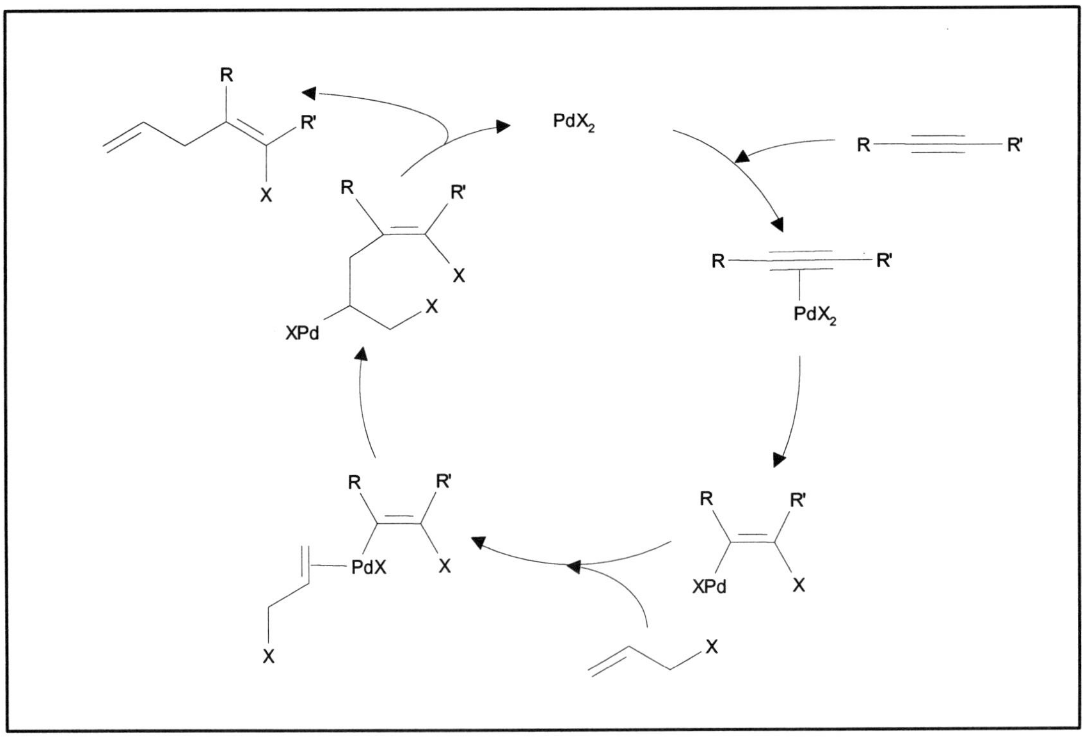

RXN29 Codimerization of Alkynes and Allyl Halides

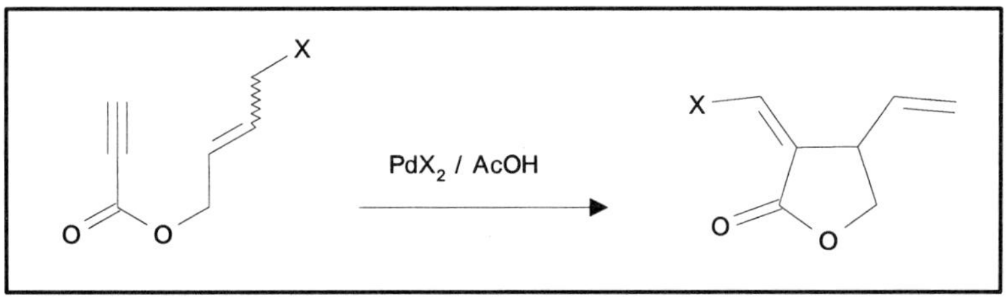

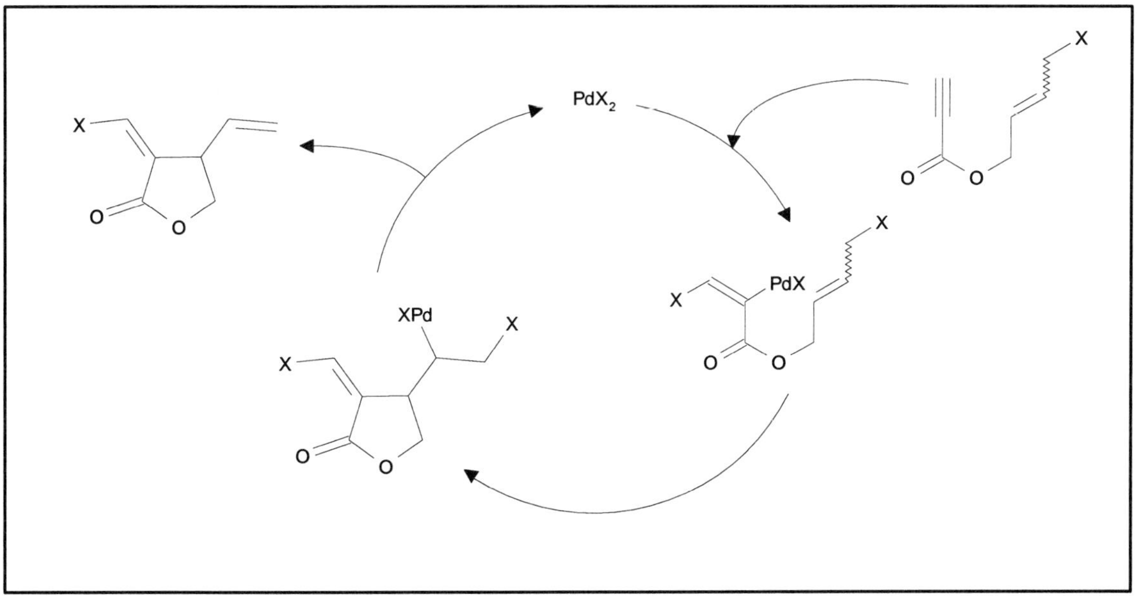

A ----> B	References
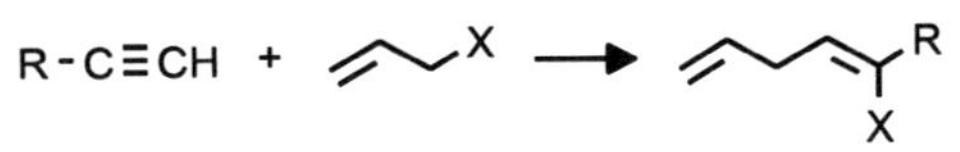	Kaneda K., *J. Org. Chem.*, 1979; 55 Camps F., *Tetrahedron Lett.*, 1988; 5811 Kosugi M., *Bull. Chem. Soc. Jpn.*, 1993; 3058
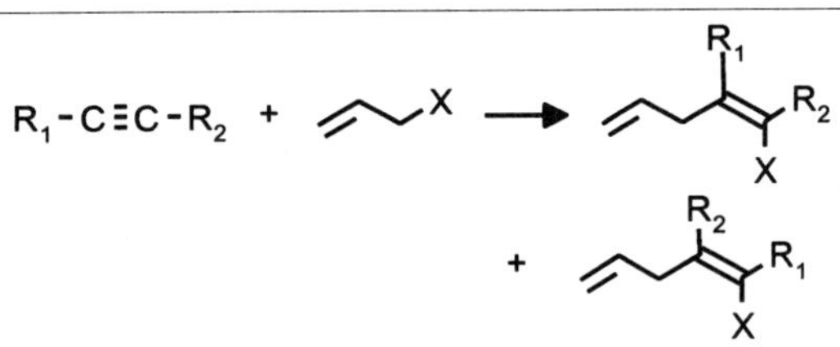	Kaneda K., *J. Org. Chem.*, 1979; 55 Yamaguchi R., *Chem. Lett.*, 1982; 1485
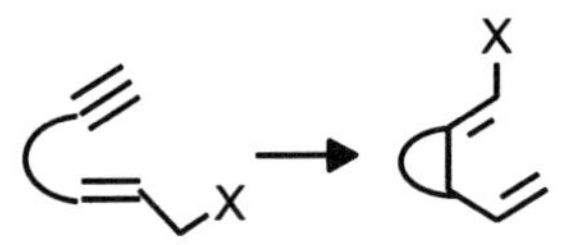	Lu X., *J. Chem. Soc., Chem. Commun.*, 1990; 733 Lu X., *J. Org. Chem.*, 1991; 5120 Lu X., *J. Org. Chem.*, 1993; 3692
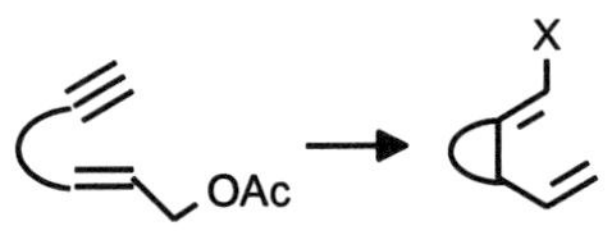	Lu X, *J. Chem. Research (S)*, 1993; 366 Lu X., *J. Org. Chem.*, 1995; 1087

RXN30 Cyclopropanation of Alkenes and 1,3-Dienes by Diazomethane

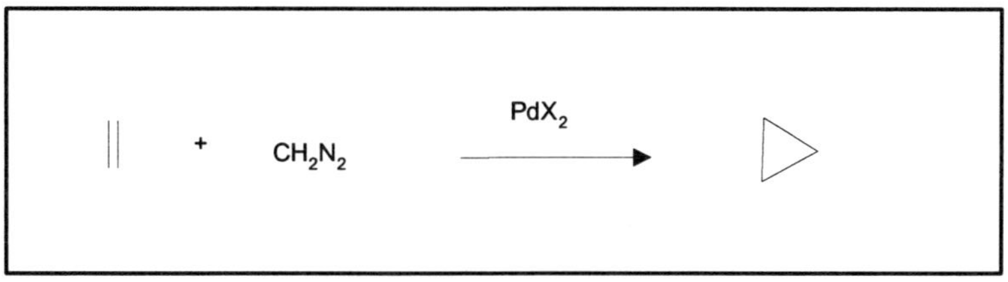

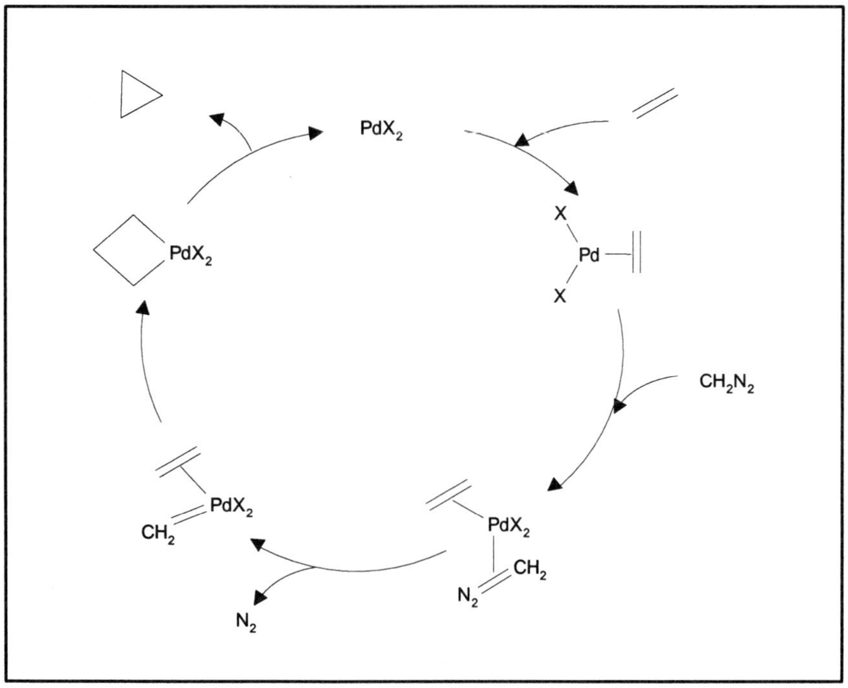

A ---> B	References
	Mende U., *Tetrahedron Lett.*, 1975; 629
	Hacksell U., *J. Chem. Soc., Perkin Trans. 1*, 1994; 461
	Tomilov Y.V., *Synthesis*, 1990; 246
	Anciaux A.J., *J. Org. Chem.*, 1980; 695 (with N_2CHCO_2Et)
	Tomilov Y.V., *Synthesis*, 1990; 246

RXN31 Rearrangement of α-Hydroxy Diazo Compounds

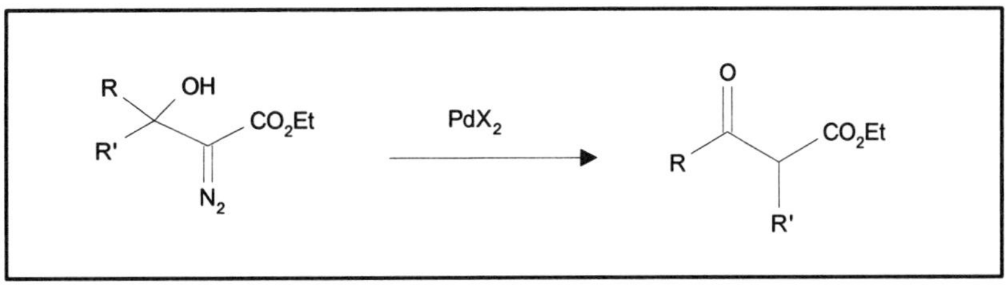

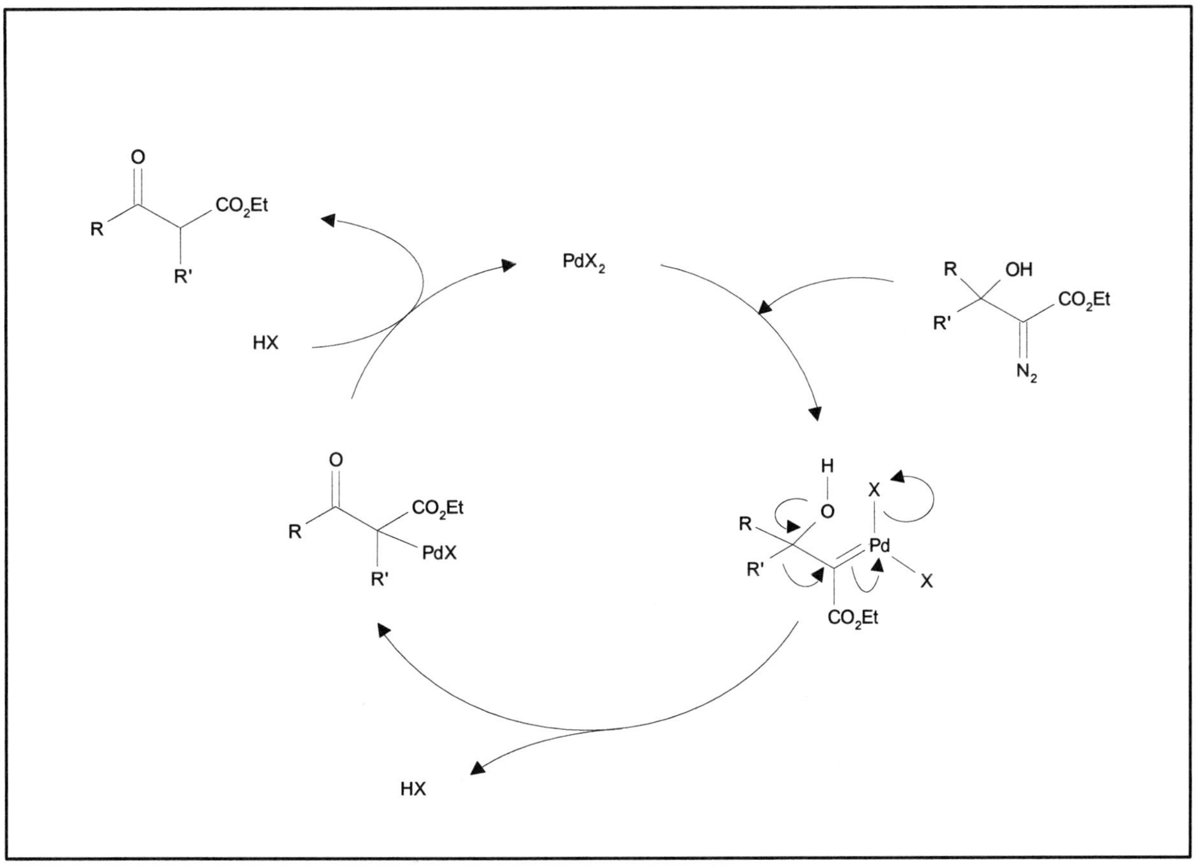

RXN31

A ---> B	References
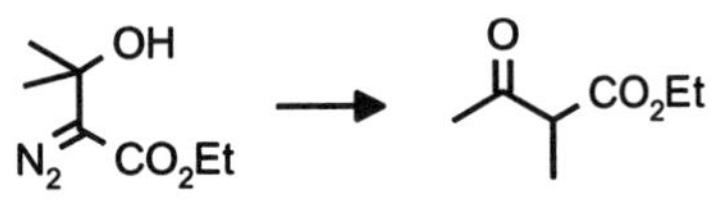	Kim S. W., *Synthesis*, 1983; 197

RXN32 Substitution, Addition and Elimination on Pro- π -Allyl Substrates

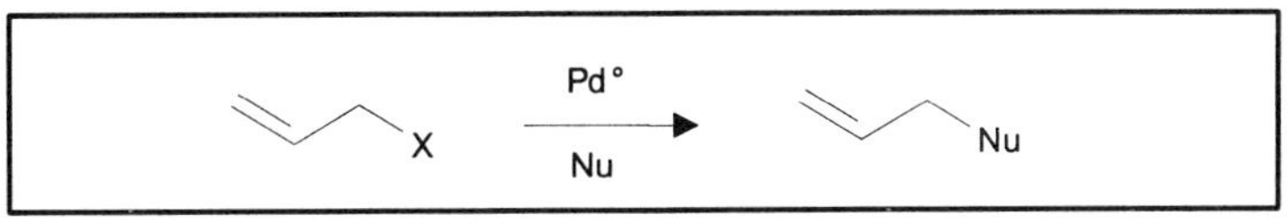

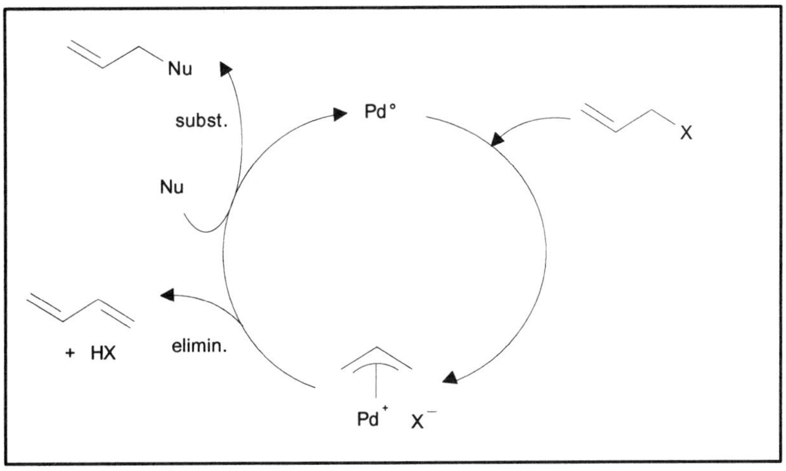

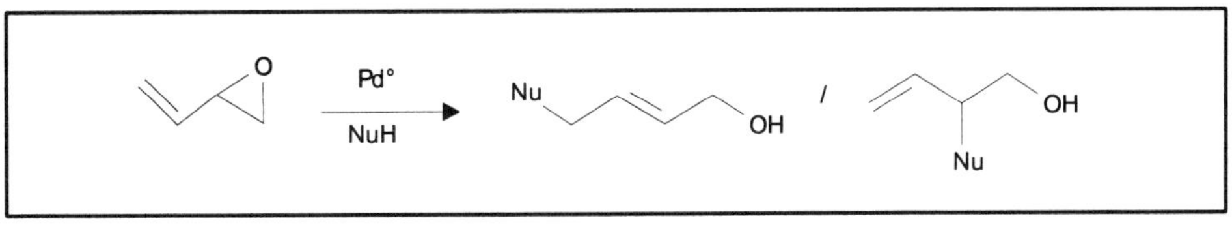

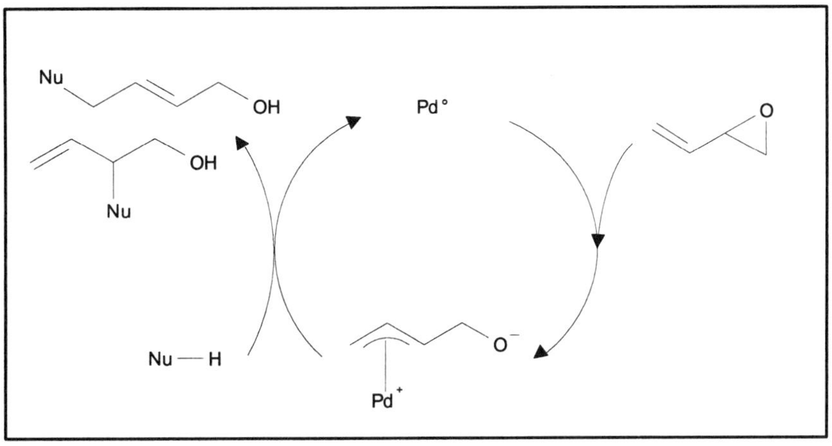

RXN32 Substitution, Addition and Elimination on Pro- π -Allyl Substrates

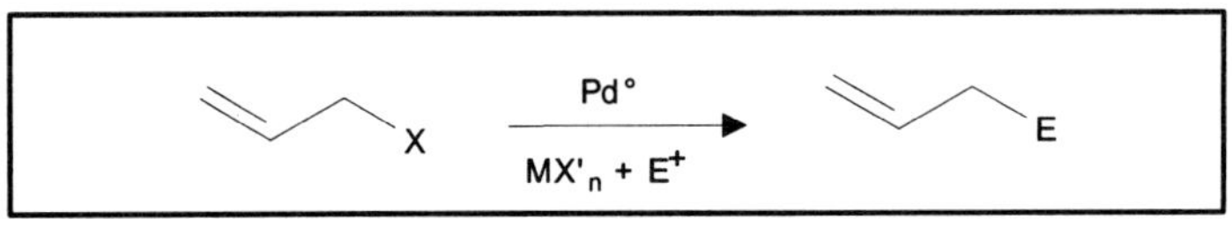

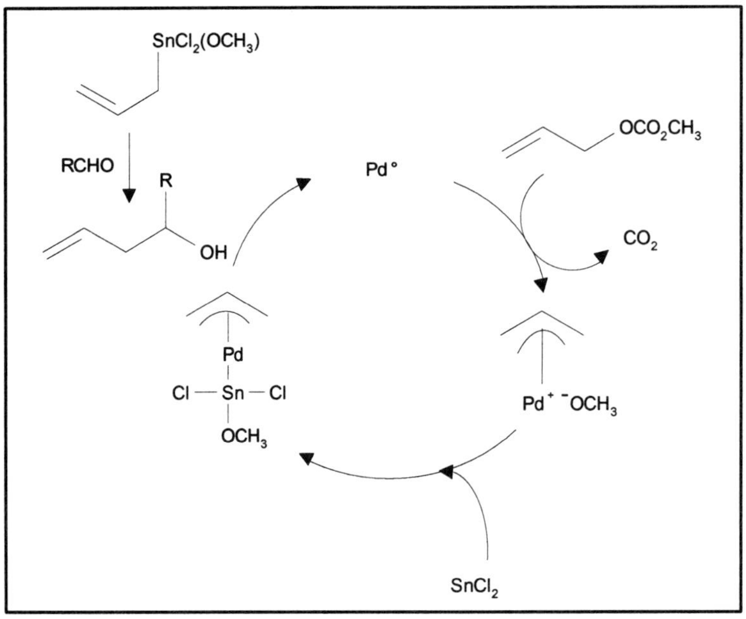

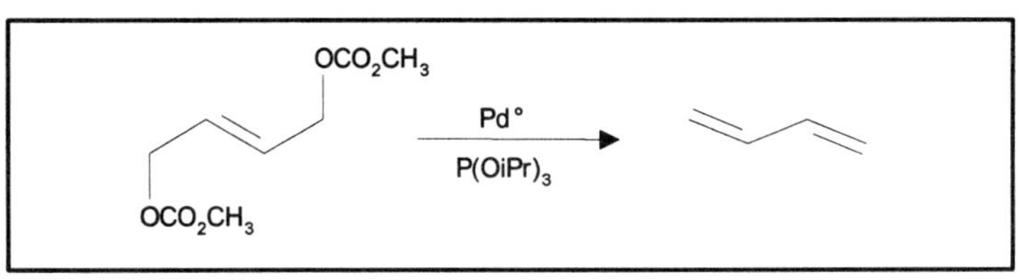

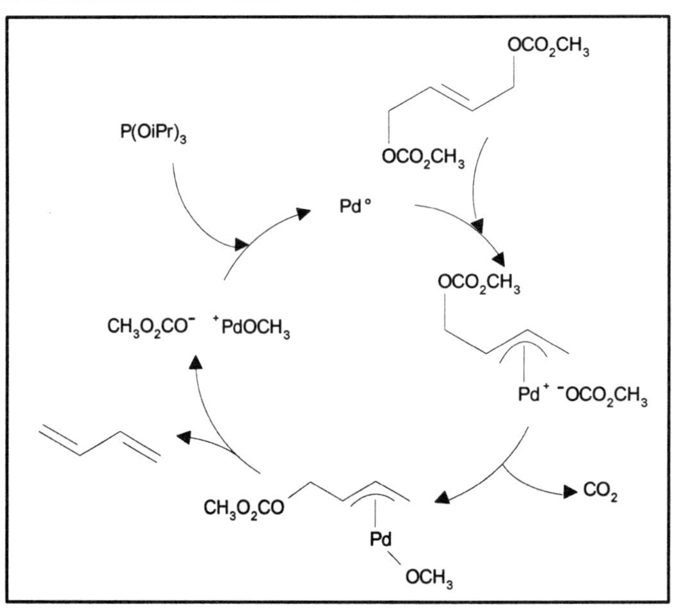

Leaving Group	References
OH	Atkins K.E., *Tetrahedron Lett.*, 1970; 3821 Bergbreiter D.E., *J. Chem. Soc., Chem. Commun.*, 1989; 883 Miura M., *J. Chem. Soc., Perkin Trans. 1*, 1992; 2833 Sakakibara M., *Tetrahedron Lett.*, 1994; 8013
OR	Klumpp G.W., *Tetrahedron Lett.*, 1988; 3579 Takahashi K., *Bull. Chem. Soc. Jpn.*, 1972; 230 Kusama T., *Chem. Pharm. Bull.*, 1992; 1718
OC_6H_5	Fiaud J.C., *J. Organomet. Chem.*, 1978; 154, 175 Takahashi K., *Bull. Chem. Soc. Jpn.*, 1972; 230 Tsuji J., *Tetrahedron Lett.*, 1978; 2075 (elimination)
OAc	Trost B.M., *J. Am. Chem. Soc.*, 1976; 630 Sinou D., *Tetrahedron Lett.*, 1991; 2025 Trost B.M., *J. Am. Chem. Soc.*, 1994; 4089 (asymmetric catalysis) Andersson P. G., *Organometallics*, 1995; 14, 1 (elimination) Trost B.M., *Tetrahedron Lett.*, 1979; 2301 (elimination)
OCOtBu	Fiaud J.C., *J. Org. Chem.*, 1990; 4840
OCOR	McCombie S.W., *J. Org. Chem.*, 1982; 587 (deallylation) Genêt J.P., *Tetrahedron*, 1994; 497 (deallylation) Genêt J.P., *Synlett.*, 1993; 680 (deallylation) Genêt J.P., *Tetrahedron Lett.*, 1994; 8783 (deallylation)
OCOAr	Fiaud J.C., *J. Org. Chem.*, 1990; 4840 Genêt J.P., *Tetrahedron Lett.*, 1995; 3007 Blart E., *Thesis, Université Paris VI*, 1993 (deallylation)
OCHO	Tsuji J., *Tetrahedron Lett.*, 1992; 2987
$OCOCH_2COCH_3$	Fiaud J.C., *Tetrahedron Lett.*, 1982; 5279
$OCOCH_2Cl$	Tanikaga R., *J. Chem. Soc., Perkin Trans. 1*, 1990; 1185
$OCOCH_2P(Ph)_2$	Kocovsky P., *J. Am. Chem. Soc.*, 1989; 4981 Kocovsky P., *Tetrahedron*, 1992; 7229
OCO_2R	Tsuji J., *J. Org. Chem.*, 1985; 1523 Tsuji J., *Tetrahedron Lett.*, 1982; 4809 Genêt J.P., *Synlett.*, 1993; 680 (deallylation) Genêt J.P., *Tetrahedron Lett.*, 1993; 4189 (deallylation) Genêt J.P., *Tetrahedron*, 1994; 497 (deallylation) Ikegami S., *Tetrahedron Lett.*, 1986; 2885 Trost B.M., *J. Am. Chem. Soc.*, 1994; 4089 (asymmetric catalysis) Kondo K., *Synlett.*, 1995; 609 (deallylation) Genêt J.P., *Tetrahedron Lett.*, 1994; 8783 (dealkylation)

Leaving Group	References
$OCO_2N{=}C(Me)_2$	Sato K., *Chem. Lett.*, 1988; 291
OCONHR	Genêt J.P., *Tetrahedron Lett.*, 1993; 4189 (deallylation) Genêt J.P., *Tetrahedron*, 1994; 497 (deallylation) Blechert S., *Synthesis*, 1995; 592 (deallylation)
$OCONR_2$	Genêt J.P., *Tetrahedron Lett.*, 1993; 4189 (deallylation) Genêt J.P., *Tetrahedron*, 1994; 497 (deallylation) Genêt J.P., *Synlett.*, 1993; 680 (deallylation) Tsuji J., *Tetrahedron Lett.*, 1985; 2449 Guibé F., *Tetrahedron Lett.*, 1986; 2365 (deallylation) Guibé F., *Tetrahedron Lett.*, 1992; 477 (deallylation) Genêt J.P., *Tetrahedron Lett.*, 1994; 8783 (deallylation)
$ON{=}C(Me)_2$	Sato K., *Chem. Lett.*, 1988; 291
OC(R)=NR'	Sato K., *Bull. Chem. Soc. Jpn.*, 1989; 239
OC(NHR)=NR'	Inoue Y., *Bull. Chem. Soc. Jpn.*, 1984; 3021
OC(SMe)=S	Bosnich B., *J. Chem. Soc., Chem. Commun.*, 1986; 146
$OPO(OR)_2$	Tanigawa Y., *Tetrahedron Lett.*, 1982; 5549 Oshima K., *Bull. Chem. Soc. Jpn.*, 1985; 1196 (umpolung)
$OP(Ph)_3^+$	Falck J.R., *Tetrahedron Lett.*, 1992; 2091
OTs	Salaün J., *Tetrahedron Lett.*, 1990; 4593
OTHP	Masuyama Y., *Tetrahedron Lett.*, 1992; 6477 (umpolung)
ONO_2	Ruzziconi R., *Tetrahedron Lett.*, 1993; 6333
$OAlMe_2$	Negishi E.I., *Tetrahedron Lett.*, 1981; 3737
OBR_2	Lu X., *J. Organomet. Chem.*, 1988; 344, 109
$OB(Ph)_3$	Kocovsky P., *Tetrahedron Lett.*, 1993; 179
$OSiMe_3$	Hayashi T., *J. Organomet. Chem.*, 1985; 285, 359
NEt_2	Atkins K.E., *Tetrahedron Lett.*, 1970; 3821 Kumobayashi H., *Chem. Lett.*, 1986; 157 (elimination)
NR_2	Guibé F., *J. Org. Chem.*, 1993; 6109

Leaving Group	References
RNH_2^+	Genêt J.P., *Tetrahedron Lett.*, 1995; 1267
R_2NH^+	Genêt J.P., *Tetrahedron Lett.*, 1995; 1267
R_3N^+	Dzhemilev U.M., *Bull. Acad. Sci. USSR, Div. Chem. Sci.*, 1987; 365 Hirao T., *J. Organomet. Chem.*, 1982; 236, 409
N(Ts)(R)-	Junk M.E., *J. Org. Chem.*, 1994; 4719
NO_2	Hegedus L.S., *J. Am. Chem. Soc.*, 1982; 3727 Ono N., *J. Chem. Soc., Chem. Commun.*, 1985; 523 Ono N., *J. Org. Chem.*, 1986; 3734
SO_2Ph	Trost B.M., *J. Am. Chem. Soc.*, 1980; 5979 Julia M., *J. Chem. Soc., Chem. Commun.*, 1994; 1905
R_2S^+	Hirao T., *J. Organomet. Chem.*, 1982; 236, 409
	Hesse M., *Synthesis*, 1992; 931
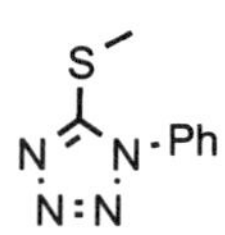	Tsuboyama K., *Chem. Pharm. Bull.*, 1990; 2357
Br	Fukumoto K., *Synlett.*, 1994; 597 Fukumoto K., *Tetrahedron*, 1995; 6927
Cl	Kurusawa H., *J. Am. Chem. Soc.*, 1990; 2813
C=C-C-L	Frost C.G., *Tetrahedron: Asymmetry*, 1992; 1089

Substrate	References
$C{=}CBr\text{-}OPO(OR)_2$	Nwokogu G.C., *Tetrahedron Lett.*, 1985; 3900
C=CF-C-OAc	Johnson W.S., *Tetrahedron Lett.*, 1992; 8001
$CF_2\text{-}C{=}C(R)\text{-}C\text{-}OAc$	Shi G.Q., *Tetrahedron Lett.*, 1995; 6305
$C{=}C(CH_2SnR_3)\text{-}C\text{-}OAc$	Trost B.M., *J. Am. Chem. Soc.*, 1985; 8277

Substrate	References
C=C(CH$_2$ZnBr)-C-OR	Klumpp G.W., *Tetrahedron Lett.*, 1988; 3579 Klumpp G.W., *Tetrahedron*, 1992; 9877
C=C(CH$_2$MgCl)-C-OR	Klumpp G.W., *Tetrahedron*, 1992; 9877 Klumpp G.W., *Tetrahedron*, 1992; 9901
C=C(CF$_3$)-CH$_2$-OR	Taguchi T., *Chem. Pharm. Bull.*, 1990; 1104 (umpolung)
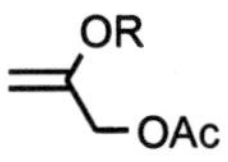	Kosugi M., *Chem. Lett.*, 1987; 1237 Kosugi M., *Bull. Chem. Soc. Jpn.*, 1989; 3383
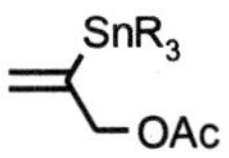	Carpita A., *Tetrahedron*, 1994; 4853
C=C(CH$_2$OAc)$_2$	Lu X., *Tetrahedron Lett.*, 1987; 6219 Malacria M., *Tetrahedron Lett.*, 1995; 6447
C=C(CH$_2$Cl)$_2$	Najera C., *Tetrahedron Lett.*, 1995; 7697
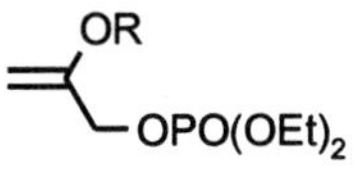	Kosugi M., *Chem. Lett.*, 1987; 1237
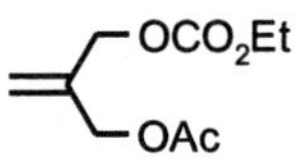	Mitsudo T., *Tetrahedron Lett.*, 1986; 5389
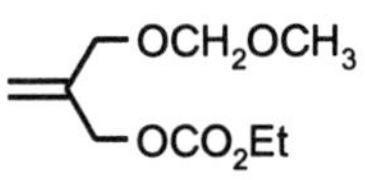	Okahara M., *Synthesis*, 1990; 32
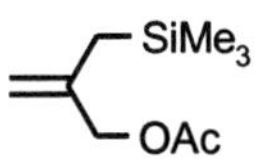	Trost B.M., *J. Am. Chem. Soc.*, 1979; 6429 Trost B.M., *J. Am. Chem. Soc.*, 1979; 6432 Trost B.M., *Angew. Chem. Int., Ed. Engl.*, 1986; 1 Trost B.M., *Angew. Chem. Int., Ed. Engl.*, 1989; 213
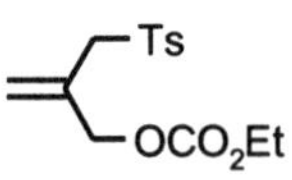	Tsuji J., *Tetrahedron Lett.*, 1984; 5183 Hayashi T., *Tetrahedron Lett.*, 1989; 375
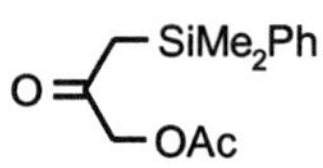	Trost B.M., *J. Am. Chem. Soc.*, 1989; 4430
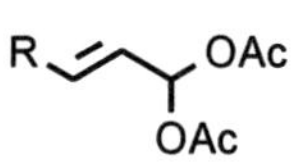	Lu X., *J. Organomet. Chem.*, 1984; 268, 185 Trost B.M., *Tetrahedron Lett.*, 1985; 131 Trost B.M., *J. Am. Chem. Soc.*, 1995; 7247 (asym. catalysis)

Substrate	References
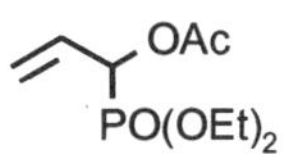	Lu X., *Tetrahedron Lett.*, 1987; 1897
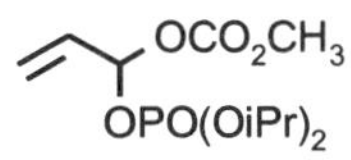	Öhler E., *Synthesis*, 1995; 539
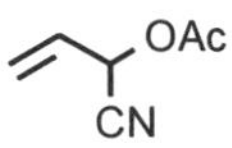	Tsuji J., *Tetrahedron Lett.*, 1981; 2573
C=C-C($SiMe_3$)-OAc	Hirao T., *Tetrahedron Lett.*, 1981; 3079
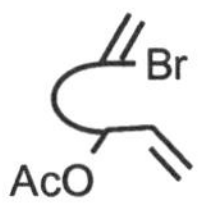	Trost B.M., *J. Am. Chem. Soc.*, 1987; 3487
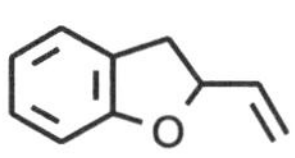	Murahashi S.I., *Bull. Chem. Soc. Jpn.*, 1986; 2191
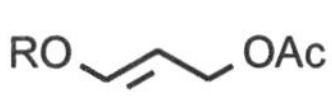	Cazes B., *Tetrahedron Lett.*, 1995; 535
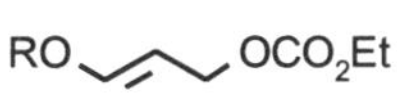	Cazes B., *Tetrahedron Lett.*, 1995; 535
RS-C=C-C-OAc	Godleski S.A., *J. Org. Chem.*, 1984; 2246
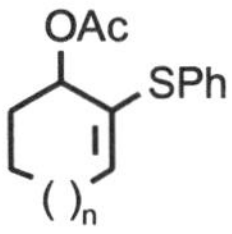	Vankar Y.D., *Tetrahedron*, 1995; 4829
Me_3Si-C=C-C-OR	Salaün J., *Synlett.*, 1994; 949
Me_3Si-C=C-C-OAc	Trost B.M., *J. Org. Chem.*, 1984; 468
Me_3Si-C=C-C-OCO_2R	Tsuji J., *Tetrahedron Lett.*, 1988; 343
R_3Si-C=C-C-OCO_2CH_3	Hayashi T., *Tetrahedron Lett.*, 1994; 4813 (asymmetric catalysis)

Substrate	References
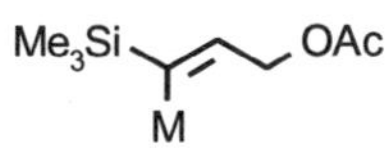	Lautens M., *Angew. Chem. Int. Ed. Engl.*, 1994; 2448
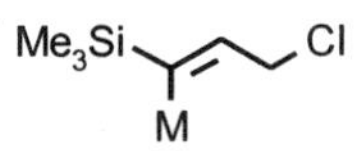	Lautens M., *Angew. Chem. Int. Ed. Engl.*, 1994; 2448
$R'O_2C$-C=C-C-OR	Tanikaga R., *J. Chem. Soc., Perkin Trans. 1*, 1990; 1185
RO_2C-C=C-C-OAc	Tsuji J., *Tetrahedron Lett.*, 1981; 2573
CF_3-C=C-C-OR	Kobayashi Y., *Chem. Pharm. Bull.*, 1988; 4209
C=C-CH$(OEt)_2$	Negishi E., *J. Org. Chem.*, 1985; 3406 Tamaru Y., *Angew. Chem., Int. Ed. Engl.*, 1995; 787
C=C-C$(OEt)_3$	Negishi E., *J. Org. Chem.*, 1985; 3406
XC-C=C-CX	Hayashi T., *J. Org. Chem.*, 1993; 6826
HO-C-C=C-C-OH	Atkins K.E., *Tetrahedron Lett.*, 1970; 3821
Cl-C-C=C-C-OAc	Backwall J.E., *Tetrahedron Lett.*, 1982; 1617
Cl-CH_2-HC=CH-CH_2-Cl	Najera C., *Tetrahedron Lett.*, 1995; 7697
RO_2CO-C-C=C-C-OCO_2R	Trost B.M., *J. Org. Chem.*, 1988; 915 (elimination) Trost B.M., *Synthesis*, 1991; 1235 (elimination) Sinou D., *Tetrahedron Lett.*, 1994; 6093 (asymmetric catalysis)
AcO-C-C=C-C-OAc	Buono G., *J. Org. Chem.*, 1995; 852 (elimination)
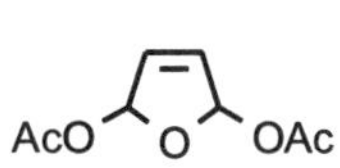	Holzapfel C.W., *Tetrahedron*, 1995; 8555
HO-C-C=C-C(OAc)	Takano S., *Tetrahedron Lett.*, 1993; 8485
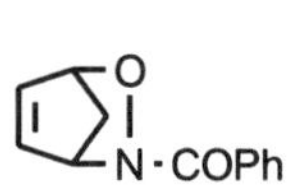	Procter G., *Tetrahedron Lett.*, 1990; 1043
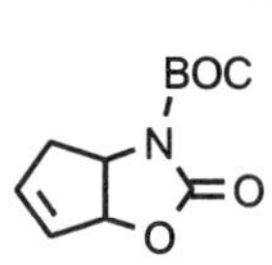	Procter G., *Tetrahedron Lett.*, 1995; 7539

Substrate	References
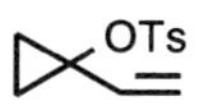	Salaün J., *Tetrahedron Lett.*, 1993; 4193
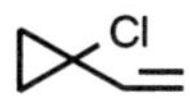	De Meijere A., *Synlett.*, 1995; 355
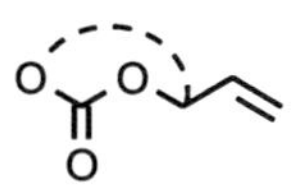	Kang S.K., *Tetrahedron: Asymmetry,* 1992; 1139 Kang S.K., *Tetrahedron Lett.*, 1994; 2357 Hanaoka M., *Tetrahedron Lett.*, 1994; 9425
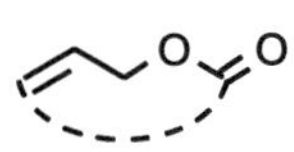	Noels A.F., *J. Org. Chem.*, 1976; 2527 (elimination) Trost B.M., *Tetrahedron*, 1981; 2999 Trost B.M., *J. Am. Chem. Soc.*, 1981; 1864 Negishi E.I., *J. Chem. Soc., Chem. Commun.*, 1982; 160 Trost B.M., *J. Am. Chem. Soc.*, 1981; 7550
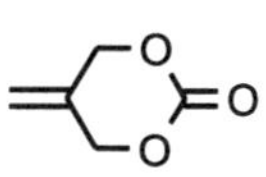	Kang S.K., *Synth. Commun.*, 1995; 1359
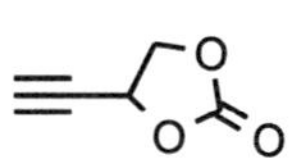	Kang S.K., *J. Chem. Soc., Perkin Trans. 1*, 1994; 237
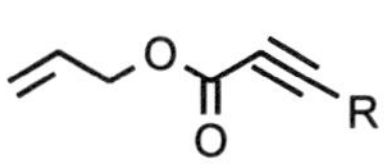	Sato F., *J. Chem. Soc., Chem. Commun.*, 1994; 279
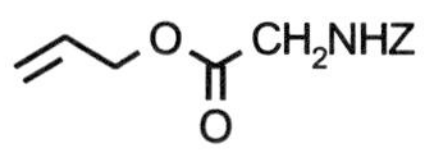	Kazmaier U., *J. Org. Chem.*, 1994; 6667
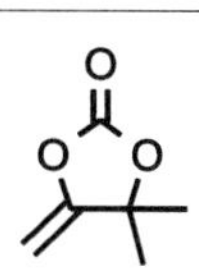	Murai S., *J. Org. Chem.*, 1993; 1173

Substrate	References
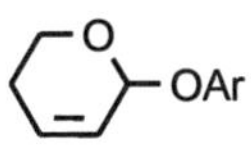	Sinou D., *J. Org. Chem.*, 1989; 1890 Sinou D., *Tetrahedron Lett.*, 1992; 2481 Sinou D., *J. Chem. Soc., Chem. Commun.*, 1995; 1103
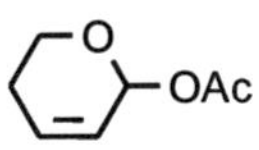	Dunkerton L.V., *J. Org. Chem.*, 1982; 2812
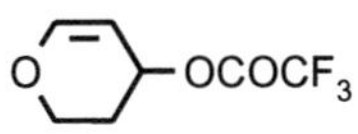	Rajanbadu T.V., *J. Org. Chem.*, 1985; 3642
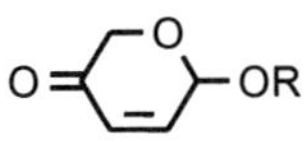	Hoffman H.M.R., *Tetrahedron Lett.*, 1989; 4489
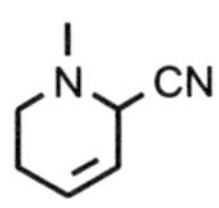	Husson H.P., *Tetrahedron Lett.*, 1982; 5055
	Tamura R., *Tetrahedron Lett.*, 1984; 4437 Tamura R., *Chem. Lett.*, 1985; 229
	Tamura R., *J. Org. Chem.*, 1987; 1640
C=C-C=C-C=C-C-OAc	Trost B.M., *J. Am. Chem. Soc.*, 1984; 7641
C=C-C(OAc)-C=C	Shibasaki M., *Tetrahedron Lett.*, 1992; 3527 Trost B.M., *J. Am. Chem. Soc.*, 1984; 7641
Ph_2N=CH-C(OAc)-CO_2R	O'Donnell M.J., *Tetrahedron Lett.*, 1990; 5135 O'Donnell M.J., *Tetrahedron Lett.*, 1994; 9383 O'Donnell M.J., *Tetrahedron Lett.*, 1995; 4205
C=C=C-CX	Cazes B., *Tetrahedron Lett.*, 1984; 203
C=C=C-C-OCO_2Me	Miyaura N., *Tetrahedron*, 1994; 7961
C=C=C(OCH_3)-C-OAc	Kleijn H., *Recl.: J. R., Neth. Chem. Soc.*, 1983; 378
C=C=C(SPh)-C-OAc	Padwa A., *Synlett.*, 1992; 869

Substrate	References
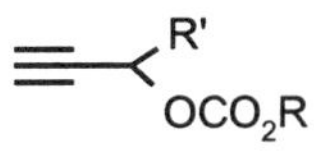	Tsuji J., *J. Am. Chem. Soc.*, 1985; 2196 Suzuki A., *Synlett.*, 1994; 149 Lu X., *J. Chem. Soc., Perkin Trans. 1*, 1992; 17 Tsuji J., *Synlett.*, 1991; 697 Mandai T., *J. Organomet. Chem.*, 1994; 473, 343
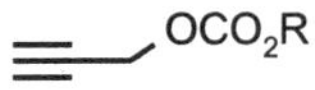	Mandai T., *Tetrahedron Lett.*, 1993; 7615 (elimination)
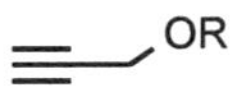	Aurrecoechea J.M., *J. Org. Chem.*, 1994; 702 Dixneuf P.H., *Synlett.*, 1994; 457
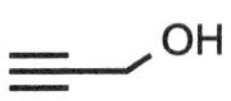	Masuyama Y., *J. Chem. Soc., Chem. Commun.*, 1993; 1219
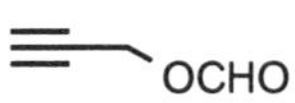	Mandai T., *Tetrahedron Lett.*, 1993; 2161 Mandai T., *J. Organomet. Chem.*, 1994; 473, 343
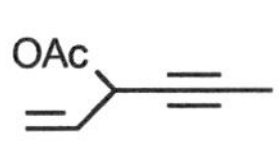	Keinan E., *J. Org. Chem.*, 1986; 4006
Ar-CH_2-OCOR	Fiaud J.C., *Tetrahedron Lett.*, 1992; 2509
Ar-CH(R)-OCO_2R'	Fiaud J.C., *Tetrahedron*, 1995; 3235 Fiaud J.C., *Tetrahedron: Asymmetry*, 1995; 1899 (asymmetric catalysis)
	Tsuji J., *Tetrahedron Lett.*, 1981; 2575 Trost B.M., *J. Am. Chem. Soc.*, 1981; 5969 Tsuji J., *Tetrahedron*, 1986; 4361 Deardorff D.R., *J. Org. Chem.*, 1988; 189 Deardorff D.R., *Org. Synth.*, 1989; 67, 114 Malacria M., *Tetrahedron Lett.*, 1995; 2487 Stille J.K., *Tetrahedron*, 1989; 979
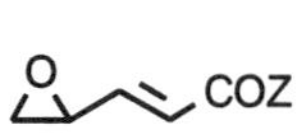	Tsuda T., *J. Org. Chem.*, 1989; 977
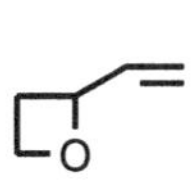	Larock R.C., *Tetrahedron Lett.*, 1989; 3487

Substrate	References
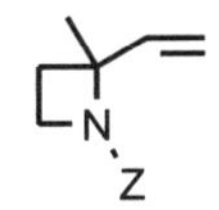	Shimizu I., *Tetrahedron*, 1995; 5331
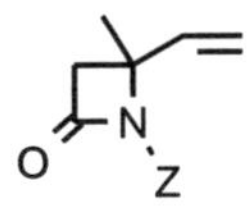	Shimizu I., *Tetrahedron*, 1995; 5331
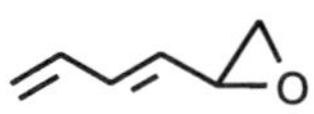	Trost B.M., *Tetrahedron Lett.*, 1986; 4949
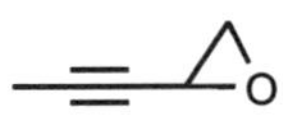	Kleijn M., *Recl.: J. R., Neth. Chem. Soc.*, 1982; 97
	Burgess K., *J. Org. Chem.*, 1987; 2046 Tsuji J., *Chem. Lett.*, 1987; 1157
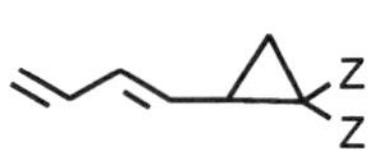	Oshima K., *Tetrahedron Lett.*, 1982; 2871
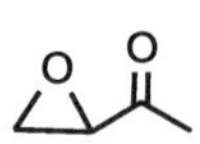	Torii S., *Chem. Lett.*, 1989; 1975

Nucleophile = Carbon	References
$CH(CO_2R)_2$	Trost B.M., *J. Am. Chem. Soc.*, 1976; 630 Fiaud J.C., *Tetrahedron Lett.*, 1980; 4437 Tanner D., *Tetrahedron Lett.*, 1994; 4631 (asymmetric cat.) Brown J.M., *Tetrahedron*, 1994; 4493 (asymmetric cat.) Williams J.M.J., *Synlett.*, 1994; 551 (asymmetric cat.) Williams J.M.J., *Tetrahedron Lett.*, 1995; 461 (asymmetric cat.) Kang J., *Tetrahedron: Asymmetry*, 1994; 1347 (asymmetric cat.) Williams J.M.J., *Tetrahedron: Asymmetry*, 1994; 1895 (asym. cat.) Helmchen G., *Tetrahedron Lett.*, 1994; 1523 (asymmetric cat.) Helmchen G., *Tetrahedron Lett.*, 1994; 8595 (asymmetric cat.) Koga K., *Tetrahedron Lett.*, 1994; 6689 (asymmetric cat.) Wimmer P., *Tetrahedron: Asymmetry*, 1995; 657 (asymmetric cat.) Lemaire M., *Tetrahedron: Asymmetry*, 1995; 1109 (asym. cat.) Pfaltz A., *Helv. Chim. Acta*, 1995; 265 (asymmetric catalysis) Pfaltz A., *Tetrahedron*, 1992; 2143 (asymmetric catalysis) Togni A., *J. Am. Chem. Soc.*, 1994; 4062 (asymmetric catalysis) Williams J.M.J., *J. Chem. Soc., Perkin Trans. 1*, 1994; 2065 (asym. cat.) Helmchen G., *J. Chem. Soc., Chem. Commun.*, 1995; 1845 (asym. cat.) Seebach D., *Helvetica Chem. Acta*, 1995; 1636 (asym. catalysis) Williams J.M.J., *Tetrahedron: Asymmetry*, 1995; 2535 (asym. catalysis)
$R\text{-}C(CO_2R')_2$	Trost B.M., *Tetrahedron Lett.*, 1993; 2271 Akermark B., *Organometallics*, 1994; 1963
$CH_2{=}C(Br)\text{-}CH_2\text{-}CH(CO_2R)_2$	Gaudin J.M., *Tetrahedron Lett.*, 1991; 6113
$RO_2C\text{-}CH\text{-}SO_2Ar$	Trost B.M., *J. Am. Chem. Soc.*, 1980; 4730 Procter G., *Tetrahedron Lett.*, 1995; 7541
$RO_2C\text{-}CH\text{-}PO(OEt)_2$	Malleron J.L., *Thesis, Université Paris Sud*, 1981 Minami T., *Tetrahedron: Asymmetry*, 1995; 2469 (asym. catalysis)
$RO_2C\text{-}CH\text{-}NO_2$	Genêt J.P., *Tetrahedron Lett.*, 1984; 3579 Genêt J.P., *Tetrahedron Lett.*, 1984; 4379
$ROC\text{-}CH\text{-}NO_2$	Hesse M., *Synthesis*, 1985; 645
$RO_2C\text{-}CH\text{-}N{=}C(Ph)_2$	Genêt J.P., *Tetrahedron Lett.*, 1986; 23

Nucleophile = Carbon	References
Camphorsultam-OC-CH-N=C(Ph)$_2$	Saläun J., *Synlett.*, 1995; 226
RO$_2$C(R')-N=C	Hayashi T., *Tetrahedron Lett.*, 1987; 4849
R-CH-CO$_2$R'	Tsuji J., *J. Org. Chem.*, 1985; 1523
RCO-CH-CO$_2$R'	Hayashi T., *J. Am. Chem. Soc.*, 1994; 4221 (asymmetric catalysis) Hayashi T., *Tetrahedron: Asymmetry*, 1995; 2495 (asym. catalysis)
CH$_3$O$_2$C-CH-CONHCH$_2$C$_6$H$_5$	Mori M., *J. Org. Chem.*, 1995; 2016
CH$_3$O$_2$C-CH-CONH-CHR$_1$R$_2$	Johnson E.P., *J. Org. Chem.*, 1995; 6595 (intramolecular)
Ph-CH-CO$_2$R	Tsuji J., *J. Org. Chem.*, 1985; 1523 Beletskaya I.P., *J. Organomet. Chem.*, 1983; 250, 551
RCO$_2$(Ph)$_2$C-	Watanabe Y., *J. Org. Chem.*, 1987; 1695
CF$_3$-CO-CH-CO$_2$Et	Shimizu I., *Synlett.*, 1992; 301
(Ph)$_2$C=N-CH-CO$_2$tBu	Williams J.M.J., *Tetrahedron: Asymmetry*, 1995; 1515 (asymmetric catalysis)
C(CO$_2$Et)$_2$-NHAc	Chauvin Y., *J. Org. Chem.*, 1979; 3063
ArSO$_2$-CH-CN	Lu X., *J. Organomet. Chem.*, 1989; 359, 139
ArSO$_2$-CH-COCH$_3$	Procter G., *Tetrahedron Lett.*, 1995; 7541
PhO$_2$S-CH-NO$_2$	Wade P.A., *J. Org. Chem.*, 1981; 765 Wade P.A., *J. Chem. Soc., Chem. Commun.*, 1980; 287 Trost B.M., *J. Am. Chem. Soc.*, 1992; 8745 Bray B.L., *Tetrahedron Lett.*, 1995; 4483
CH(SO$_2$Ph)$_2$	Trost B.M., *J. Am. Chem. Soc.*, 1980; 4730
RCHNO$_2$	Wade P.A., *J. Org. Chem.*, 1982; 365 Deardorff D.R., *Tetrahedron Lett.*, 1989; 6625 Piotrowska H., *Monatsh. Chem.*, 1982; 1221
R$_2$C=N-CH-PO(OEt)$_2$	Genêt J.P., *Tetrahedron Lett.*, 1988; 4559 Williams J.M.J., *Tetrahedron: Asymmetry*, 1995; 679 (asymmetric cat.)
(EtO)$_2$PO-CH-PO(OEt)$_2$	Magnin D.R., *Synlett.*, 1993; 933
CH(CN)$_2$	Lu X., *J. Organomet. Chem.*, 1989; 359, 139

Nucleophile = Carbon	References
NC-C(Ph)-OTHP	Tsuji J., *J. Org. Chem.*, 1985; 1523
NC-CH-N=C(Ph)$_2$	Genêt J.P., *Tetrahedron Lett.*, 1986; 23 Salaün J., *Tetrahedron Lett.*, 1995; 2979
R'-CO-CH-CO-R	Fiaud J.C., *J. Organomet. Chem.*, 1978; 154, 175 Hayashi T., *J. Org. Chem.*, 1988; 113
MeO$_2$C-CH-CO-CH$_2$COMe	Moreno-Manas M., *Tetrahedron Lett.*, 1989; 3105
MeO$_2$C-CH-CO-CH$_2$CO$_2$Me	Mori M., *J. Org. Chem.*, 1995; 2016
RO$_2$C-CH-CO-CH$_2$-OMe	Greeves N., *Synthesis*, 1993; 1109
MeO$_2$C-CHCO-CH$_2$CH$_2$C(CO$_2$Bu)$_2$	Malacria M., *Tetrahedron Lett.*, 1995; 6447
RO$_2$C-CH-CO-CH$_2$-SPh	Greeves N., *Synthesis*, 1993; 1109
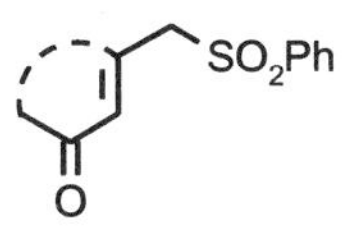	Trost B.M., *J. Am. Chem. Soc.*, 1980; 5979
C=C-CH-SO$_2$Ph	Tsuji J., *J. Org. Chem.*, 1985; 1523
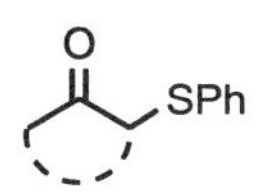	Trost B.M., *J. Am. Chem. Soc.*, 1981; 5969
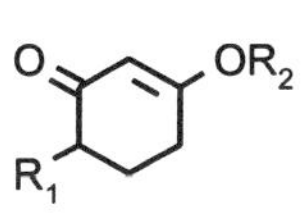	Fukumoto K., *Tetrahedron*, 1995; 6927
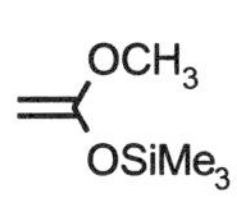	Musco A., *J. Org. Chem.*, 1991; 3924
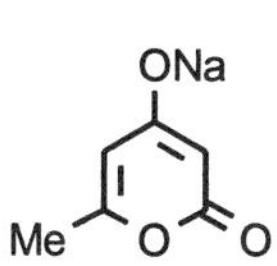	Moreno-Manas M., *Tetrahedron Lett.*, 1989; 3109 Moreno-Manas M., *Tetrahedron*, 1992; 1695
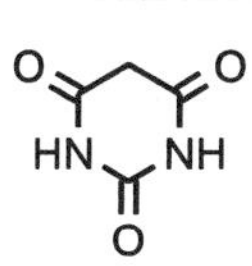	Moreno-Manas M., *Tetrahedron*, 1988; 7205 Guibé F., *J. Org. Chem.*, 1993; 6109

Nucleophile = Carbon	References
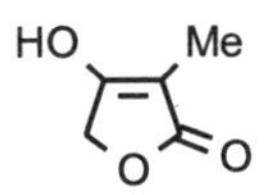	Moreno-Manas M., *Tetrahedron*, 1988; 7205
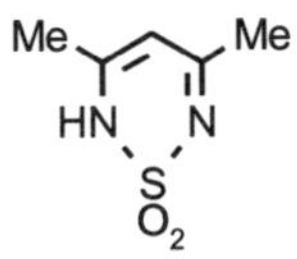	Moreno-Manas M., *Tetrahedron*, 1988; 7205
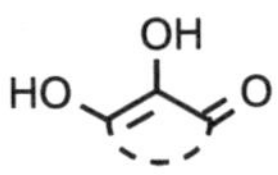	Moreno-Manas M., *J. Org. Chem.*, 1990; 4925
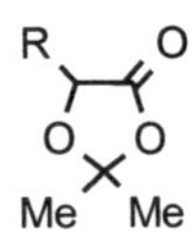	Kellog R.M., *Recl.: J. R., Neth. Chem. Soc.*, 1992; 129
C=C-NR_2	Tsuji J., *Bull. Chem. Soc. Jpn.*, 1973; 1896 Hiroi K., *J. Org. Chem.*, 1994; 203 Hiroi K., *Tetrahedron Lett.*, 1995; 7251
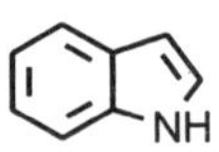	Dzhemilev U.M., *Bull. Acad. Sci. USSR*, 1987; 365
CN	Tsuji Y., *J. Org. Chem.*, 1993; 16
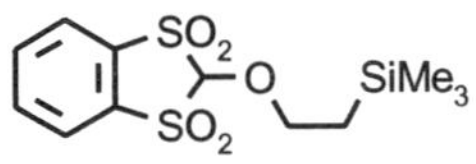	Trost B.M., *J. Am. Chem. Soc.*, 1984; 2469
C=C(OEt)-SnR_3	Migita T., *Chem. Lett.*, 1987; 1237
R_2C=C-O(-)	Tsuji J., *Tetrahedron Lett.*, 1983; 1793

Nucleophile = R-Metal	References
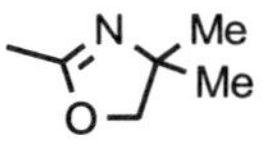 / Sn	Kosugi M., *Chem. Lett.*, 1988; 1351
CH-CO-R' / B	Ono N., *Bull. Chem. Soc. Jpn.*, 1985; 1863 Negishi E.I., *J. Org. Chem.*, 1982; 3188
CH-CO-R' / Li	Fiaud J.C., *J. Chem. Soc., Chem. Commun.*, 1981; 1159
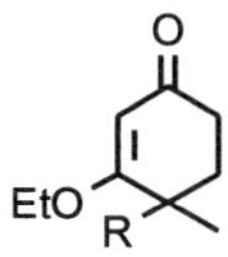 / Li	Fukumoto K., *Synlett.*, 1994; 597
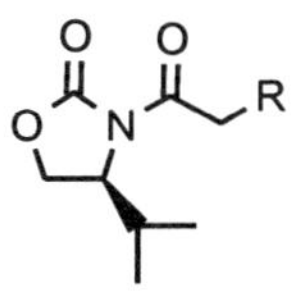 / Li	Kanno H., *Tetrahedron Lett.*, 1995; 5375
CH-CO-R' / Si	Tsuji J., *Chem. Lett.*, 1983; 1325
CH-CO-R' / Sn	Beletskaya I.P., *J. Organomet. Chem.*, 1983; 250, 551 Trost B.M., *Tetrahedron Lett.*, 1980; 2591 Shi G.Q., *Tetrahedron Lett.*, 1995; 6305
CH-CO-R' / Zn	Negishi E.I., *J. Org. Chem.*, 1983; 4098
Br-CH_2-CO_2R' / Zn	Boldrini G.P., *Tetrahedron Lett.*, 1986; 4223
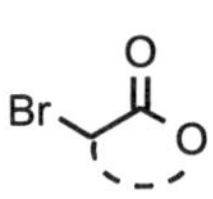 / Zn	Boldrini G.P., *Tetrahedron Lett.*, 1986; 4223
C=C / Al	Negishi E.I., *J. Chem. Soc., Chem. Commun.*, 1982; 160 Negishi E.I., *J. Org. Chem.*, 1985; 3406 O' Donnell M. J., *Tetrahedron Lett.*, 1994; 9383
C=C / B	Miyaura N., *Tetrahedron*, 1994; 7961
C=C / Sn	Beletskaya I.P., *J. Organomet. Chem.*, 1983; 250, 551 Farina V., *Tetrahedron Lett.*, 1988; 5739
C=C / Si	Hiyama T., *Tetrahedron Lett.*, 1995; 1539
C=C($SiMe_3$)-MgBr	Buono G., *Tetrahedron Lett.*, 1990; 77

Nucleophile = R-Metal	References
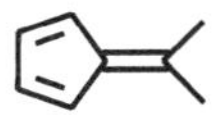 / Li	Soderberg B.C., *Tetrahedron*, 1994; 61
C=C-C=C / Sn	Stille J.K., *Tetrahedron*, 1989; 979
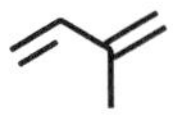 / Sn	Stille J.K., *Tetrahedron*, 1989; 979
C=C-C- / Si	Yamamoto K., *Synlett.*, 1994; 134
C=C-C- / Sn	Trost B.M., *Tetrahedron Lett.*, 1980; 2595 Stille J.K., *Tetrahedron Lett.*, 1980; 2599
C=C-C- / Zn	Inomata K., *Chem. Lett.*, 1985; 315
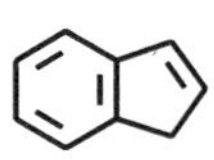 / Li	Fiaud J.C., *J. Organomet. Chem.*, 1985; 291, 393
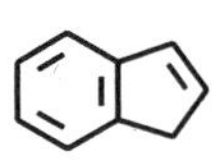 / Si	Trost B.M., *Tetrahedron Lett.*, 1980; 2595
 / Na	Fiaud J.C., *Tetrahedron Lett.*, 1980; 4437
H_2C=CH-CH_2-CH=CH_2 / Li	Malleron J.L., *Thesis, Université Paris Sud*, 1981
C=C=C / Sn	Keinan E., *J. Org. Chem.*, 1983; 5302
 / Zn	Keinan E., *J. Am. Chem. Soc.*, 1994; 11151 (elimination)
Ph / Cd	Negishi E.I., *Tetrahedron Lett.*, 1981; 3737
Ph / Mg	Negishi E.I., *Tetrahedron Lett.*, 1981; 3737
Ph / Zn	Negishi E.I., *Tetrahedron Lett.*, 1981; 3737
Ph / Zr	Negishi E.I., *Tetrahedron Lett.*, 1981; 3737
Ar / B	Fiaud J.C., *Tetrahedron Lett.*, 1990; 7453 Kurusawa H., *J. Am. Chem. Soc.*, 1990; 2813

Nucleophile = R-Metal	References
Ar / Mg	Kumada M., *J. Chem. Soc., Chem. Commun.*, 1981; 313 Hayashi T., *J. Organomet. Chem.*, 1985; 285, 359 Sinou D., *J. Chem. Soc., Chem. Commun.*, 1995; 1103
Ar / Sn	Hegedus L.S., *J. Org. Chem.*, 1990; 3019 Stille J.K., *Tetrahedron*, 1989; 979
Ar / Zn	Negishi E.I., *J. Chem. Soc., Chem. Commun.*, 1982; 160
Het / Sn	Murakami Y., *Tetrahedron Lett.*, 1985; 6457
$C_6H_5CH_2$ / Mg	Sinou D., *J. Chem. Soc., Chem. Commun.*, 1995; 1103
CH_3 / Mg	Sato F., *J. Chem. Soc., Chem. Commun.*, 1993; 1595
R / B	Miyaura., *Tetrahedron*, 1994; 7961
R / Si	Kumada M., *Tetrahedron Lett.*, 1978; 2161
R / Zn	Negishi E.I., *J. Org. Chem.*, 1982; 4161 (elimination)
EtZn	Oppolzer W., *Tetrahedron Lett.*, 1994; 7939 (Zinc-Ene reaction)

Nucleophile = Heteroatom	References
N_3	Williams J.M.J., *Tetrahedron Lett.*, 1993; 6619 Murahashi S.I., *J. Org. Chem.*, 1989; 3292 Sinou D., *Tetrahedron Lett.*, 1990; 527 Waegell B., *Tetrahedron Lett.*, 1988; 4851
RNH	Trost B.M., *J. Org. Chem.*, 1979; 3451
$NHCH_2Ph$	Pfaltz A., *Tetrahedron: Asymmetry*, 1994; 573 (asymmetric catalysis)
$N(R)CH_2Ph$	Achiwa K., *Tetrahedron: Asymmetry*, 1995; 51 (asymmetric catalysis)
NHPh	Takahashi K., *Bull. Chem. Soc. Jpn.*, 1972; 230
HN-CH-NH	Tsuda T., *J. Org. Chem.*, 1990; 3388
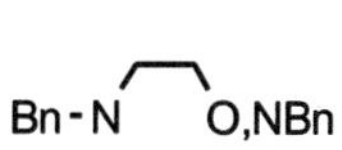	Achiwa K., *Tetrahedron: Asymmetry*, 1995; 1021 (asymmetric catalysis)

Nucleophile = Heteroatom	References
	Muzart J., *Tetrahedron Lett.*, 1995; 5527
NR_2	Dzhemilev U.M., *Bull. Acad. Sci. USSR, Div. Chim. Sci.*, 1987; 365 Genêt J.P., *Tetrahedron*, 1994; 505
NMe_2	Blart E., *Thesis, Université Paris VI*, 1993
NEt_2	Atkins K.E., *Tetrahedron Lett.*, 1970; 3821
RR'N	Taylor R.J.K., *Synthesis*, 1989; 767 Sato F., J. *Chem. Soc., Chem. Commun.*, 1994; 279 (deallylation)
$Me_2N(R)^+$	Pfeffer M., *Tetrahedron Lett.*, 1994; 2877
PhCONHNH	Pfaltz A., *Tetrahedron: Asymmetry*, 1994; 573 (asymmetric catalysis)
Ar-N-Ac	Lebedev S.A., *Zh. Org. Khim.*, 1988; 1112 (CA, 110, 134848f)
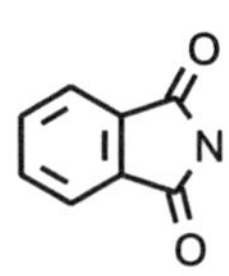	Trost B.M., *J. Am. Chem. Soc.*, 1994; 4089 Bray B.L., *Tetrahedron Lett.*, 1995; 4483
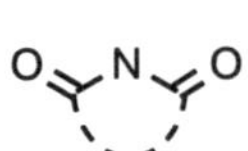	Inoue Y., *Bull. Chem. Soc. Jpn.*, 1984; 3021 Deardorff D.R., *J. Org. Chem.*, 1989; 2759 Blart E., *Thesis, Université Paris VI*, 1993 Tsuda T., *J. Org. Chem.*, 1989; 977 Williams J.M.J., *Tetrahedron Lett.*, 1993; 6619 Hesse M., *Helv. Chim. Acta*, 1994; 819
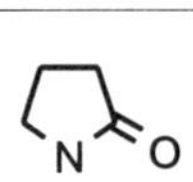	Suzuki A., *J. Organomet. Chem.*, 1982; 233, C13
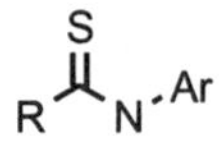	Suzuki A., *J. Organomet. Chem.*, 1982; 233, C13
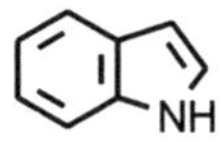	Billups W.E., *Synth. Commun.*, 1980; 147
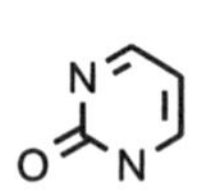	Trost B.M., *J. Am. Chem. Soc.*, 1988; 621

Nucleophile = Heteroatom	References
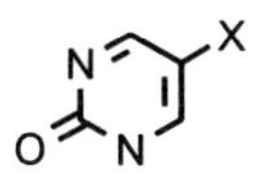	Benneche T., *Acta Chem. Scand.*, 1989; 251
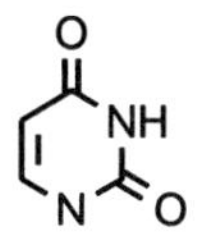	Nokami J., *Chem. Lett.*, 1994; 1071 Sinou D., *Tetrahedron Lett.*, 1994; 7085
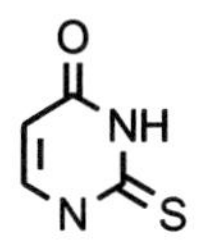	Sinou D., *Tetrahedron Lett.*, 1994; 7085
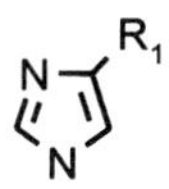	Moreno-Manas M., *J. Het. Chem.*, 1995; 1325
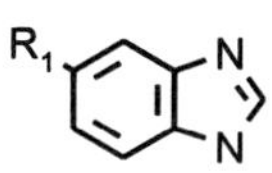	Moreno-Manas M., *J. Het. Chem.*, 1995; 1325
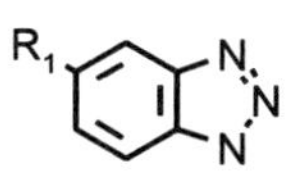	Moreno-Manas M., *J. Het. Chem.*, 1995; 1325
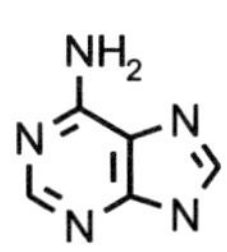	Trost B.M., *J. Am. Chem. Soc.*, 1988; 621
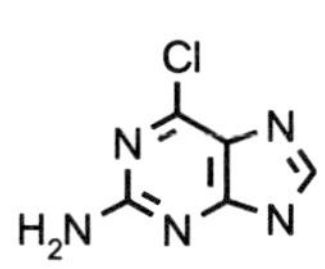	Peel M.R., *J. Org. Chem.*, 1991; 4990
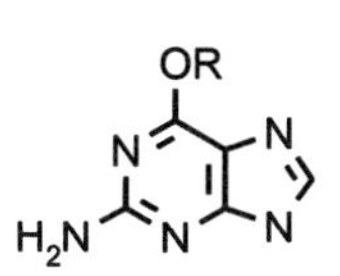	Benneche T., *Tetrahedron Lett.*, 1992; 1085
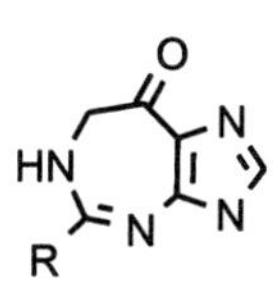	Lindell S.D., *Tetrahedron*, 1994; 6695
$N(BOC)_2$	Akermark B., *J. Org. Chem.*, 1988; 3845 Williams J.M.J., *Tetrahedron Lett.*, 1993; 6619 Pfaltz A., *Tetrahedron: Asymmetry*, 1994; 573 (asymmetric catalysis)

Nucleophile = Heteroatom	References
BOC-O-N-BOC	Genêt J.P., *Tetrahedron Lett.*, 1993; 1159 Genêt J.P., *Synlett.*, 1992; 715 Kanzler S., *Synthesis*, 1995; 539
Ac-O-N-Ac	Kanzler S., *Synthesis*, 1995; 539
$(EtO)_2PO$-N-BOC	Hutchins R.O., *J. Org. Chem.*, 1994; 4007
H-N-SO_2Ar	Backwall J.E., *Tetrahedron Lett.*, 1985; 1749 Tsuda T., *J. Org. Chem.*, 1989; 977 Pfaltz A., *Tetrahedron: Asymmetry*, 1994; 573 (asymmetric catalysis)
R-N-SO_2Ar	Dzhemilev U.M., *Zh. Org. Khim.*, 1987; 826 (CA, 109, 189933q) Trost B.M., *J. Am. Chem. Soc.*, 1992; 9327 (asym. catalysis.; intramolecular)
R-N-CO_2R'	Hayashi T., *Tetrahedron Lett.*, 1987; 4837 Trost B.M., *J. Am. Chem. Soc.*, 1987; 3792 Yamamoto K., *Chem. Lett.*, 1989; 2123 Tamaru Y., *J. Org. Chem.*, 1994; 1465 Tamaru Y., *J. Chem. Soc., Chem. Commun.*, 1992; 1498 Trost B.M., *Angew. Chem. Int. Ed. Engl.*, 1992; 229 (asymmetric cat.)
Ts-N-CO_2R'	Trost B.M., *Angew. Chem. Int., Ed. Engl.*, 1992; 228
R-N-OH	Murahashi S.I., *Tetrahedron Lett.*, 1988; 2973
$OSiR_3$	Trost B.M., *Tetrahedron Lett.*, 1993; 1421
OR	Keinan E., *J. Org. Chem.*, 1985; 3558 Trost B.M., *Tetrahedron Lett.*, 1988; 2927 Trost B.M., *Tetrahedron Lett.*, 1988; 2931 Klumpp G.W., *Tetrahedron Lett.*, 1989; 4863 Klumpp G.W., *Tetrahedron Lett.*, 1989; 5497 Sinou D., *Tetrahedron Lett.*, 1989; 4669 Sinou D., *Tetrahedron Lett.*, 1995; 1251
OPh	Keinan E., *J. Org. Chem.*, 1985; 3558 Genêt J.P., *J. Organomet. Chem.*, 1987; 326, C23
OAr	Larock R.C., *Tetrahedron Lett.*, 1991; 6315 Sinou D., *Synlett.*, 1992; 725 Holzapfel C.W., *Tetrahedron*, 1995; 8555
OAc	Kusama T., *Chem. Pharm. Bull.*, 1992; 1718
OCOR	Trost B.M., *J. Am. Chem. Soc.*, 1994; 10320 (asymmetric catalysis)

Nucleophile = Heteroatom	References
OCO_2R	Mc Ghee W.D., *Organometallics*, 1993; 1429
$OCO_2(-)$	Fujinami T., *Chem. Lett.*, 1985; 199
$OCONR_2$	Mc Ghee W.D., *Organometallics*, 1993; 1429
R-N=C-O	Miller M.J., *Tetrahedron Lett.*, 1995; 2913
SR	Bosnich B., *J. Chem. Soc., Chem. Commun.*, 1986; 146 Trost B.M., *Tetrahedron Lett.*, 1986; 4141
SPh	Trost B.M., *Tetrahedron Lett.*, 1986; 4141 Deardorff D.R., *J. Org. Chem.*, 1989; 2759
S-Het	Sinou D., *Tetrahedron Lett.*, 1992; 8099 Sinou D., *Tetrahedron*, 1994; 10321
SO_2Ar	Julia M., *J. Organomet. Chem.*, 1982; 235, 113 Gais H.J., *Tetrahedron: Asymmetry*, 1995; 643 (asymmetric catalysis) Trost B.M., *J. Am. Chem. Soc.*, 1995; 9662 (asymmetric catalysis) Boldrini G.P., *J. Organomet. Chem.*, 1984; 268, 97
$S{=}P(Ph)_2$	Fiaud J.C., *J. Chem. Soc., Chem. Commun.*, 1983; 1055
PPh_3	Kinoshita H., *Bull. Chem. Soc. Jpn.*, 1984; 3013
$SiMe_3$	Suzuki H., *Bull. Chem. Soc. Jpn.*, 1984; 607 Trost B.M., *J. Am. Chem. Soc.*, 1983; 4494 Tsuji Y., *J. Org. Chem.*, 1993; 3607
$SnBu_3$	Trost B.M., *J. Am. Chem. Soc.*, 1984; 6835 Oshima K., *Bull. Chem. Soc. Jpn.*, 1985; 1196
PR_3	Okukado N., *Chem. Lett.*, 1988; 1449
SePh	Fukuzawa S.I., *Chem. Lett.*, 1990; 927
$OP(OR)_2$	Moreno-Manas M., *Synth. Commun.*, 1992; 2219

Nucleophile = Hydrogen	References
HCO_2NH_4	Tsuji J., *Chem. Lett.*, 1984; 1017 Tsuji J., *J. Org. Chem.*, 1985; 3416
$HCO_2H\text{-}NEt_3$	Shimizu I., *Chem. Lett.*, 1989; 1457 Yamaguchi M., *Tetrahedron Lett.*, 1990; 3913 Tsuji J., *J. Org. Chem.*, 1989; 5395
$LiHBEt_3$	Inomata K., *Chem. Lett.*, 1985; 451 Inomata K., *Chem. Lett.*, 1987; 707

Nucleophile = Hydrogen	References
$NaBH_3CN$	Hutchins R.O., *Tetrahedron Lett.*, 1980; 27
$NaBH_4$	Hutchins R.O., *Tetrahedron Lett.*, 1980; 27 Zhu J., *Tetrahedron Lett.*, 1994; 4349 Zhu J., *Tetrahedron Lett.*, 1995; 3129
R_3SiH	Keinan E., *Israel J. Chem.*, 1984; 82 Guibé F., *Tetrahedron Lett.*, 1995; 5741
R_2SiH_2	Keinan E., *Israel J. Chem.*, 1984; 82
Bu_3SnH	Guibé F., *Tetrahedron Lett.*, 1986; 2365 Keinan E., *Tetrahedron Lett.*, 1982; 241 Speckamp W.N., *J. Org. Chem.*, 1995; 1733 (deallylation)

Electrophile / Metal	References (Umpolung)
ArCHO / Sn	Takahara J.P., *J. Am. Chem. Soc.*, 1992; 2577
RCHO / $SnCl_2$	Masuyama Y., *J. Am. Chem. Soc.*, 1988; 4473 Masuyama Y., *Chem. Lett.*, 1989; 1647 Masuyama Y., *Tetrahedron Lett.*, 1992; 6477 Iwata C., *J. Chem. Soc., Chem. Commun.*, 1994; 1835 (intramolecular)
RCHO / SmI_2	Inanaga J., *Tetrahedron Lett.*, 1986; 1195 Taguchi T., *Chem. Pharm. Bull.*, 1990; 1104
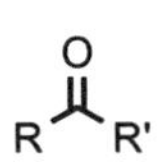 / SmI_2	Mikami K., *Tetrahedron Lett.*, 1995; 907
RCHO / Zn	Masuyama Y., *J. Org. Chem.*, 1987; 3702
RCHO / $ZnEt_2$	Julia M., *J. Chem. Soc., Chem. Commun.*, 1994; 1905 Tamaru Y., *Angew. Chem., Int. Ed. Engl.*, 1995; 787 Tamaru Y., *Tetrahedron Lett.*, 1993; 7619
RCHO / electron	Wang Z., *J. Chem. Soc., Chem. Commun.*, 1989; 356
R_3SnX / SmI_2	Inanaga J., *Tetrahedron Lett.*, 1987; 215
ROH / SmI_2	Inanaga J., *Tetrahedron Lett.*, 1986; 601 Inanaga J., *Tetrahedron Lett.*, 1987; 215 Mikami K., *Tetrahedron Lett.*, 1995; 907

RXN33 [3,3]-Sigmatropic Rearrangement and [1,3]-Shift on Allylic Derivatives

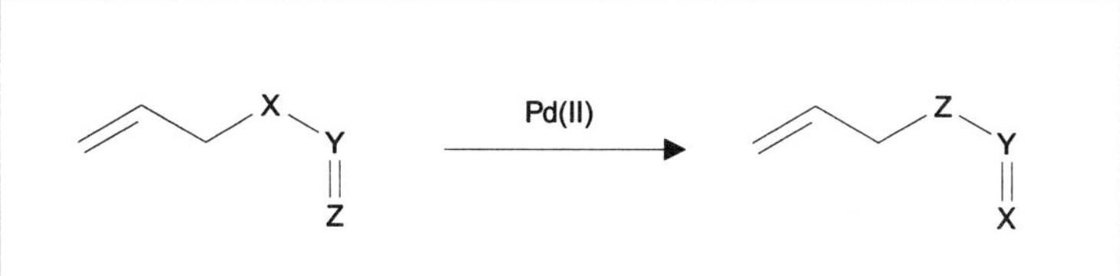

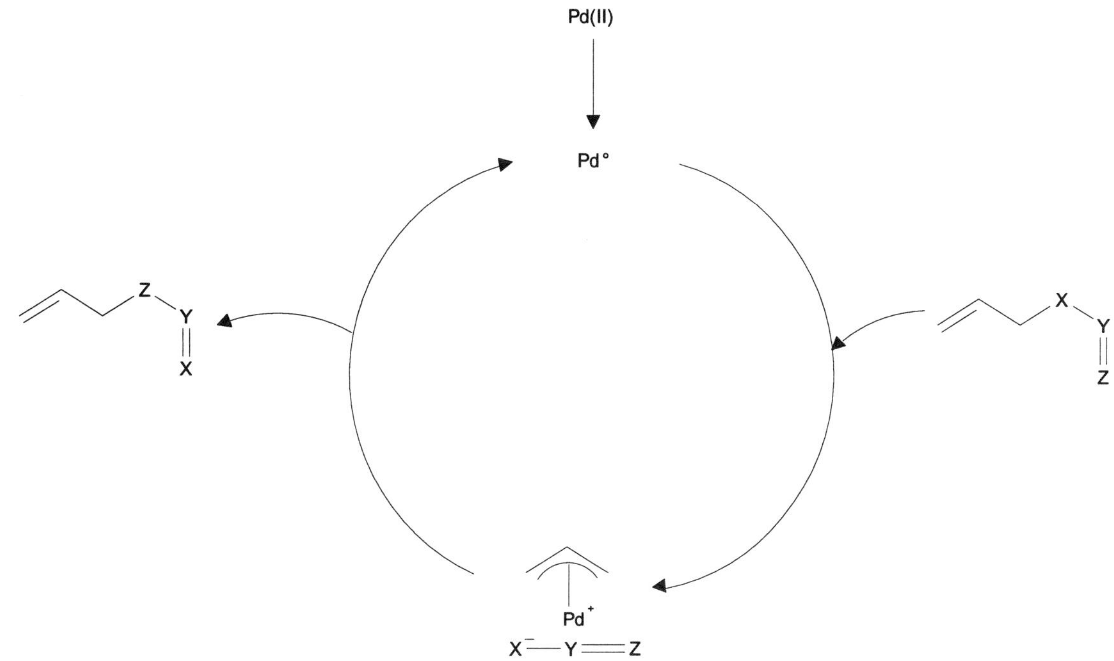

[1,3]- Shift

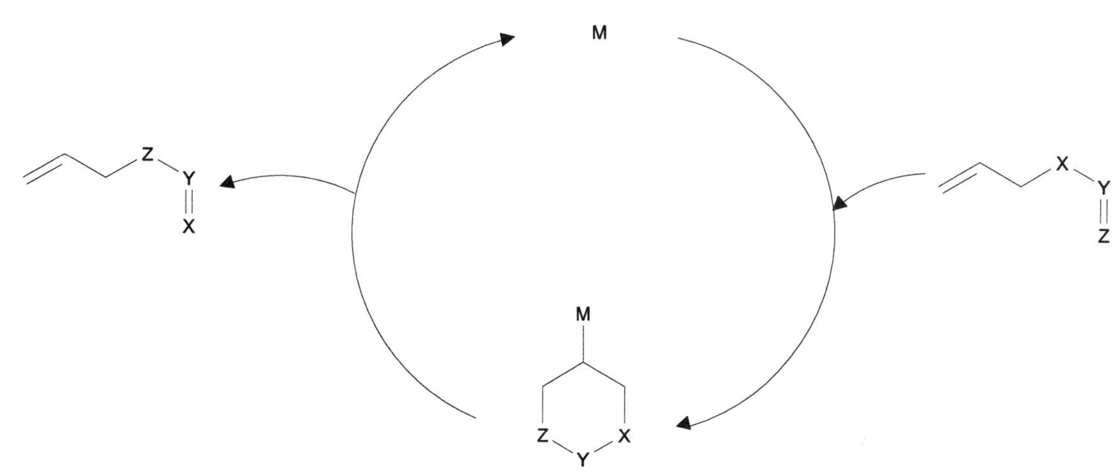

[3,3]- Sigmatropic rearrangement

A ---> B	References
	Bickelhaupt F., *Tetrahedron Lett.*, 1986; 6267 Nakai T., *Tetrahedron Lett.*, 1987; 5879 Hayashi T., *Synth. Commun.*, 1989; 2109 Nakai T., *Synlett.*, 1995; 447 Trost B.M., *J. Am. Chem. Soc.*, 1980; 2840
	Ohshima M., *Chem. Lett.*, 1984; 1535
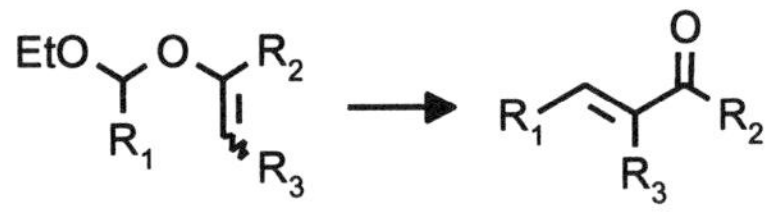	Suzuki H., *Chem. Lett.*, 1981; 1361 (-EtOH)
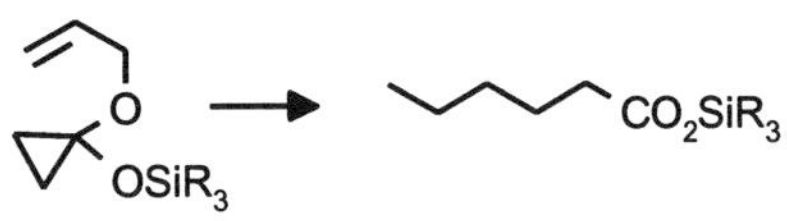	Tamaru Y., *Tetrahedron Lett.*, 1992; 789 (after hydrogenation)
	Overman L.E., *Tetrahedron Lett.*, 1979; 321 Overman L.E., *Angew. Chem. Int. Ed. Engl.*, 1984; 579 Warren S., *J. Chem. Soc., Perkin Trans. 1*, 1993; 2913 Grieco P.A., *J. Am. Chem. Soc.*, 1980; 7587 Yoshida Z., *Tetrahedron*, 1984; 1791 Ohfune Y., *Tetrahedron Lett.*, 1985; 83 Keinan E., *Synthesis*, 1982; 687
	Golding B.T., *J. Chem. Soc., Chem. Commun.*, 1981; 1030 Golding B.T., *J. Chem. Soc., Perkin Trans. 1*, 1988; 2061
	Saito S., *Tetrahedron Lett.*, 1988; 1157
	Ikariya T., *Chem. Lett.*, 1982; 1815 Metz P., *Tetrahedron*, 1992; 1071 Salaün J., *Tetrahedron Lett.*, 1995; 2975 Bosnich B., *J. Am. Chem. Soc.*, 1985; 2058
	Grigg R., *Tetrahedron Lett.*, 1991; 279
	Yamada Y., *Tetrahedron Lett.*, 1979; 5015 Yoshida Z., *J. Org. Chem.*, 1983; 1293

A ---> B	References
	Bosnich B., *Organometallics*, 1986; 1533
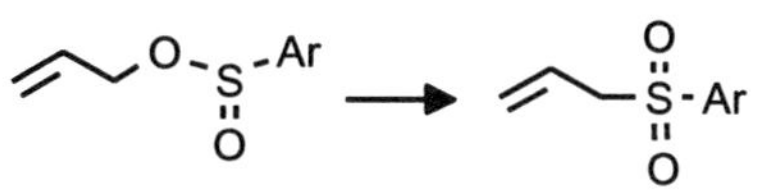	Hiroi K., *Chem. Lett.*, 1986; 617
	Overman L.E., *Angew. Chem. Int. Ed. Engl.*, 1984; 579 Bellus D., *Synthesis*, 1993; 729
	Falck-Pedersen M.L., *Acta. Chem. Scand.*, 1993; 72
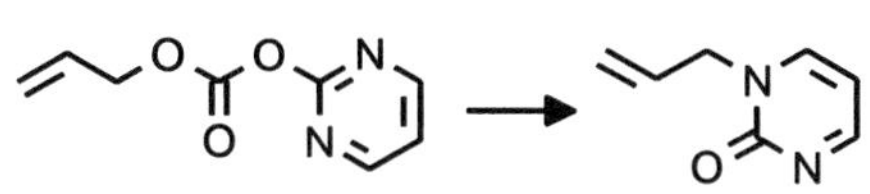	Falck-Pedersen M.L., *Acta. Chem. Scand.*, 1993; 63
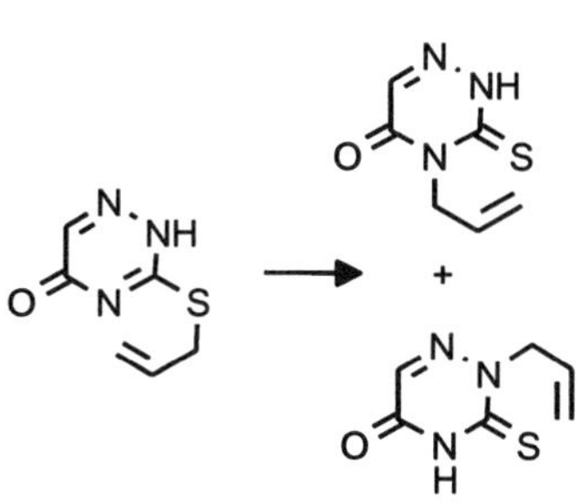	Sanemitsu, *J. Org. Chem.*, 1983; 4585
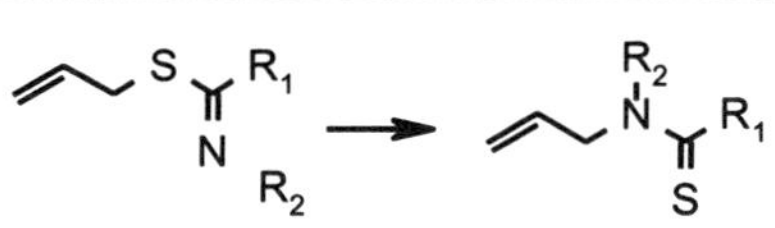	Yoshida Z., *J. Org. Chem.*, 1980; 5221 Yoshida Z., *Tetrahedron Lett.*, 1981; 4245
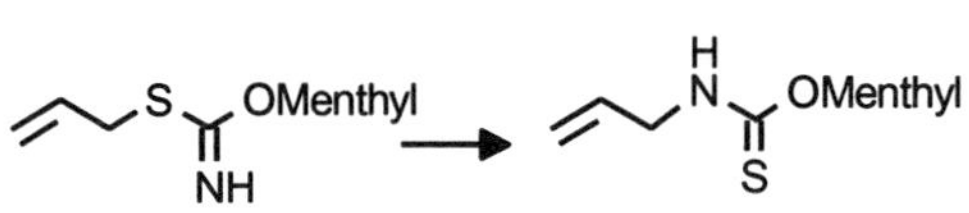	Overman L.E., *Angew. Chem. Int. Ed. Engl.*, 1984; 579
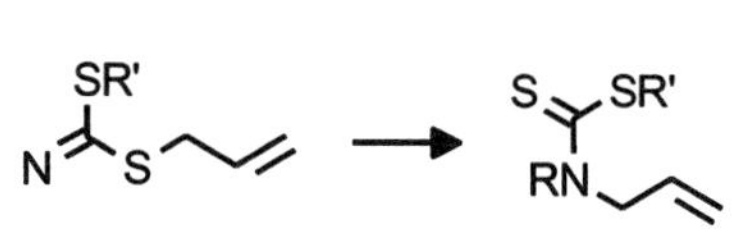	Garin J., *Synthesis*, 1991; 147
	Tamaru Y., *J. Org. Chem.*, 1990; 1823

A ---> B	References
	Inomata K., *Chem. Lett.*, 1985; 931
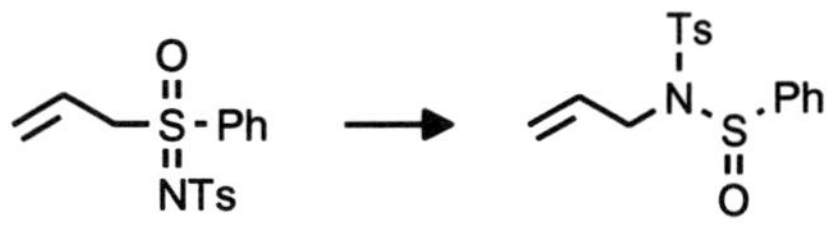	Pyne S.G., *Tetrahedron Lett.*, 1995; 3029
	Murahashi S-I., *Tetrahedron Lett.*, 1985; 5563

RXN34 1,3-Diene Monoepoxide Rearrangement

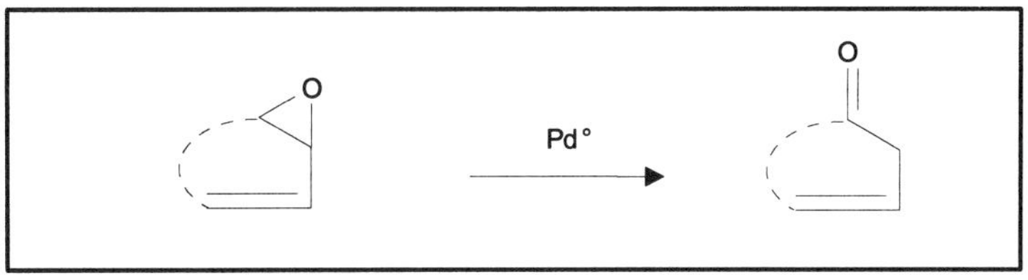

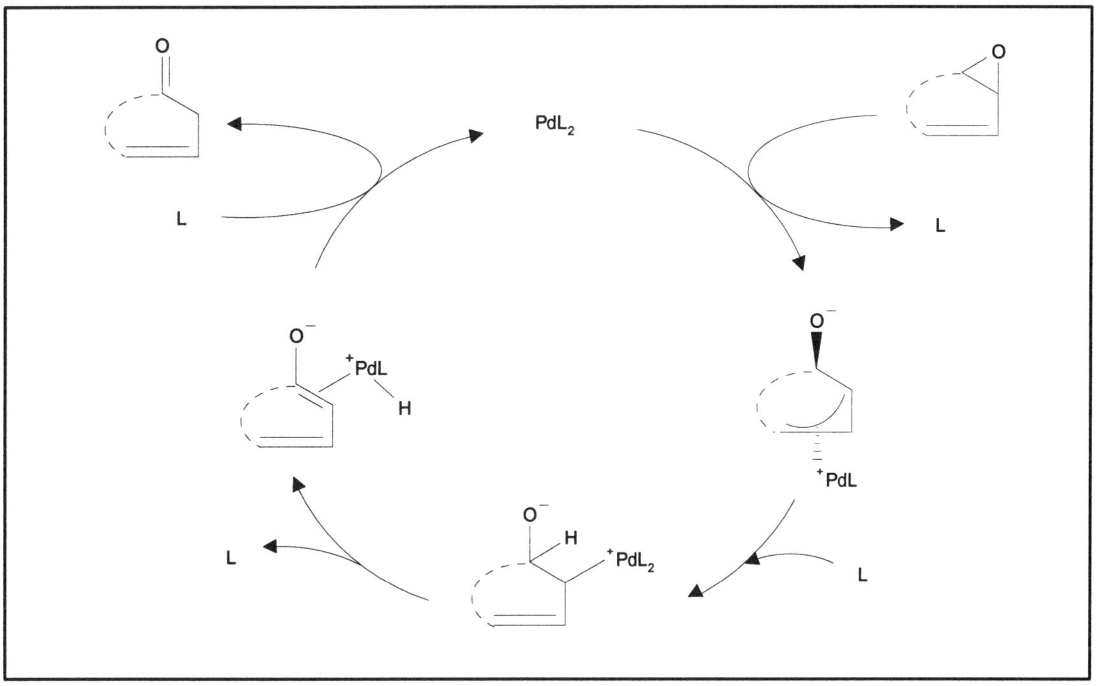

A ---> B	References
	Noyori R., *J. Am. Chem. Soc.*, 1979; 1623 Piccolo O., *J. Mol. Catal.*, 1990; 61, L1
	Nakatami Y., *Agr. Biol. Chem.*, 1975; 2431
	Malacria M., *Tetrahedron Lett.*, 1992; 3859
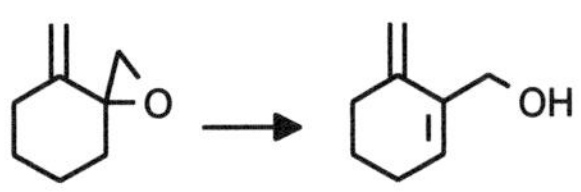	Noyori R., *J. Am. Chem. Soc.*, 1979; 1623
	Zbiral E., *Synthesis*, 1988; 623 Noyori R., *J. Am. Chem. Soc.*, 1980; 2095
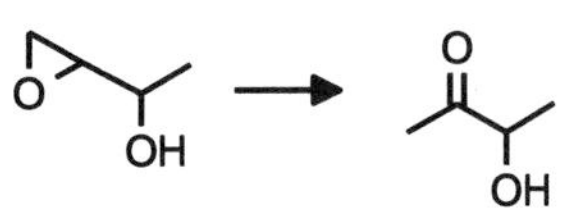	Vankar Y.D., *Synth. Commun.*, 1986; 1621
	Kulawiec R. J., *J. Org. Chem.*, 1994; 7195

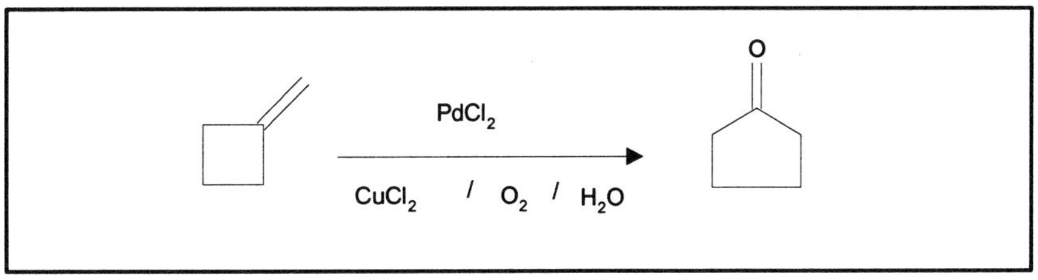
PdCl2
CuCl2 / O2 / H2O
O

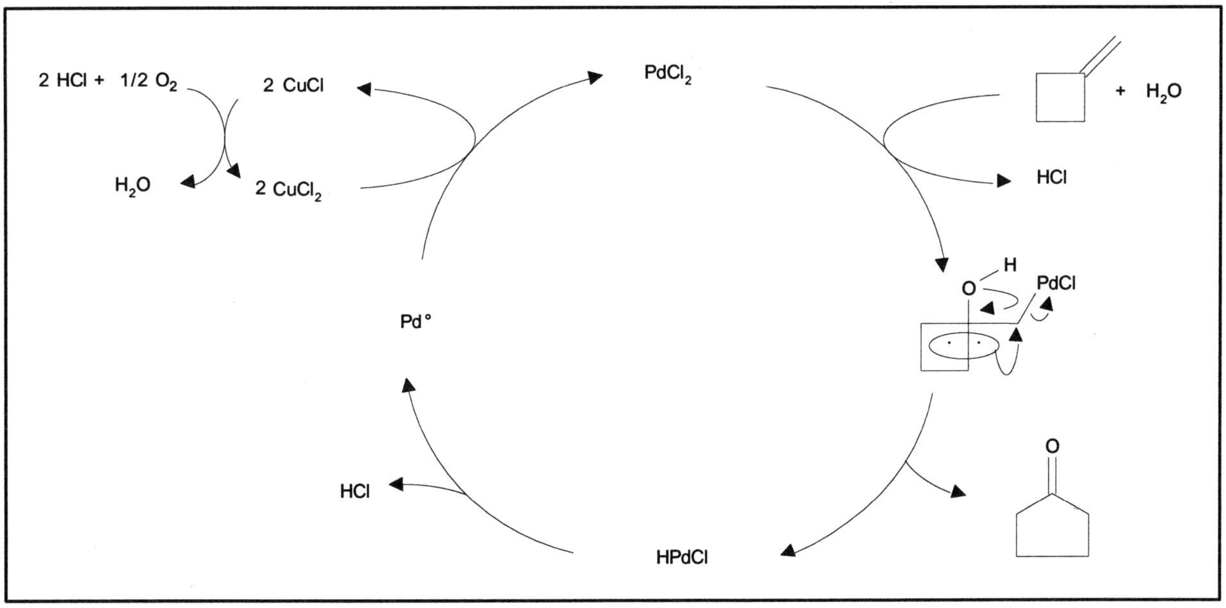
2 HCl + 1/2 O2
2 CuCl
H2O
2 CuCl2
PdCl2
+ H2O
HCl
H
O
PdCl
Pd°
HCl
HPdCl
O

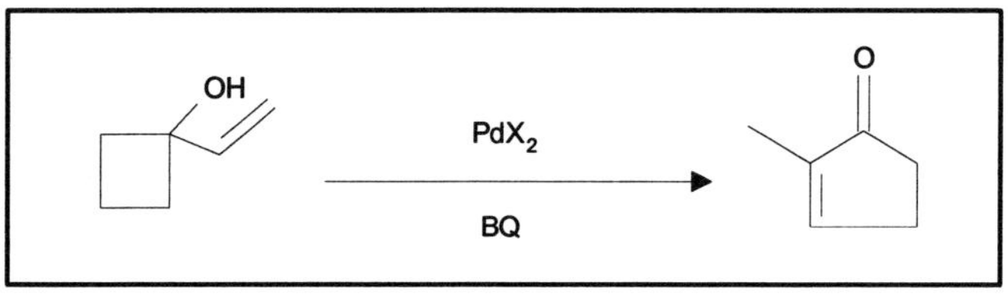

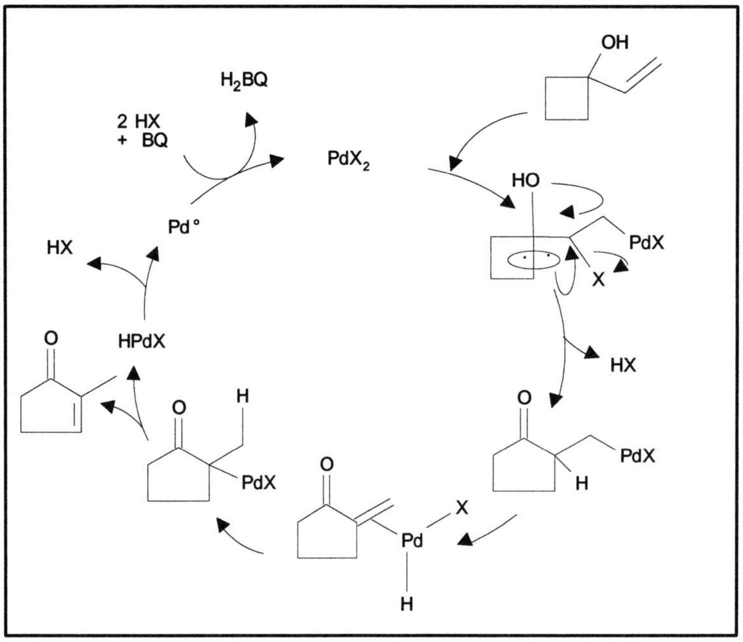

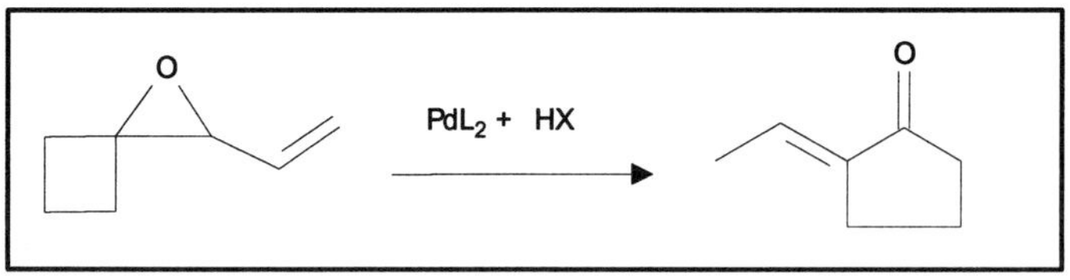

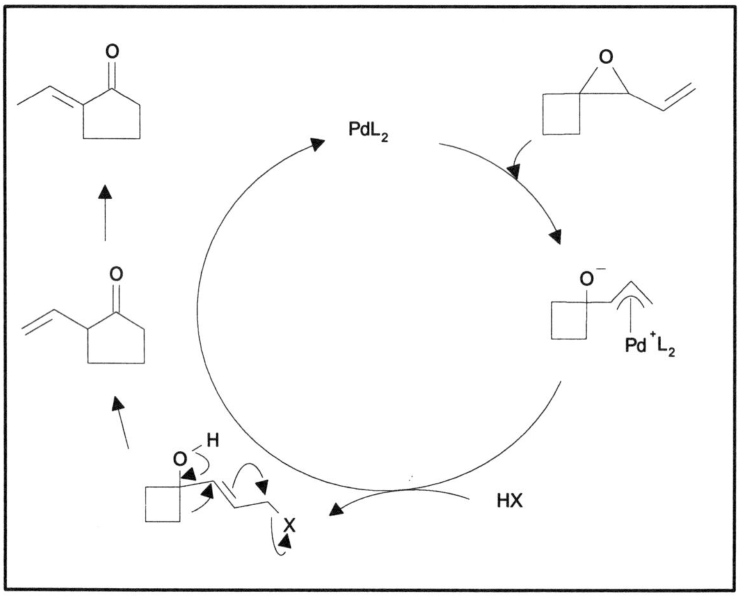

 J.-L. Malleron et al

A ----> B	References
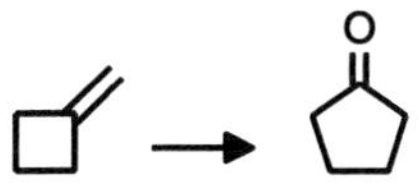	Grigg R., *J. Chem. Soc., Chem. Commun.*, 1977; 583
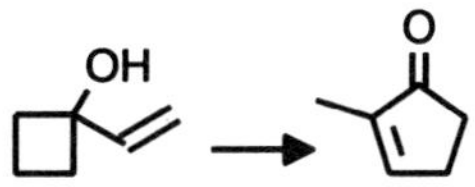	Clark G.R., *Tetrahedron Lett.*, 1985; 2503
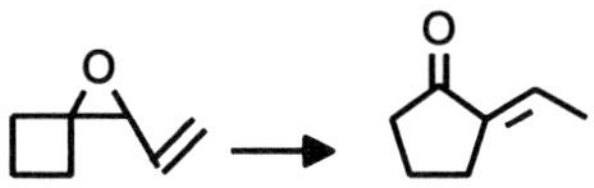	Kim S., *Tetrahedron Lett.*, 1991; 3395

RXN36 [3+2], [3+4], [3+6], [1+2] Cycloadditions

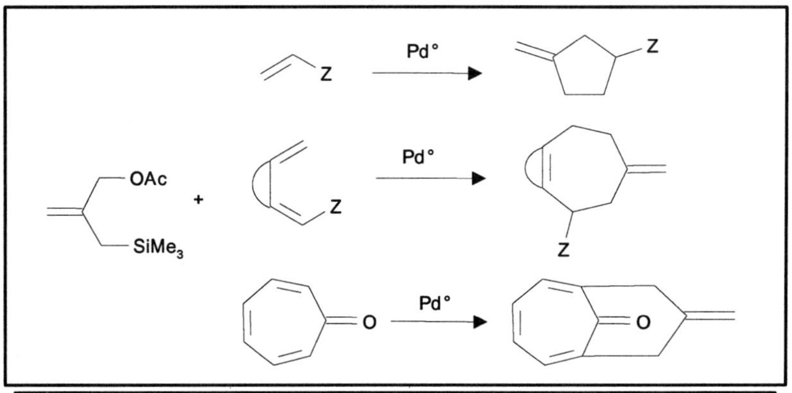

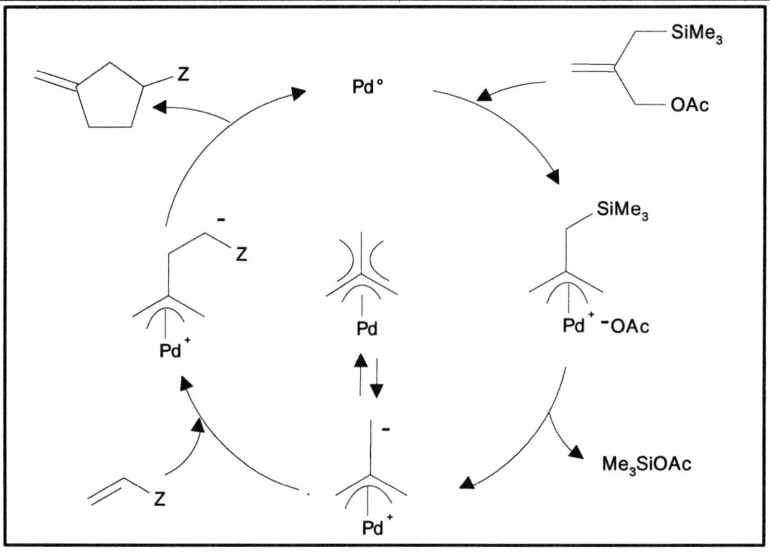

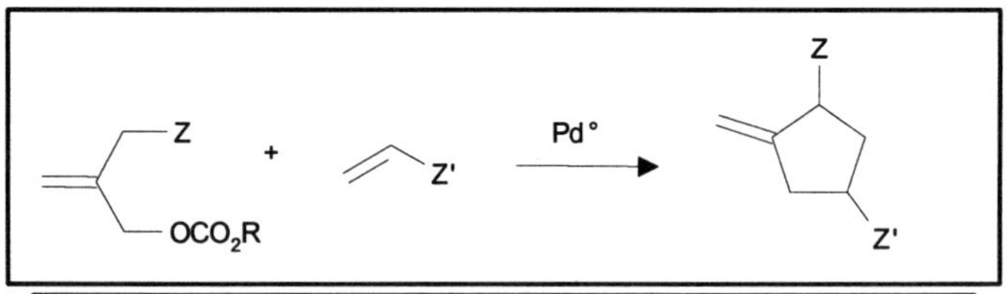

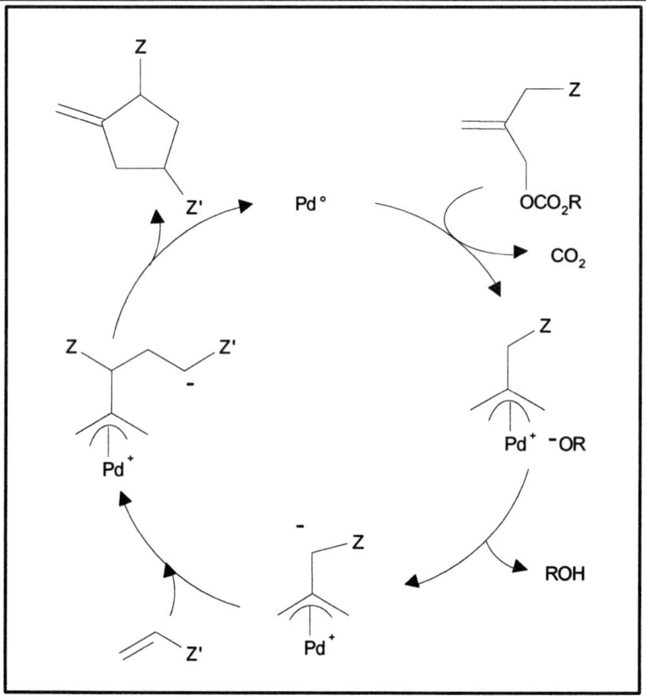

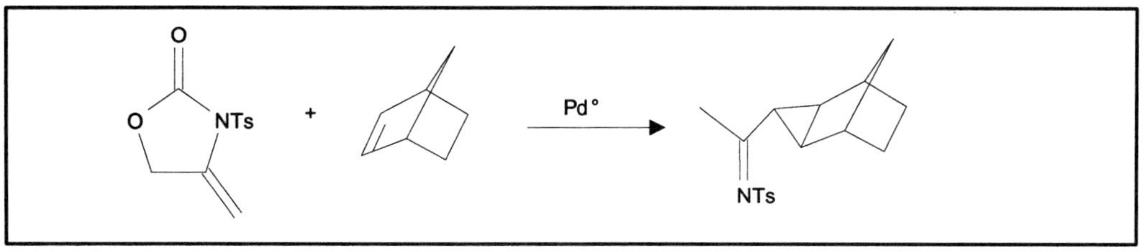
O
NTs
O
+
Pd °
NTs

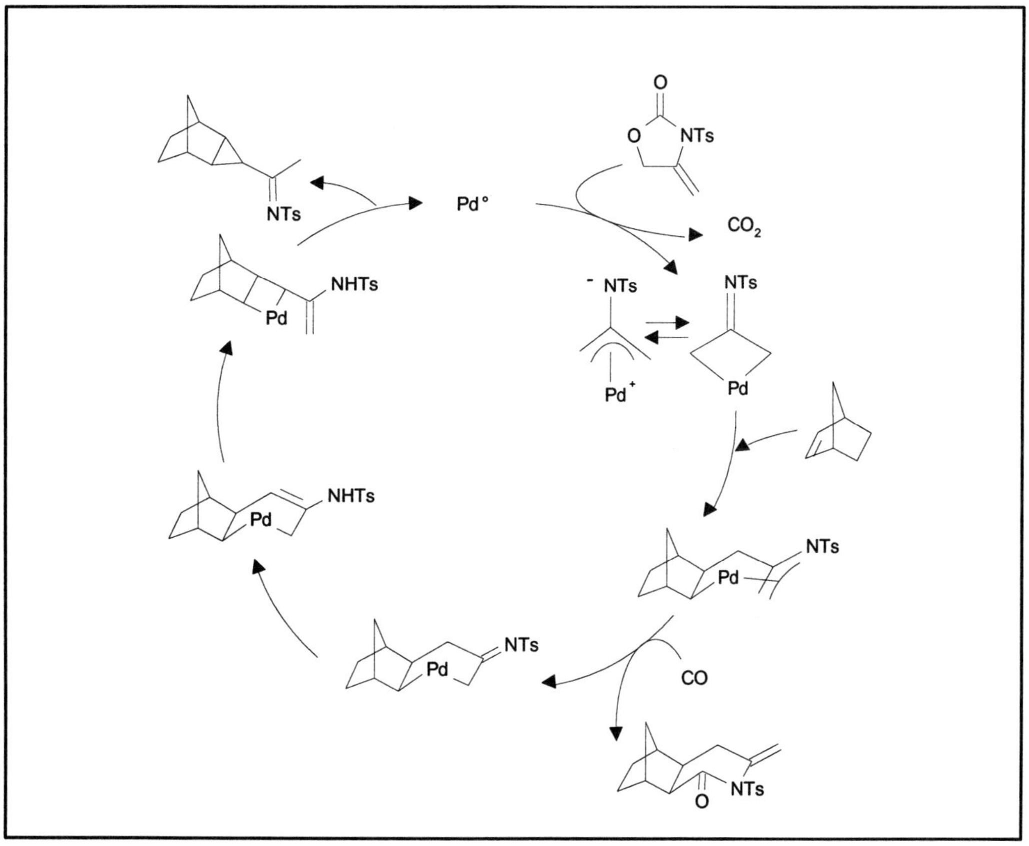
NTs
Pd °
O
O
NTs
CO2
NHTs
Pd
- NTs
NTs
Pd+
Pd
NHTs
Pd
NTs
Pd
NTs
Pd
CO
O
NTs

[3+2] Cycloaddition	References
C=C-Z	Trost B.M., *J. Am. Chem. Soc.*, 1979; 6429 Trost B.M., *J. Am. Chem. Soc.*, 1992; 7903 Trost B.M., *Pure Appl. Chem.*, 1988; 1615 Trost B.M., *Angew. Chem. Int. Ed. Engl.*, 1986; 1 Tsuji J., *Tetrahedron Lett.*, 1984; 5183 Hayashi T., *Tetrahedron Lett.*, 1989; 375
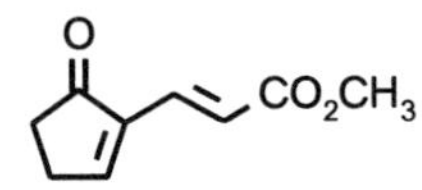	Trost B.M., *Chem. Lett.*, 1994; 2245
R"-(R')C=N-R	Jones M.D., *J. Chem. Soc., Chem. Commun.*, 1986; 1201
Z-C=N-Ar	Trost B.M., *J. Am. Chem. Soc.*, 1993; 6636
Ar-C=N-Ts	Trost B.M., *J. Am. Chem. Soc.*, 1993; 6636 Trost B.M., *Tetrahedron Lett.*, 1995; 2917
R-C=N-Ts	Trost B.M., *J. Am. Chem. Soc.*, 1993; 6636
Ar-N=C=O	Murai S., *J. Org. Chem.*, 1993; 1173
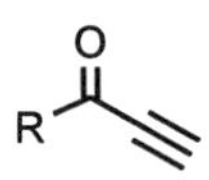	Trost B.M., *Tetrahedron Lett.*, 1993; 7183
intramolecular	Trost B.M., *J. Org. Chem.*, 1992; 686
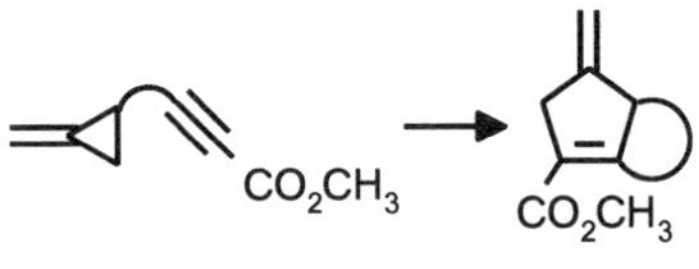	Motherwell W.B., *Tetrahedron Lett.*, 1989; 7107 Lautens M., *J. Am. Chem. Soc.*, 1994; 8821 Motherwell W.B., *Tetrahedron*, 1995; 3303
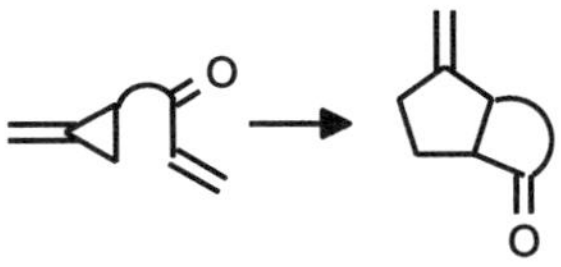	Motherwell W.B., *Tetrahedron*, 1995; 3289

[3+4] Cycloaddition	References
	Trost B.M., *J. Am. Chem. Soc.*, 1987; 3483
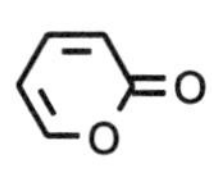	Trost B.M., *Angew. Chem. Int. Ed. Engl.*, 1989; 213

[3+4] Cycloaddition	References
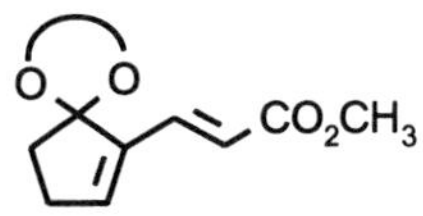	Trost B.M., *Chem. Lett.*, 1994; 2245

[3+6] Cycloaddition	References
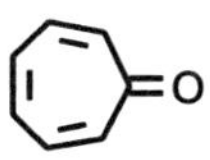	Trost B.M., *J. Am. Chem. Soc.*, 1987; 615

[1+2] Cycloaddition	References
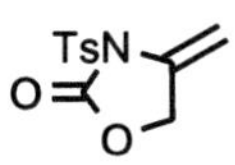 / C=C	Murai S., *J. Am. Chem. Soc.*, 1990; 9646
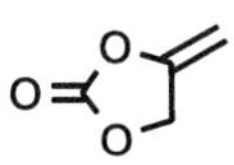 / C=C	Murai S., *J. Org. Chem.*, 1993; 1173
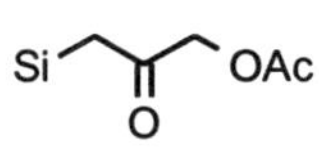 / C=C	Trost B.M., *J. Am. Chem. Soc.*, 1989; 4430
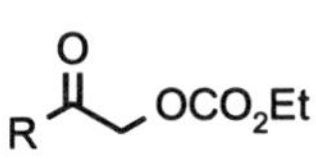 / C=C	Murai S., *J. Org. Chem.*, 1993; 9

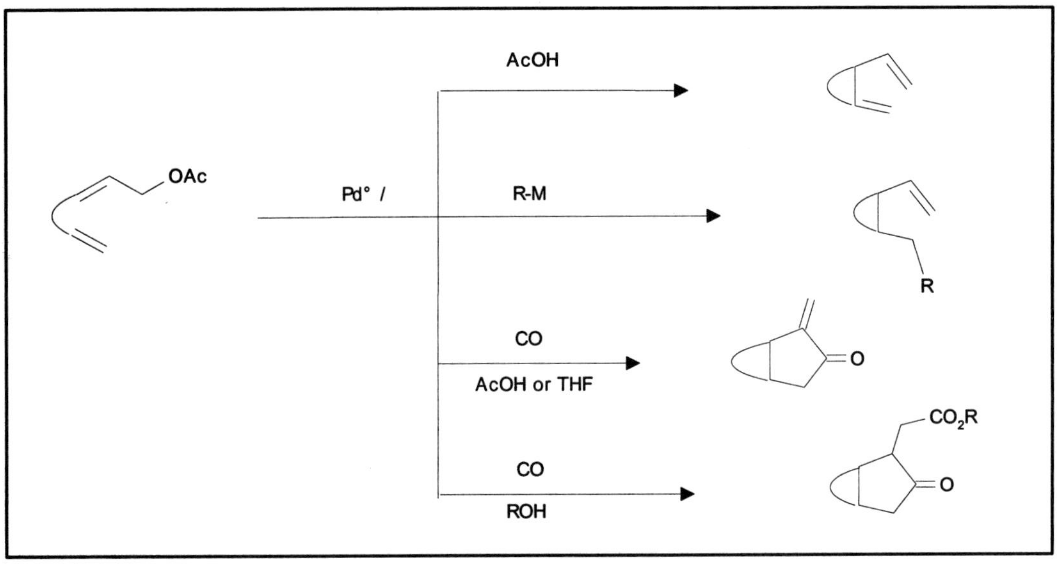
AcOH
OAc
Pd° /
R-M
R
CO
AcOH or THF
O
CO_2R
CO
ROH
O

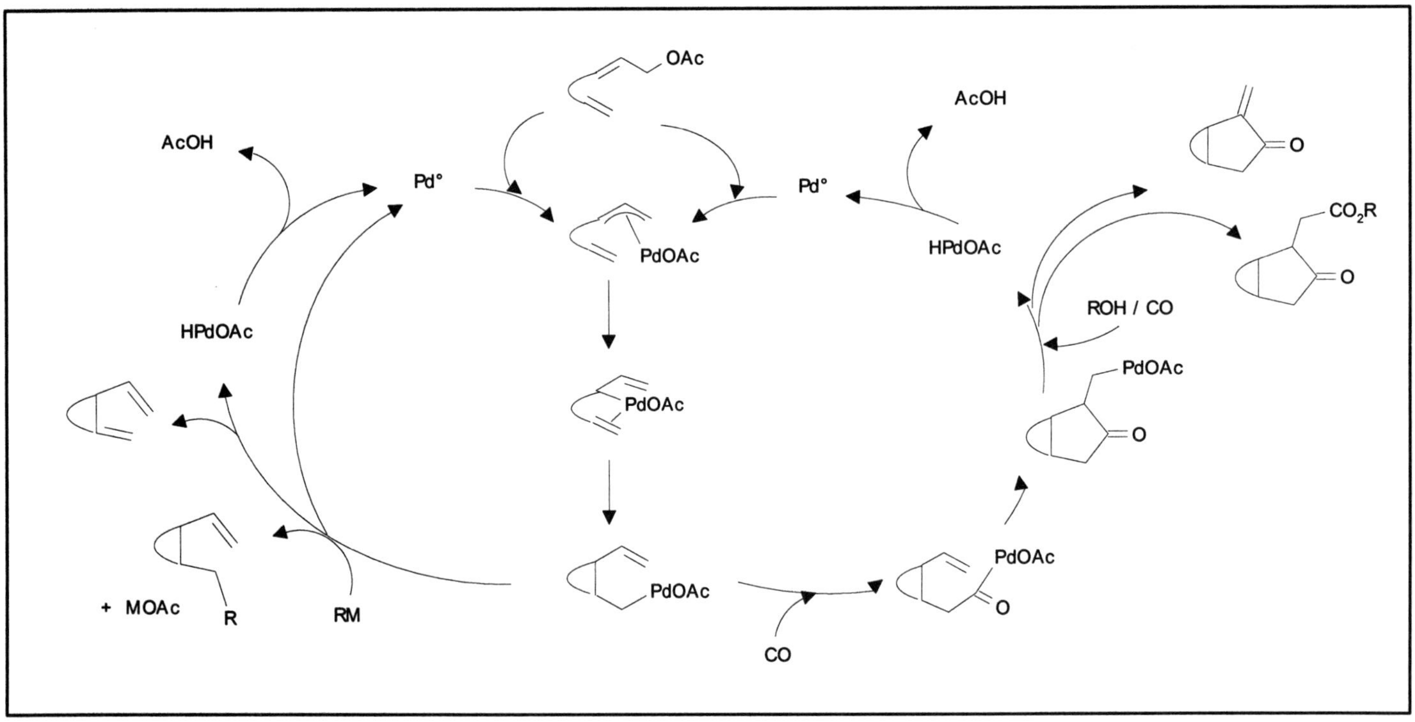
OAc
AcOH
AcOH
Pd°
Pd°
PdOAc
HPdOAc
O
CO_2R
O
HPdOAc
ROH / CO
PdOAc
PdOAc
O
PdOAc
PdOAc
O
+ MOAc
R
RM
CO

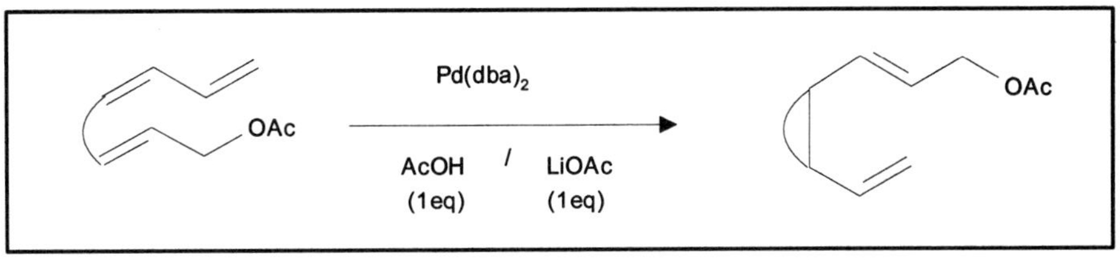
Pd(dba)$_2$
OAc
AcOH / LiOAc
(1eq) (1eq)
OAc

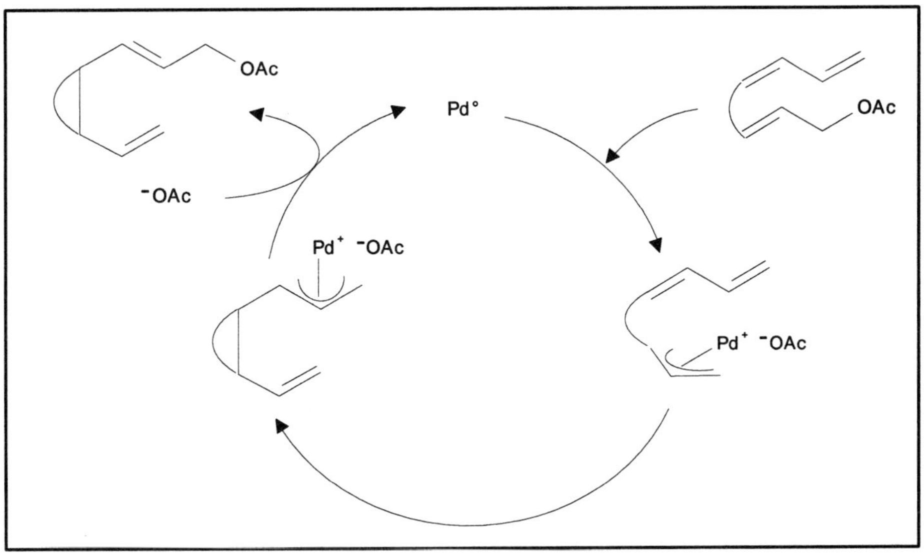
OAc
Pd°
OAc
$^-$OAc
Pd$^+$ $^-$OAc
Pd$^+$ $^-$OAc

A ---> B	References
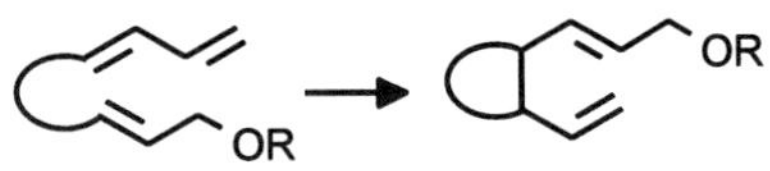	Trost B.M., *J. Am. Chem. Soc.*, 1988; 8239
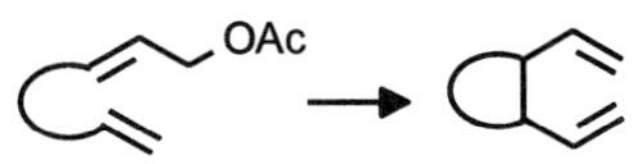	Negishi E.I., *Tetrahedron Lett.*, 1989; 291 Oppolzer W., *Angew. Chem. Int. Ed. Engl.*, 1989; 28, 38
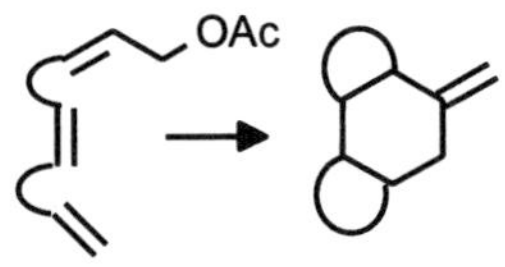	Oppolzer W., *J. Org. Chem.*, 1991; 6256 (tandem reaction)
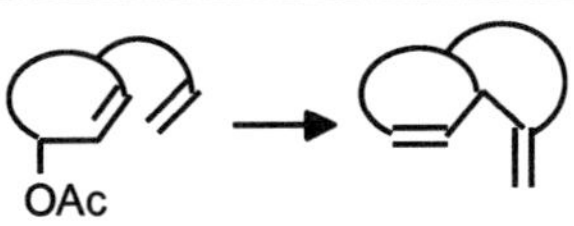	Oppolzer W., *Tetrahedron Lett.*, 1988; 4705
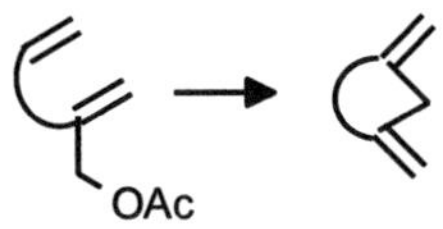	Oppolzer W., *Tetrahedron Lett.*, 1988; 5529
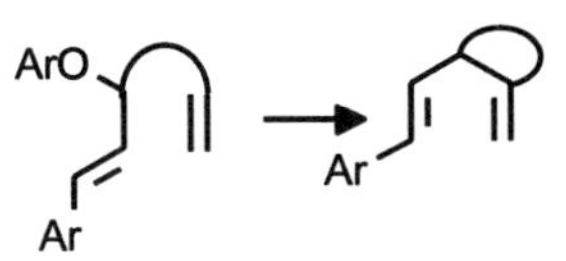	Oppolzer W., *Helv. Chim. Acta*, 1989; 14
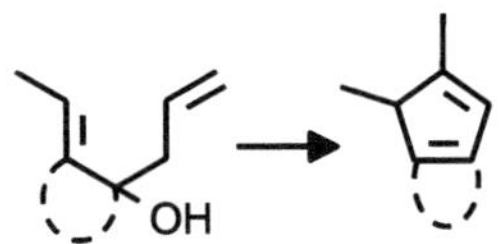	Santelli M., *Tetrahedron Lett.*, 1991; 4501
	Trost B.M., *J. Org. Chem.*, 1989; 2271
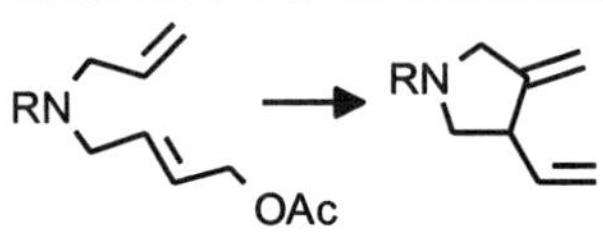	Oppolzer W., *Tetrahedron Lett.*, 1988; 4709
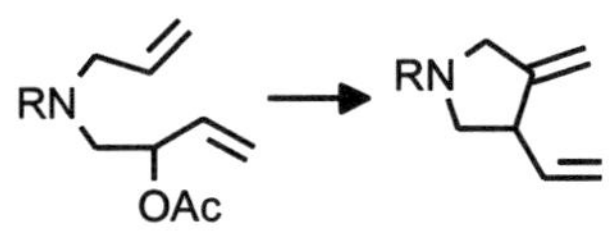	Oppolzer W., *Tetrahedron Lett.*, 1988; 4709
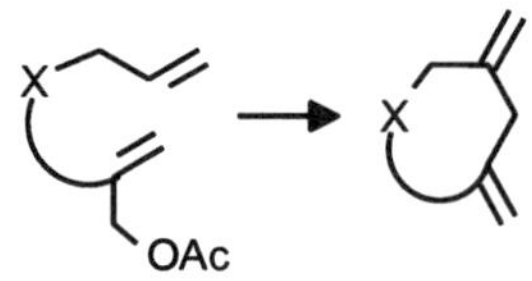	Oppolzer W., *Tetrahedron Lett.*, 1988; 5529
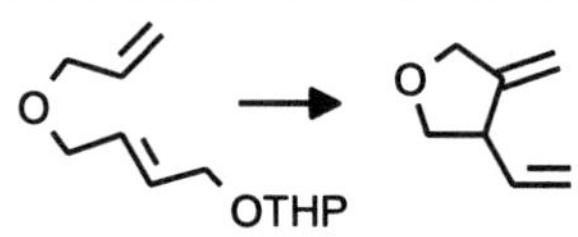	Oppolzer W., *Tetrahedron Lett.*, 1988; 4709

Tandem reaction	References
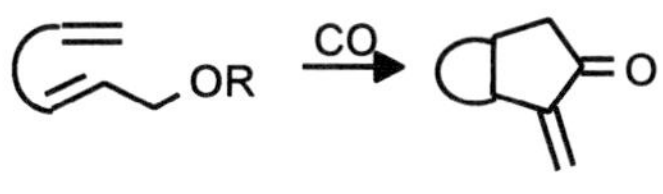	Oppolzer W., *J. Am. Chem. Soc.*, 1991; 9660 Yamamoto K., *Tetrahedron*, 1994; 5705
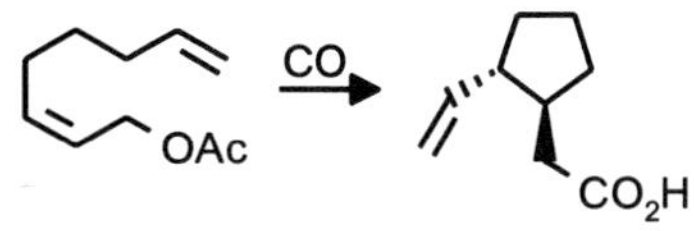	Yamamoto K., *Chem. Lett.*, 1989; 955
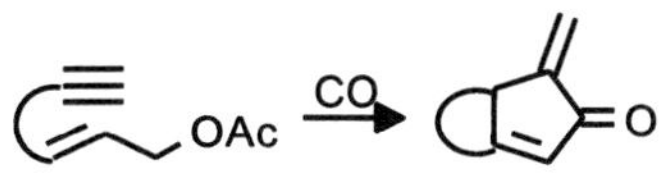	Heathcock C.H., *J. Org. Chem.*, 1993; 560
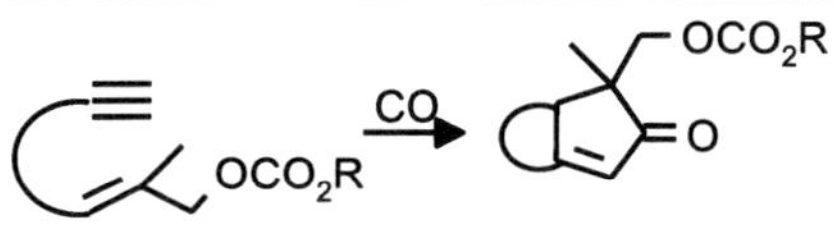	Oppolzer W., *Tetrahedron*, 1994; 415
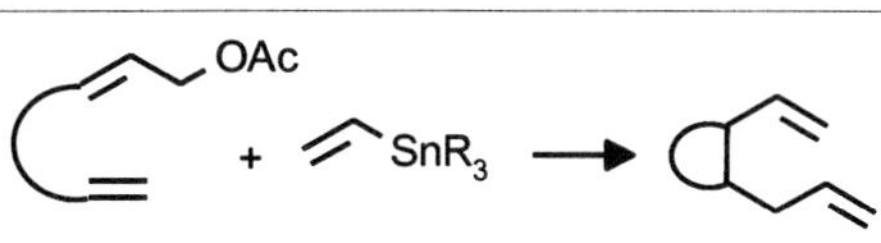	Oppolzer W., *Helv. Chim. Acta*, 1993; 1266

RXN38 Cyclization of Hexatrienolate Derivatives

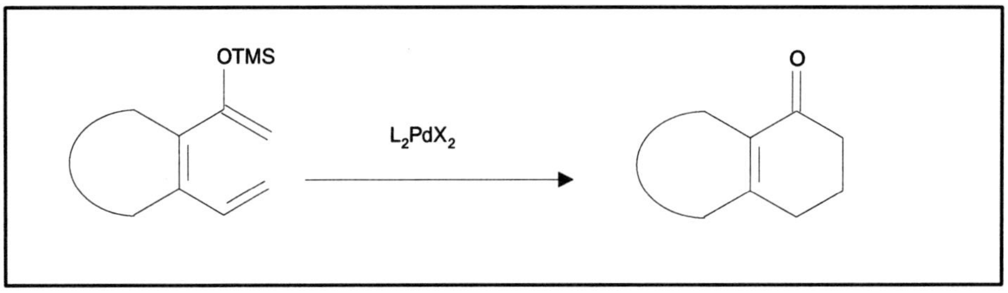

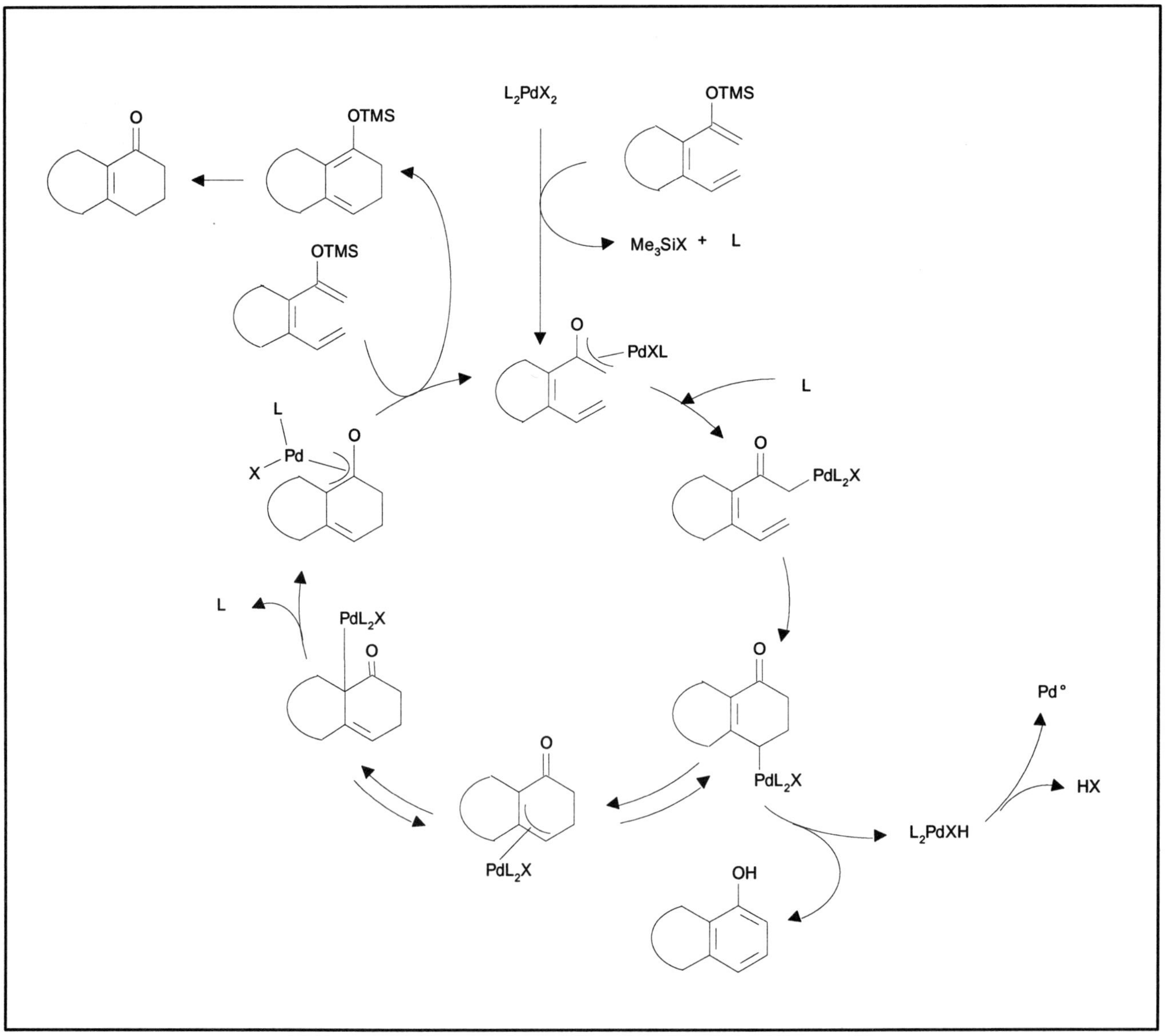

A ---> B	References
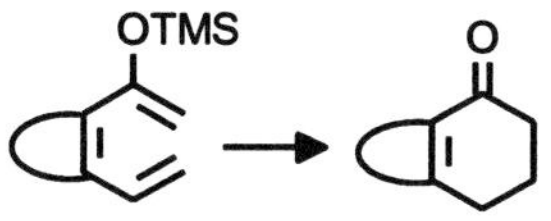	Scott W.J., *J. Am. Chem. Soc.*, 1991; 4903

RXN39 Amination or Amidation of Alkenes

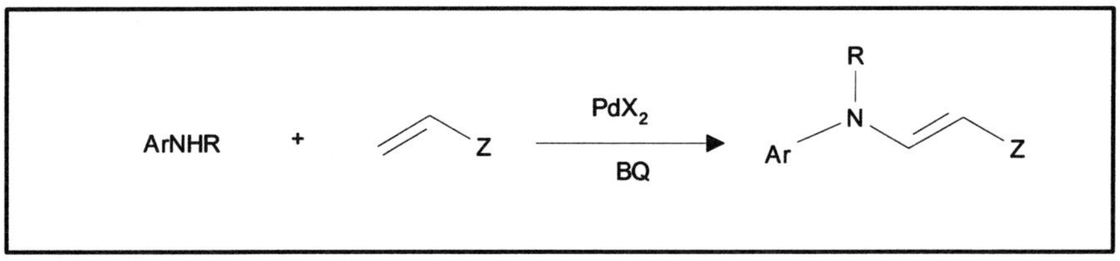

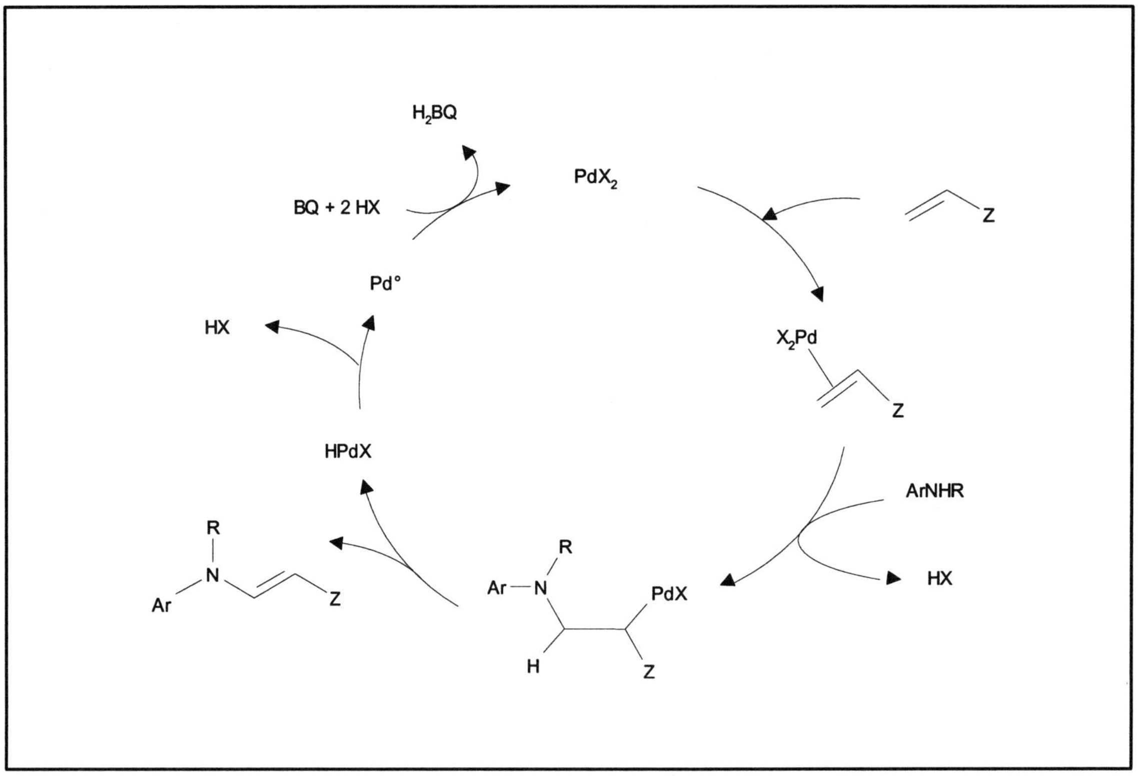

A ---> B	References
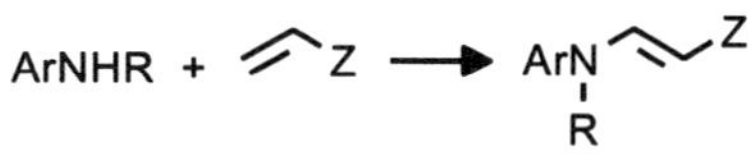	Hegedus L.S., *J. Org. Chem.*, 1981; 2561
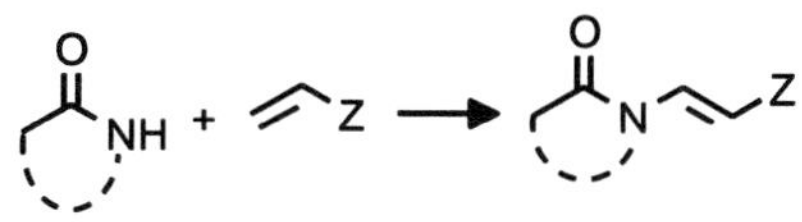	Murahashi S.I., *Tetrahedron Lett.*, 1992; 6643

RXN40 Alkoxylation of Alkenes and Alkynes

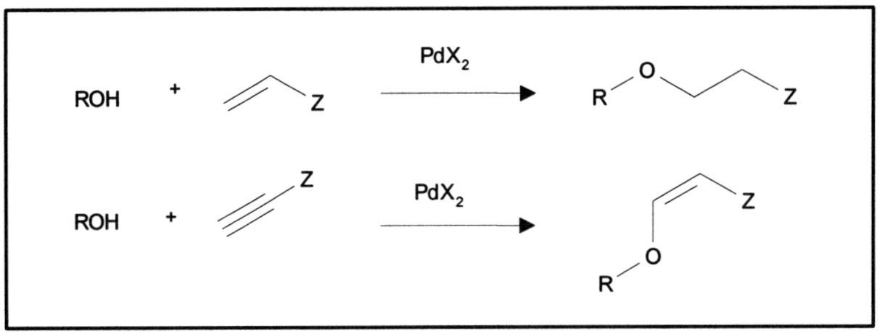

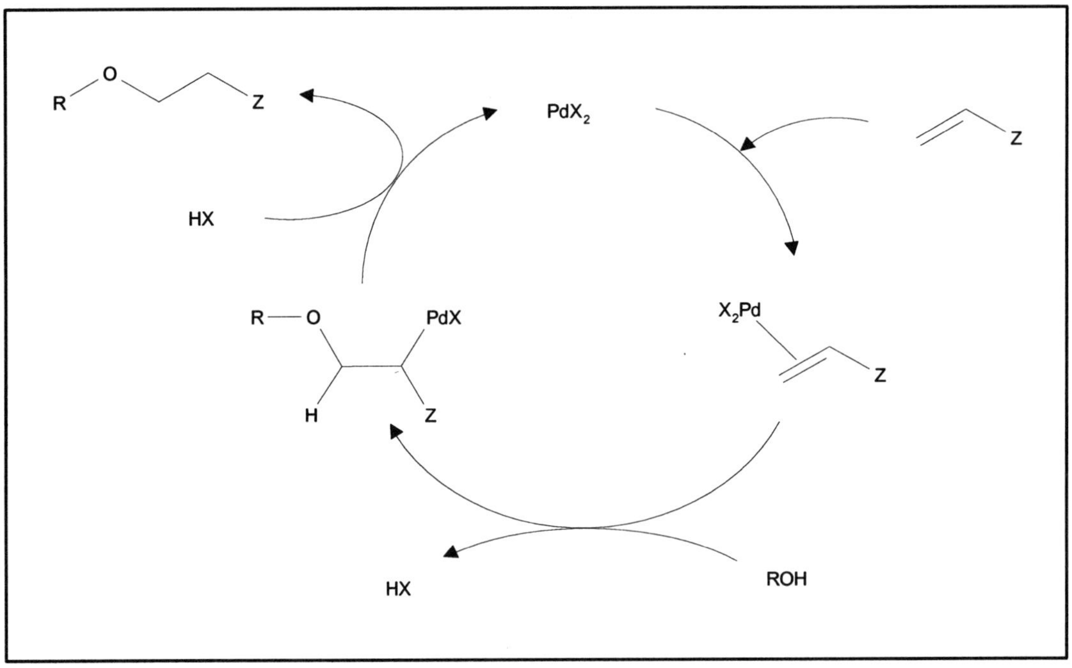

A ---> B	References
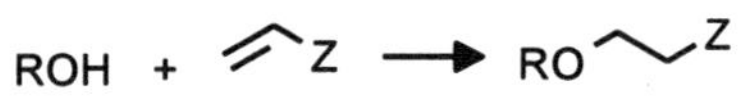	Hosokawa T., *Chem Lett.*, 1989; 2001
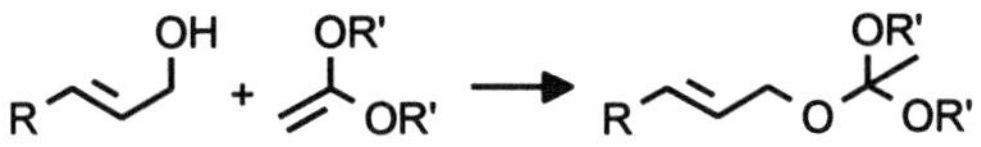	Ohshima M., *Chem. Lett.*, 1984; 1535 (tandem reaction)
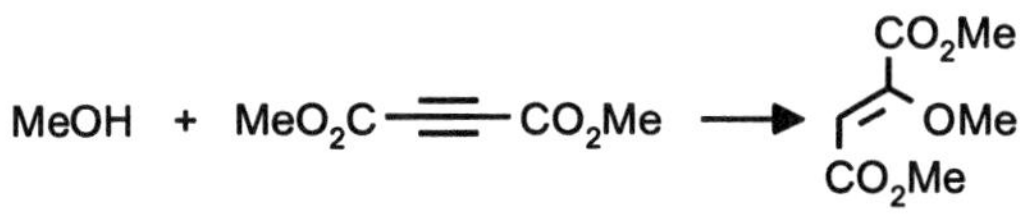	Tani K., *Chem. Lett.*, 1994; 1283
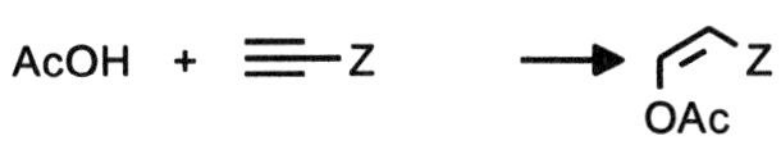	Lu X., *Tetrahedron Lett.*, 1992; 7205 Midai M., *Tetrahedron Lett.*, 1995; 5585
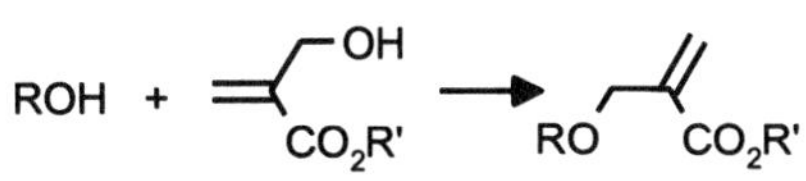	Murahashi S.I., *J. Organomet. Chem.*, 1994; 470, 253 (- XPdOH)

RXN41 Acetalization of Alkenes

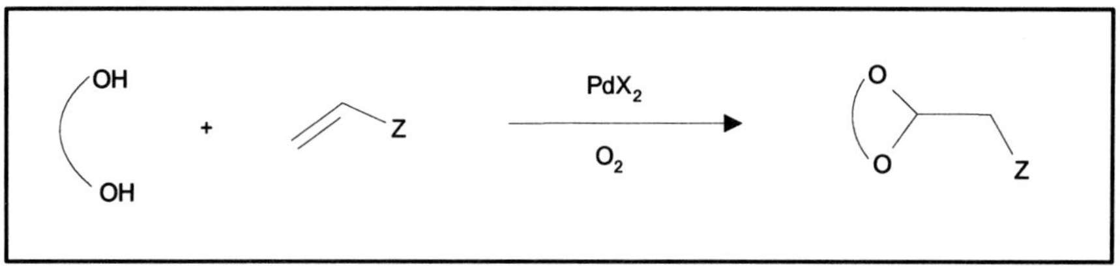

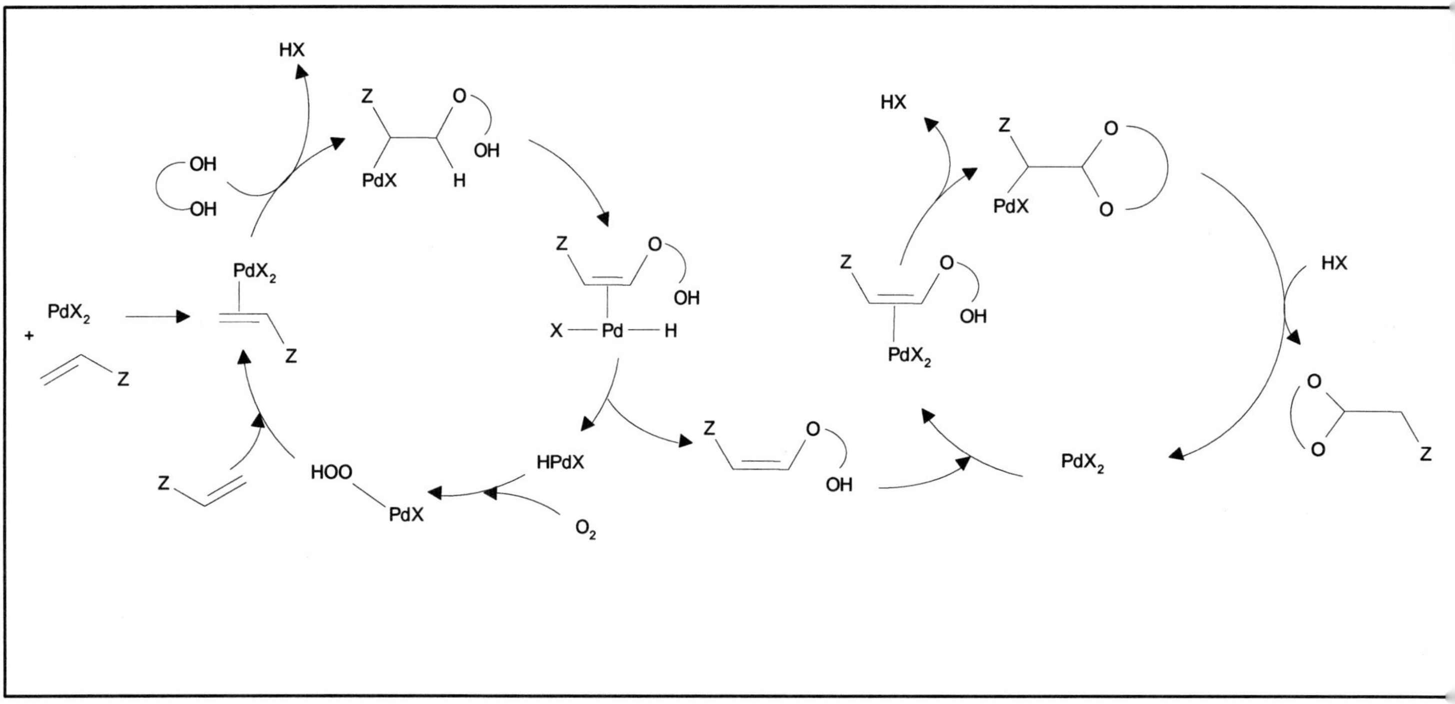

A ---> B	References
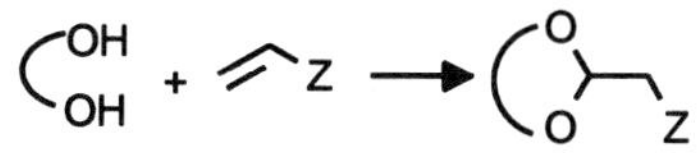	Murahashi S-I., *Acc. Chem. Res.*, 1990; 49 Murahashi S-I., *J. Org. Chem.*, 1987; 1758 Murahashi S-I., *Heterocycles*, 1992; 33, 1079
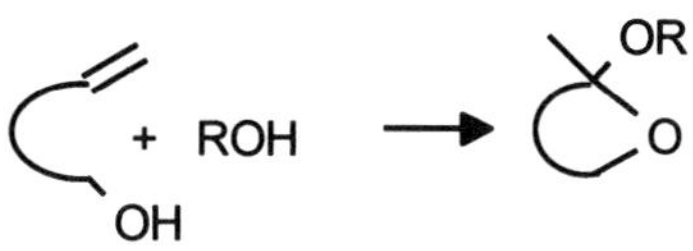	Hosokawa T., *Chem. Lett.*, 1990; 1387
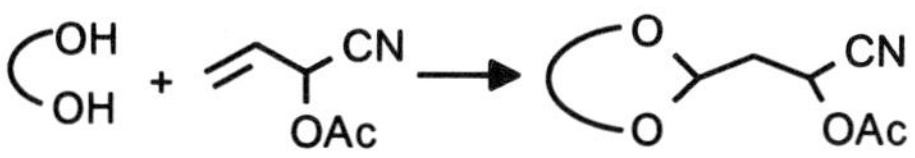	Murahashi S-I., *Synthesis*, 1992; 558
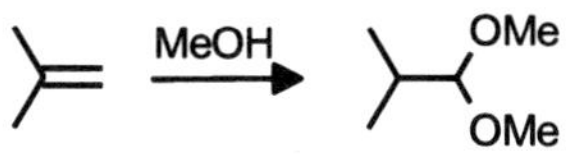	Hosokawa T., *J. Chem. Soc., Chem. Commun.*, 1993; 117
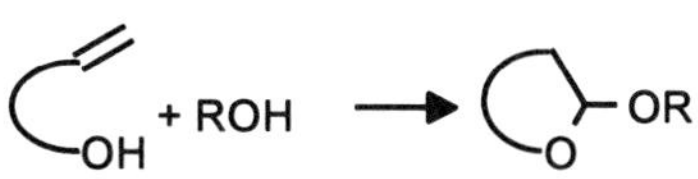	Feringa B.L., *Tetrahedron Lett.*, 1994; 455
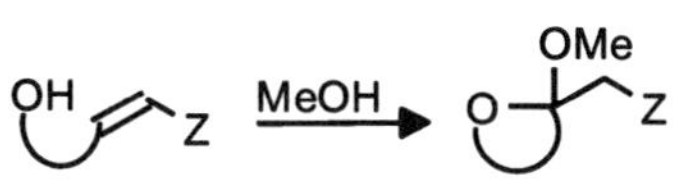	Sturgess M.A., *Tetrahedron Lett.*, 1992; 7739
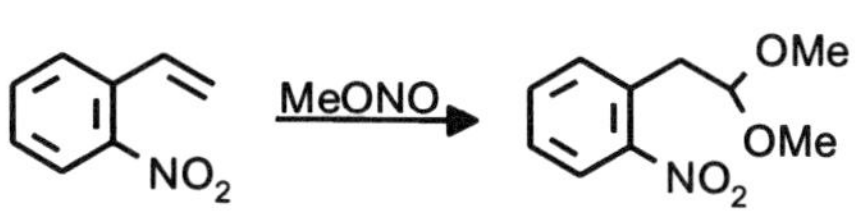	Izumi T., *J. Heterocycl. Chem.*, 1992; 1625

RXN42 Allylic Acyloxylation of Cycloalkenes

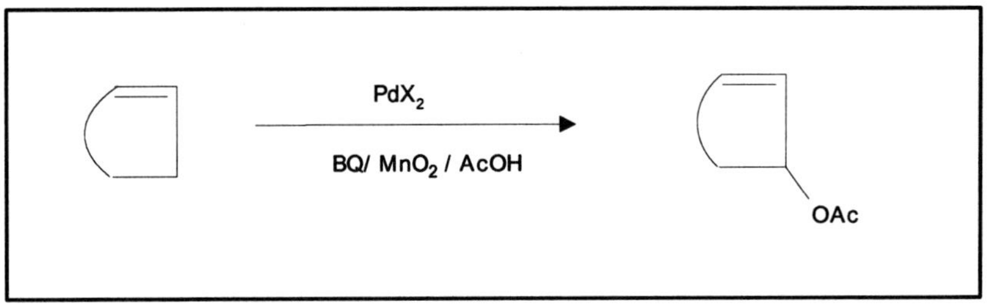

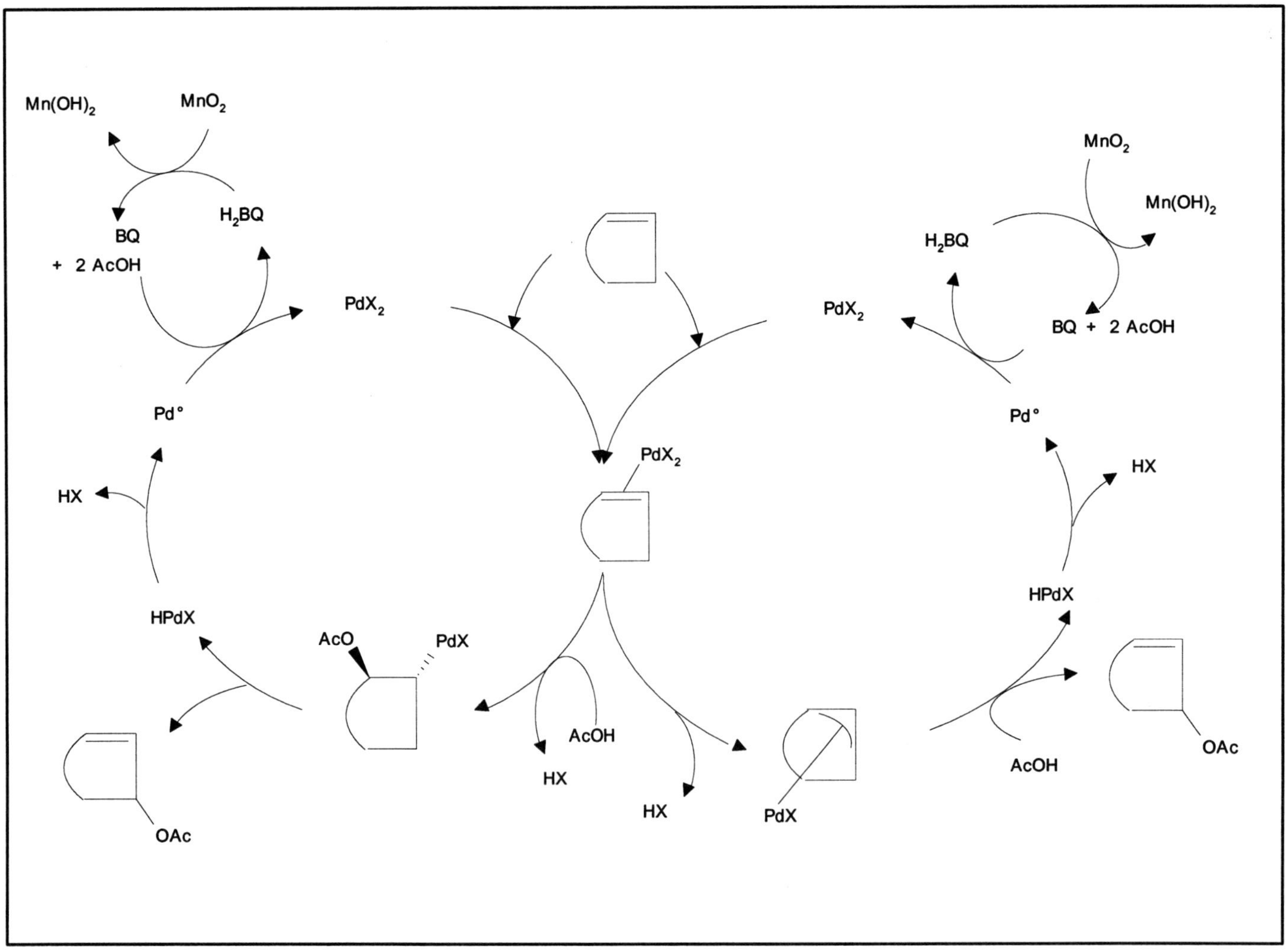

A ---> B	References
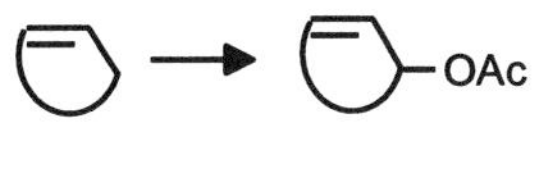	Bäckvall J.E., *J. Am. Chem. Soc.*, 1990; 5160 Akermark B., *J. Org. Chem.*, 1990; 975 Akermark B., *Org. Synth.*, 1990; 68, 109 Bäckvall J.E., *J. Chem. Soc., Chem. Commun.*, 1994; 265 Akermark B., *Tetrahedron Lett.*, 1993; 2523
	Akermark B., *J. Org. Chem.*, 1994; 5729
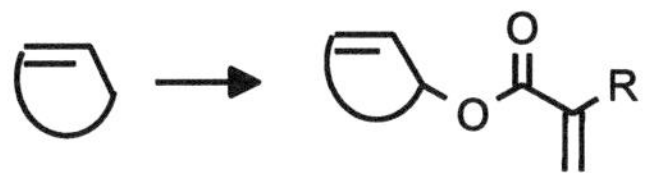	Waegell B., *J. Chem. Soc., Chem. Commun.*, 1994; 2589

RXN42

RXN43 Tandem Acyloxylation-Cyclization of 1,5-Dienes

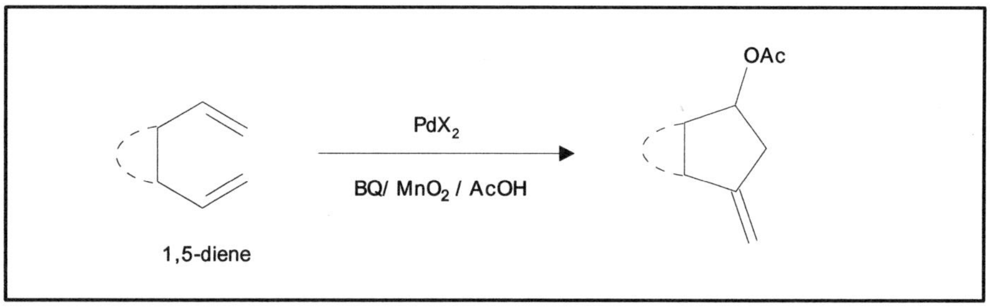

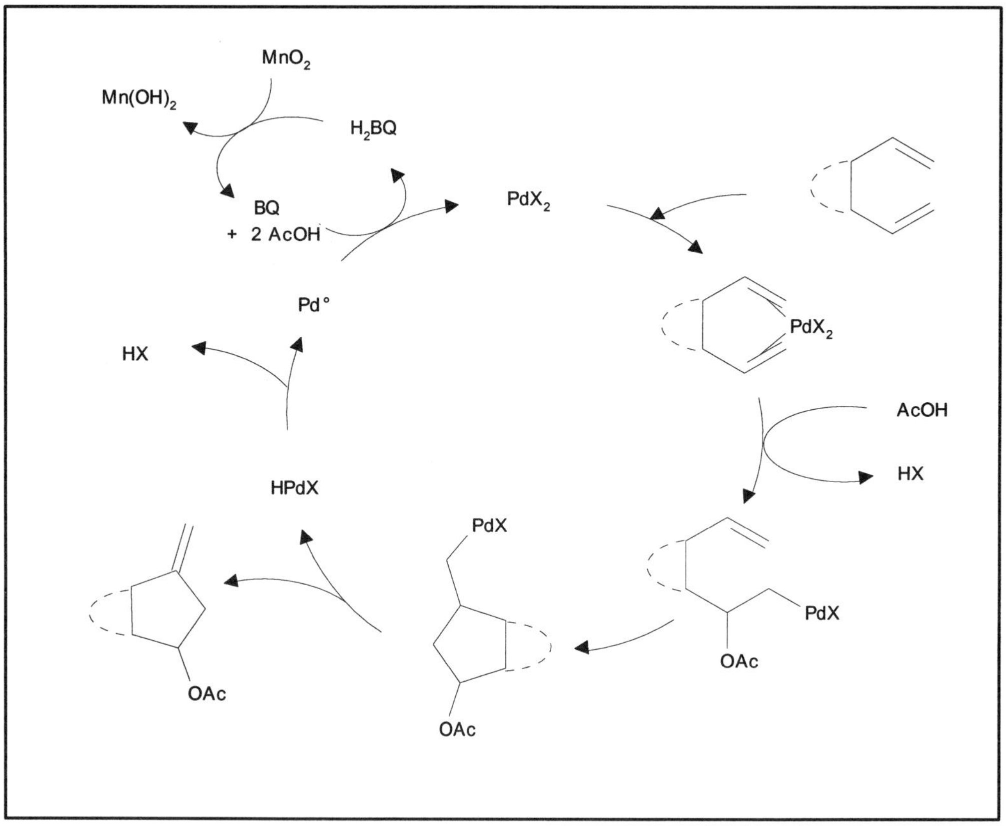

A ---> B	References
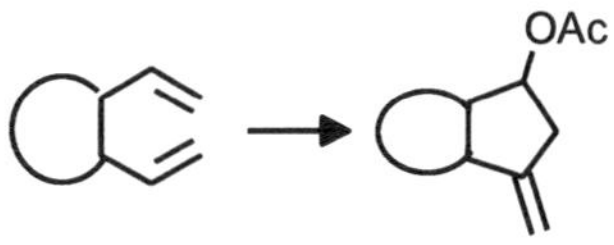	Moberg C., *J. Chem. Soc., Chem. Commun.*, 1986; 518 Moberg C., *Acta Chem. Scand.*, 1991; 77
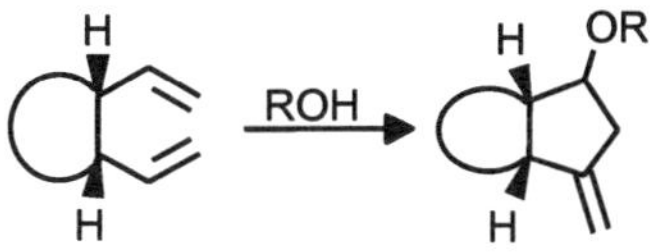	Moberg C., *J. Chem. Soc., Chem. Commun.*, 1988; 1516
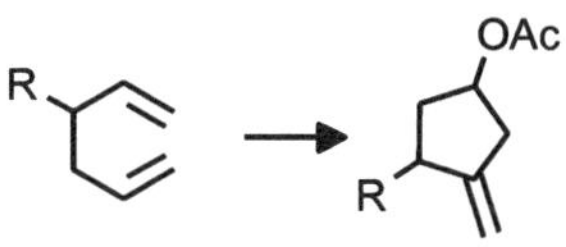	Moberg C., *J. Org. Chem.*, 1989; 4914 Moberg C., *Tetrahedron Lett.*, 1988; 5973

RXN43

RXN44 Tandem Acyloxychlorination-Cyclization of 1,6-Dienes

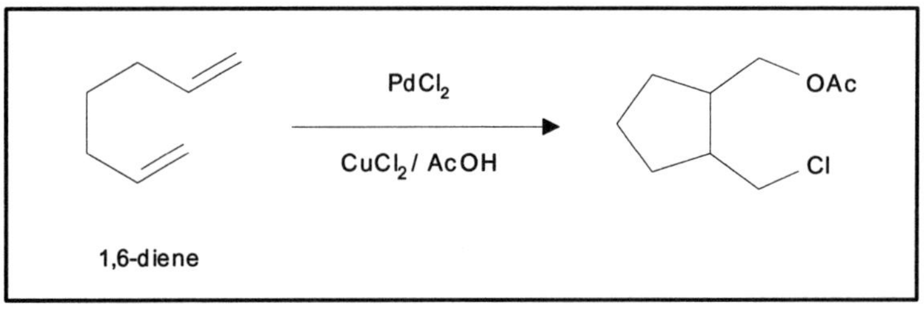

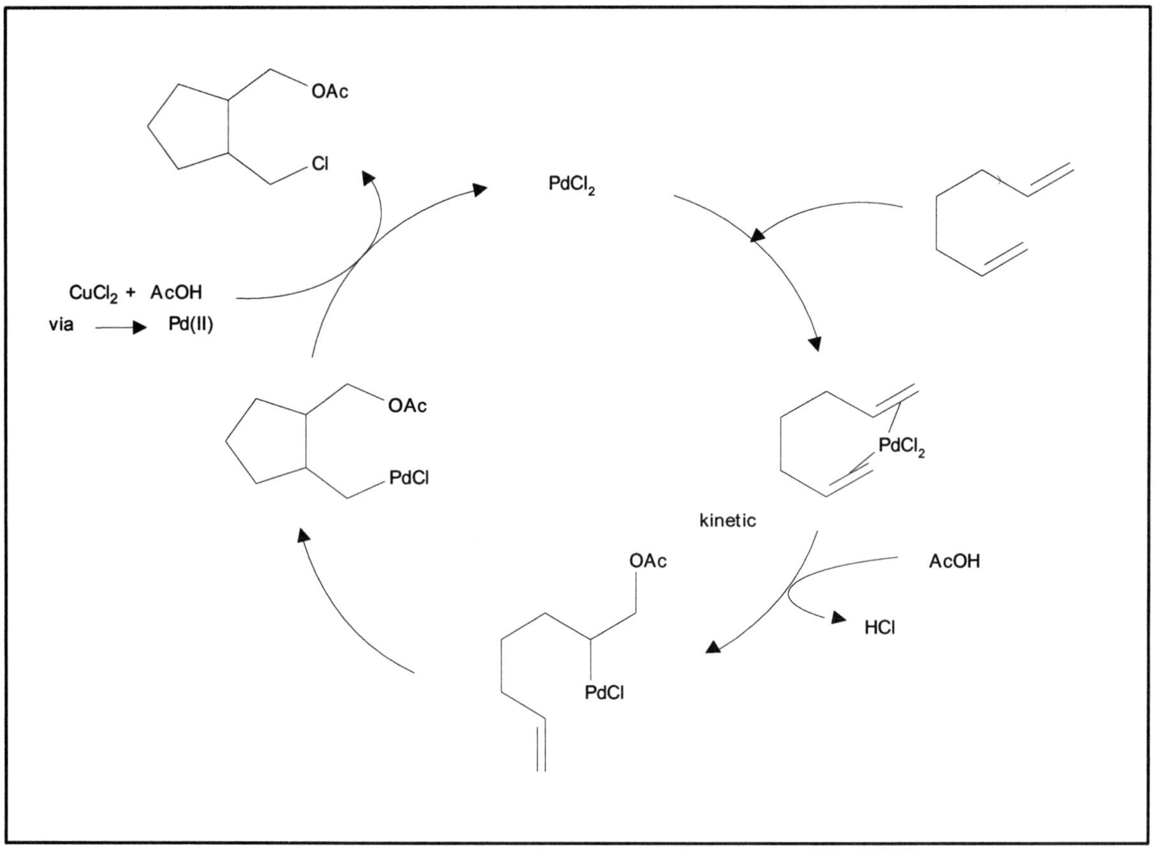

 J.-L. Malleron et al

A ---> B	References
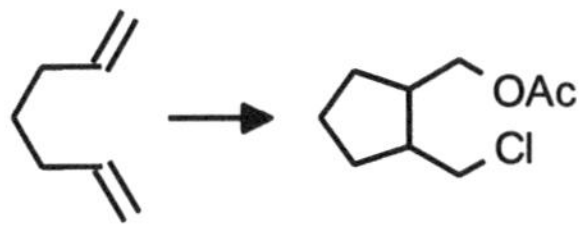	Trost B.M., *J. Chem. Soc., Chem. Commun.*, 1985; 1084
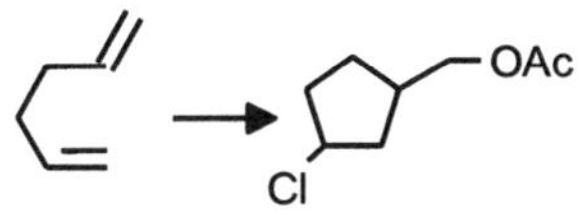	Tenaglia A., *J. Chem. Soc., Chem. Commun.*, 1993; 420

RXN44

RXN45 1,4-Acyloxychlorination of 1,3-Dienes

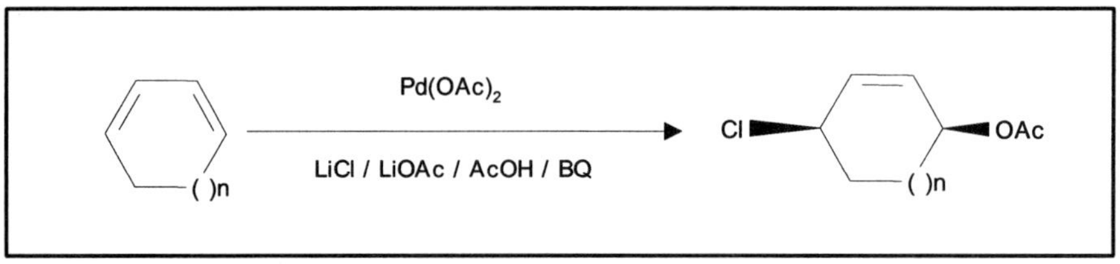

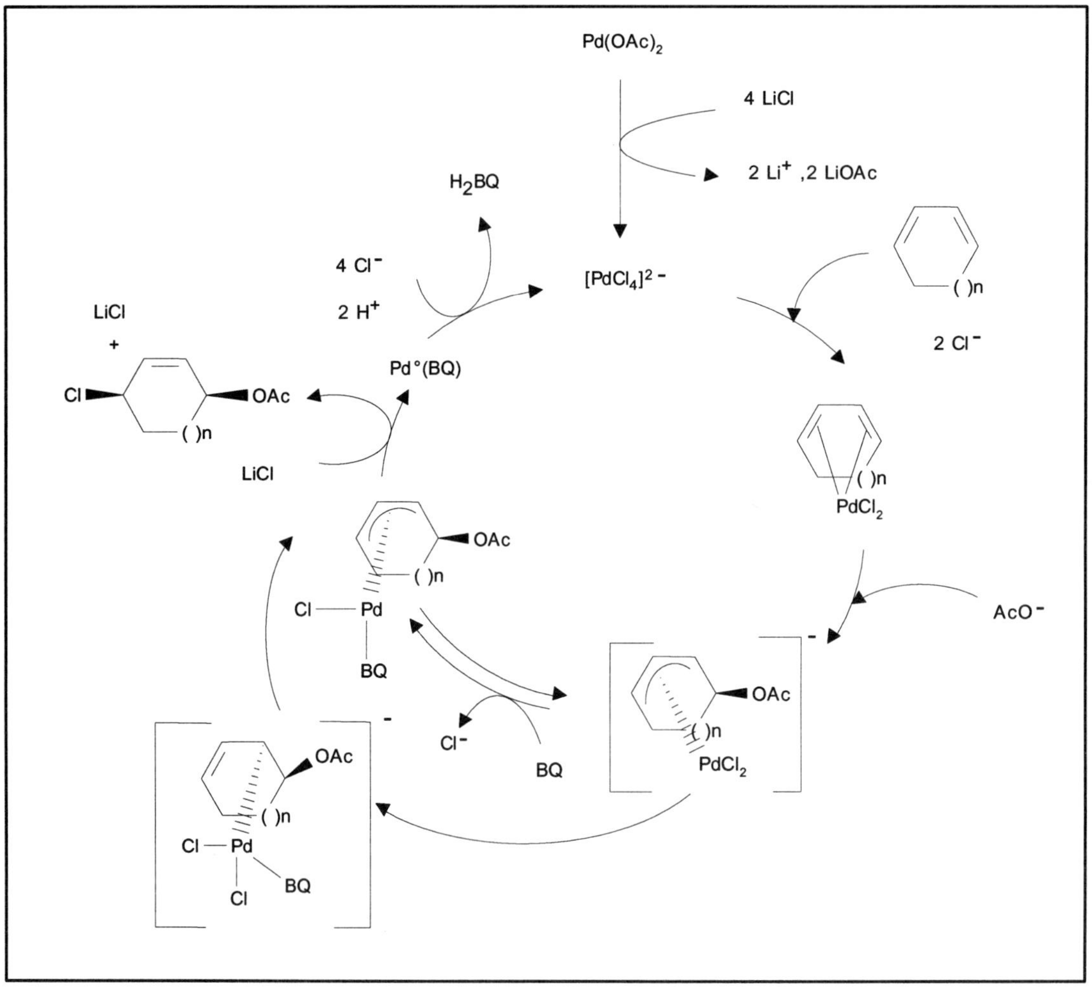

A ---> B	References
	Bäckvall J.E., *Acta Chem. Scand.*, 1990; 492 Bäckvall J.E., *J. Chem. Soc., Chem. Commun.*, 1987; 1236
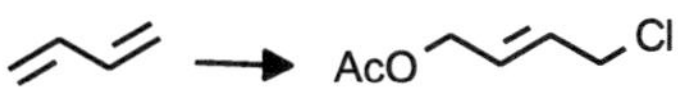	Bäckvall J.E., *Org. Synth.*, 1989; 67, 105 Bäckvall J.E., *Tetrahedron Lett.*, 1982; 1617
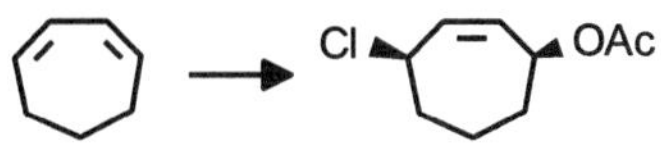	Bäckvall J.E., *J. Am. Chem. Soc.*, 1985; 3676
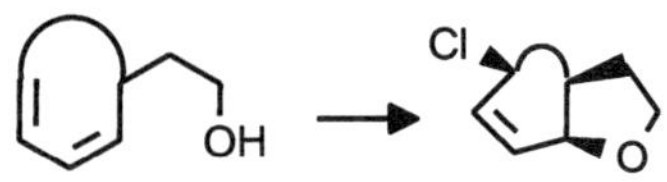	Bäckvall J.E., *J. Am. Chem. Soc.*, 1992; 6374
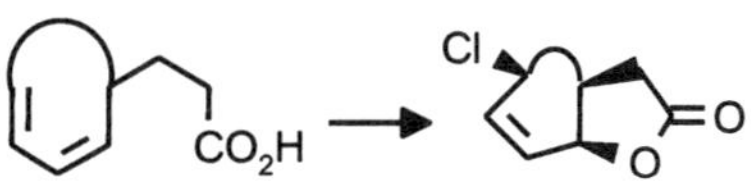	Bäckvall J.E., *Tetrahedron Lett.*, 1989; 137 Bäckvall J.E., *J. Org. Chem.*, 1993; 5445
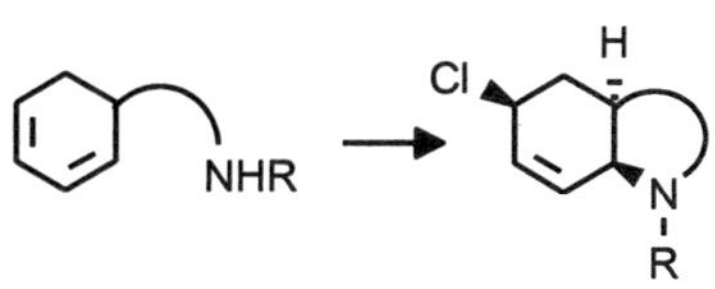	Bäckvall J.E., *J. Am. Chem. Soc.*, 1990; 3683
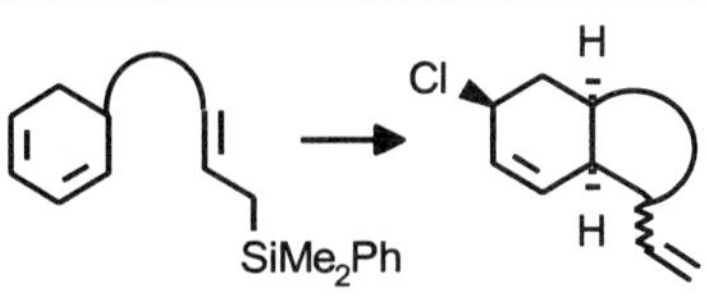	Bäckvall J.E., *J. Am. Chem. Soc.*, 1995; 560

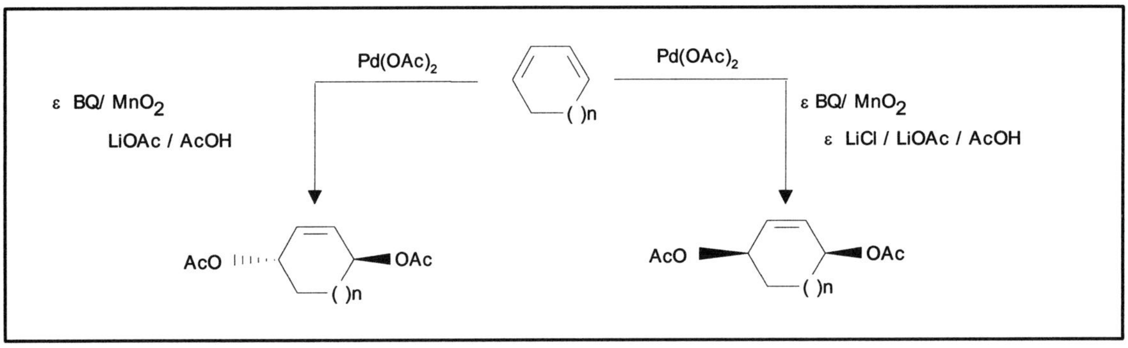
Pd(OAc)2
Pd(OAc)2
ε BQ/ MnO2
LiOAc / AcOH
ε BQ/ MnO2
ε LiCl / LiOAc / AcOH
( )n
AcO
OAc
( )n
AcO
OAc
( )n

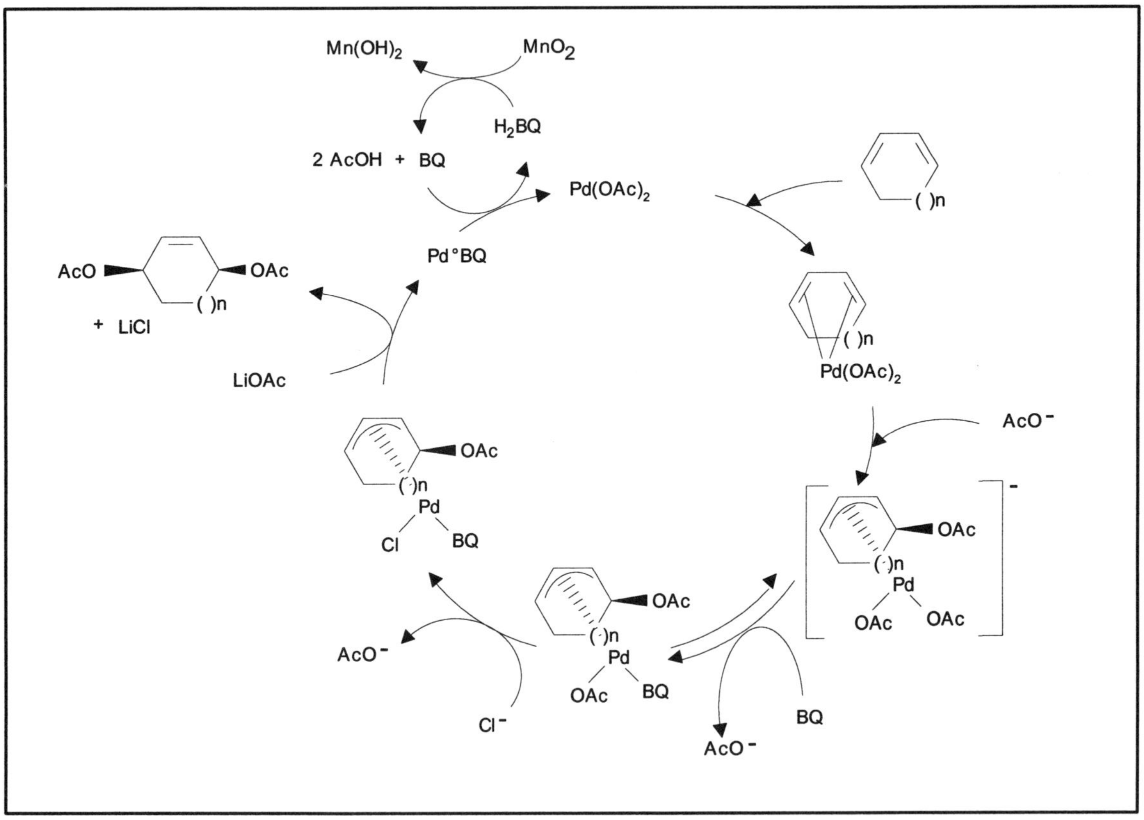
Mn(OH)2
MnO2
H2BQ
2 AcOH + BQ
Pd(OAc)2
( )n
Pd°BQ
AcO
OAc
( )n
+ LiCl
LiOAc
( )n
Pd(OAc)2
AcO⁻
OAc
( )n
Pd
Cl
BQ
OAc
( )n
Pd
OAc
OAc
OAc
( )n
Pd
OAc
BQ
AcO⁻
Cl⁻
BQ
AcO⁻

A ---> B	References
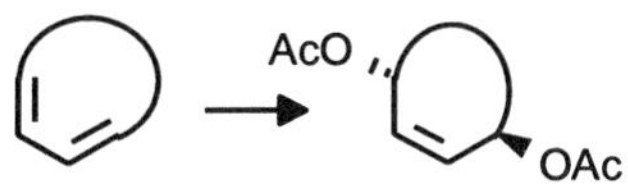	Bäckvall J.E., *J. Am. Chem. Soc.*, 1990; 5160 Bäckvall J.E., *Acta Chem. Scand.*, 1990; 492 Bäckvall J.E., *J. Org. Chem.*, 1984; 4619 Bäckvall J.E., *J. Chem. Soc., Chem. Commun.*, 1987; 1236 Bäckvall J.E., *J. Am. Chem. Soc.*, 1981; 4959 Bäckvall J.E., *J. Am. Chem. Soc.*, 1987; 4750
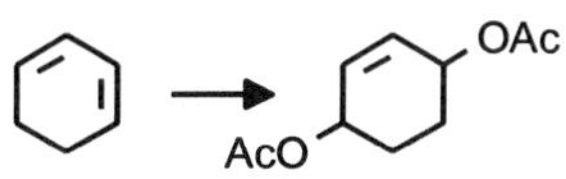	Akermark B., *J. Org. Chem.*, 1994; 5729
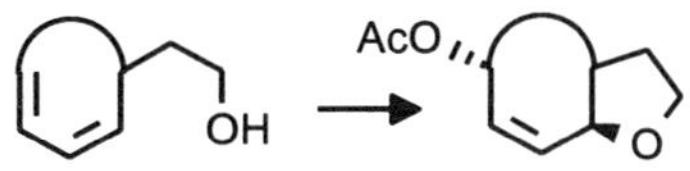	Bäckvall J.E., *J. Am. Chem. Soc.*, 1992; 6374 Andersson P.G., *Tetrahedron Lett.*, 1995; 5397
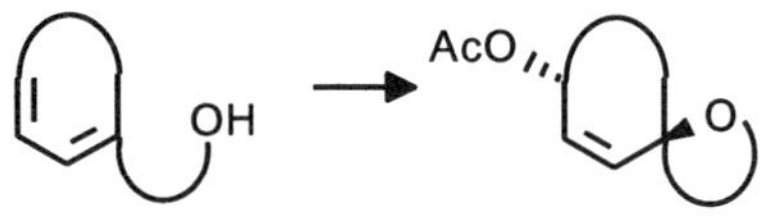	Bäckvall J.E., *J. Org. Chem.*, 1991; 2274
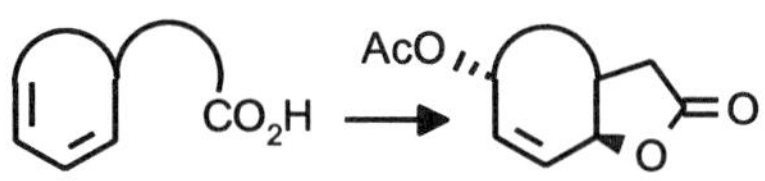	Bäckvall J.E., *Tetrahedron Lett.*, 1989; 137 Bäckvall J.E., *J. Org. Chem.*, 1993; 5445
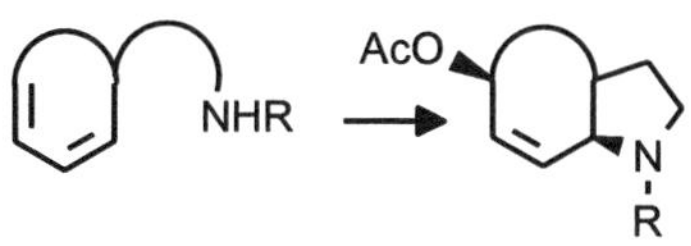	Bäckvall J.E., *J. Am. Chem. Soc.*, 1990; 3683
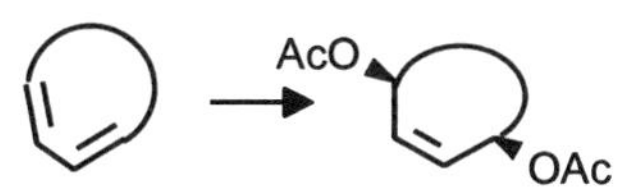	Bäckvall J.E., *J. Am. Chem. Soc.*, 1981; 4959 Bäckvall J.E., *Acta Chem. Scand.*, 1990; 492 Bäckvall J.E., *J. Org. Chem.*, 1984; 4619 Johnson C.R., *J. Org. Chem.*, 1995; 615
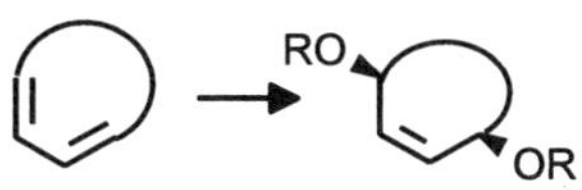	Bäckvall J.E., *J. Org. Chem.*, 1988; 5695
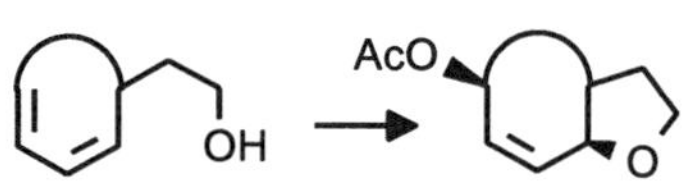	Bäckvall J.E., *J. Am. Chem. Soc.*, 1992; 6374 Andersson P.G., *Tetrahedron Lett.*, 1995; 5397
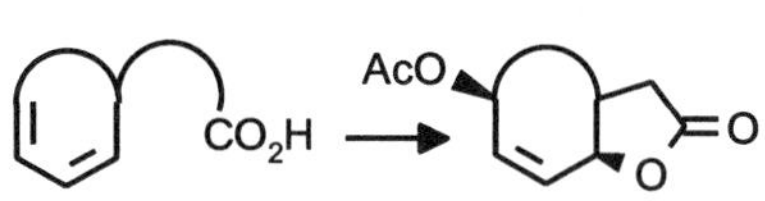	Bäckvall J.E., *Tetrahedron Lett.*, 1989; 137 Bäckvall J.E., *J. Org. Chem.*, 1993; 5445
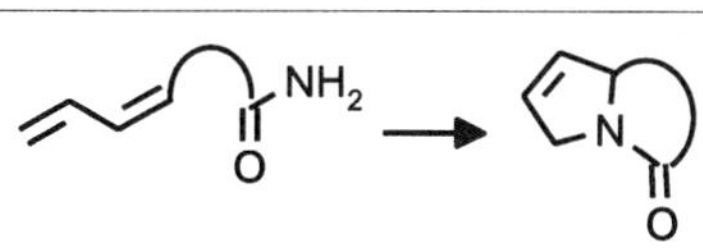	Bäckvall J.E., *J. Am. Chem. Soc.*, 1992; 8696

RXN47 Intramolecular Amination, Alkoxylation or Acyloxylation of Alkynes

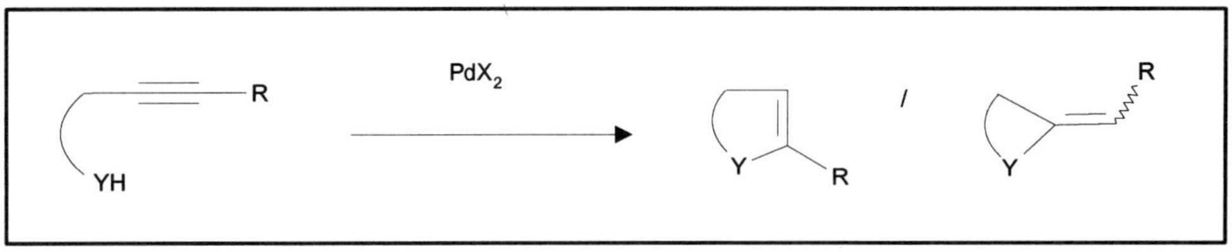

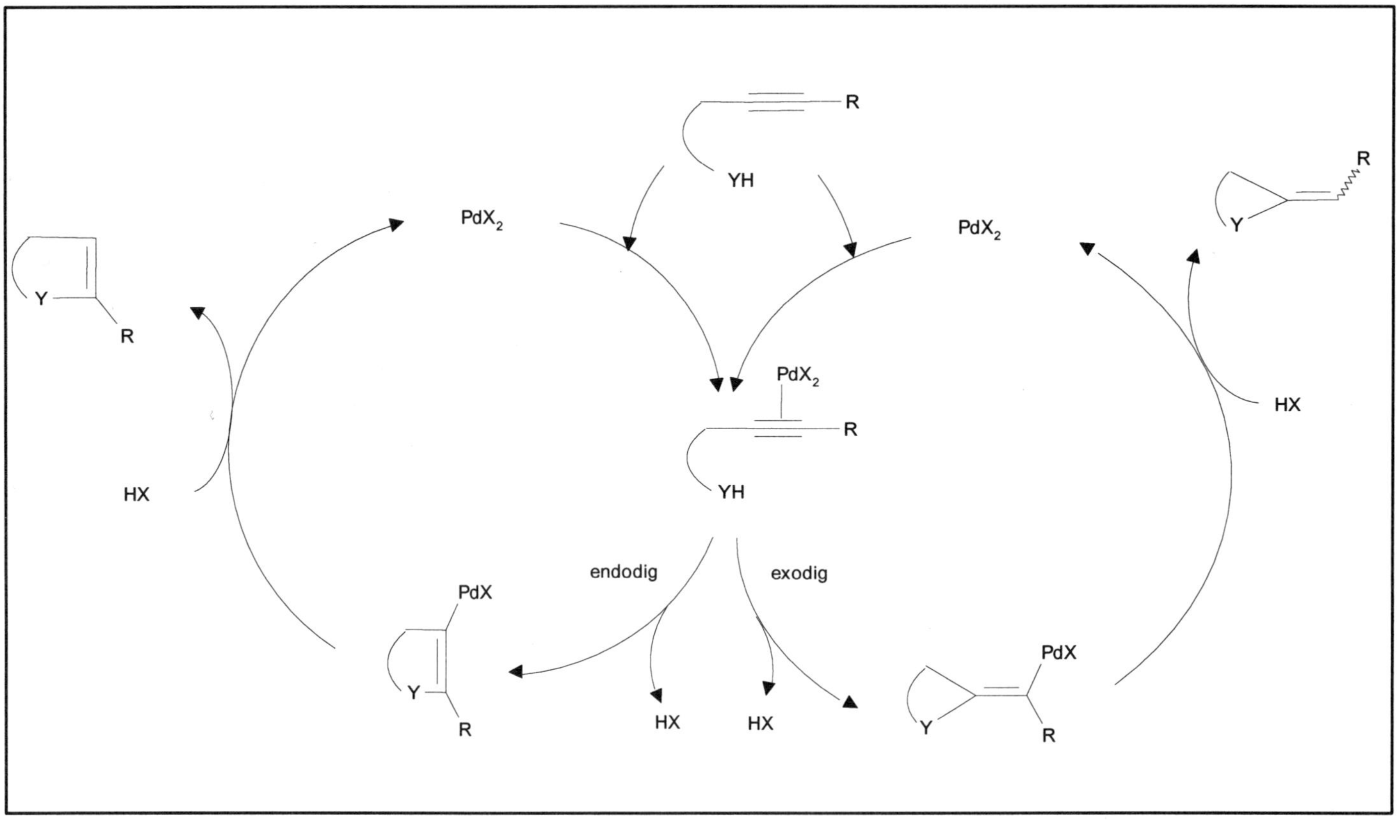

Reagent	YH	Product	Catalyst	References
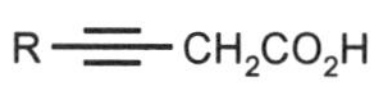	CO_2H	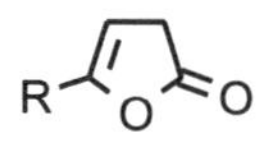	$PdCl_2L_2$	Utimoto K., *Tetrahedron Lett.*, 1984; 5323
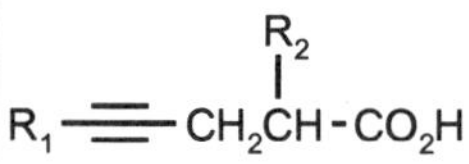	CO_2H	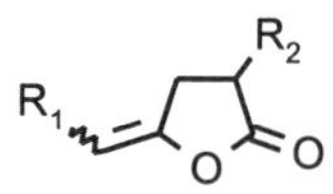	$PdCl_2L_2$	Utimoto K., *Tetrahedron Lett.*, 1984; 5323
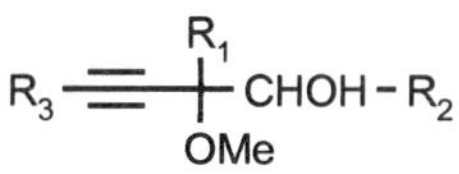	OH	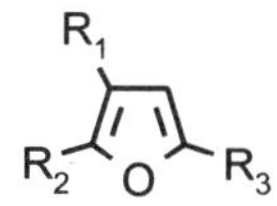	$PdCl_2L_2$	Utimoto K., *Tetrahedron*, 1985; 3655
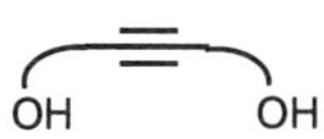	OH	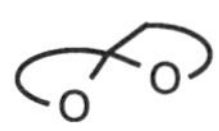	$PdCl_2L_2$	Utimoto K., *Pure Appl. Chem.*, 1983; 1845
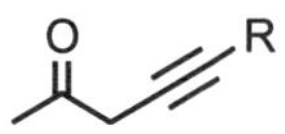	OH	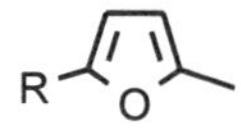	$PdCl_2L_2$	Utimoto K., *J. Org. Chem.*, 1991; 5816 Murahashi S-I., *Heterocycles*, 1992; 33, 1079
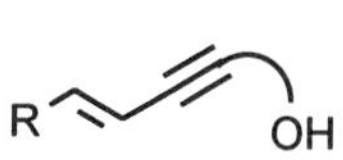	OH	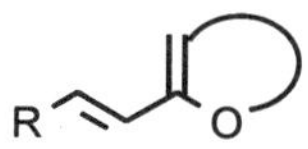	$PdCl_2L_2$	Cacchi S., *Tetrahedron*, 1993; 4955
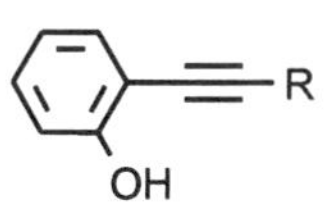	OH	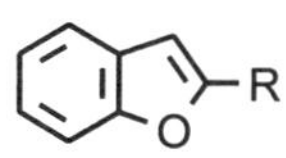	$PdCl_2L_2$	Villemin D., *Heterocycles*, 1989; 29, 1255
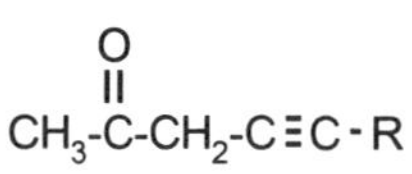	OH	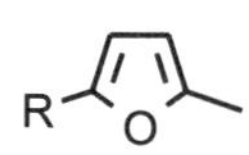	$PdCl_2L_2$	Utimoto K., *J. Org. Chem.*, 1991; 5816
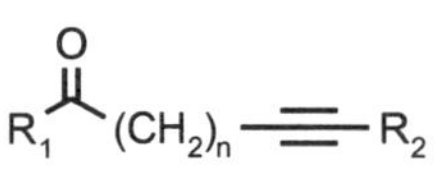	C=O	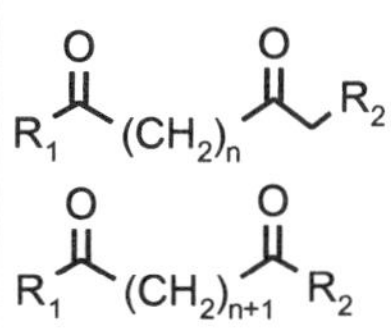	$PdCl_2L_2$	Utimoto K., *Tetrahedron Lett.*, 1987; 3127
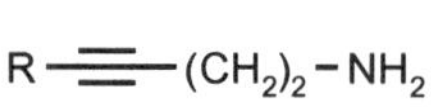	NH_2	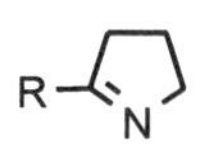	$PdCl_2L_2$	Utimoto K., *Pure Appl. Chem.*, 1983; 1845 Utimoto K., *J. Org. Chem.*, 1991; 5812
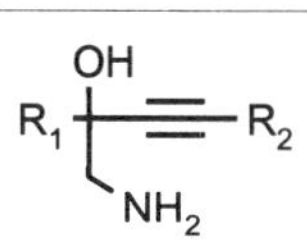	NH_2	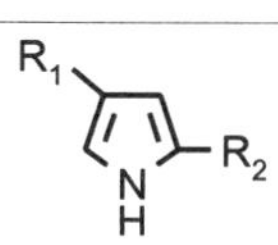	$PdCl_2L_2$	Utimoto K., *Tetrahedron Lett.*, 1981; 4277
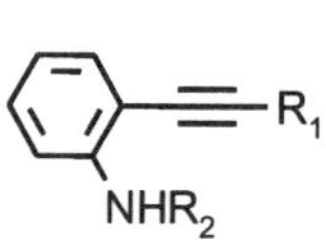	NH_2	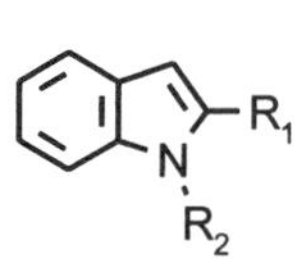	$PdCl_2$	Utimoko K., *Tetrahedron Lett.*, 1988; 1799
	NH_2		$PdCl_2L_2$	Cacchi S., *Tetrahedron Lett.*, 1989; 2581
	NH_2		$PdCl_2$	Cacchi S., *J. Organomet. Chem.*, 1994; 475, 289
	NH_2, NHAc		$PdCl_2$	Utimoko K., *Tetrahedron Lett.*, 1988; 1799
	NHAc		$PdCl_2L_2$	Stille J.K., *J. Org. Chem.*, 1989; 5856
	$NHSO_2Me$		$PdCl_2L_2$	Yamanaka H., *Chem. Pharm. Bull.*, 1988; 1305

RXN47

RXN48 Tandem Intramolecular Amination, Alkoxylation or Acyloxylation-Allylation of Alkynes and Allenes

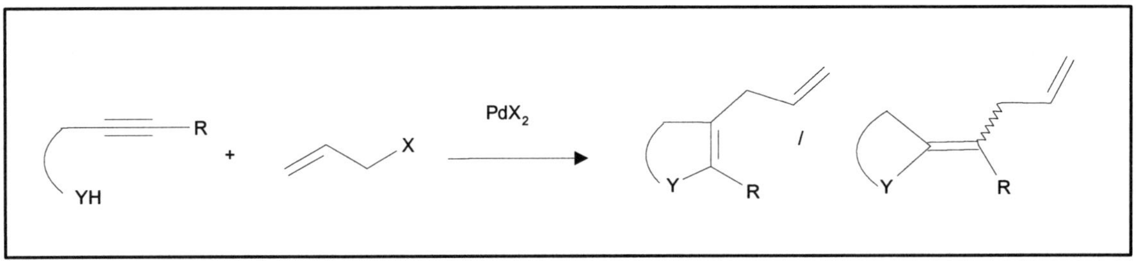

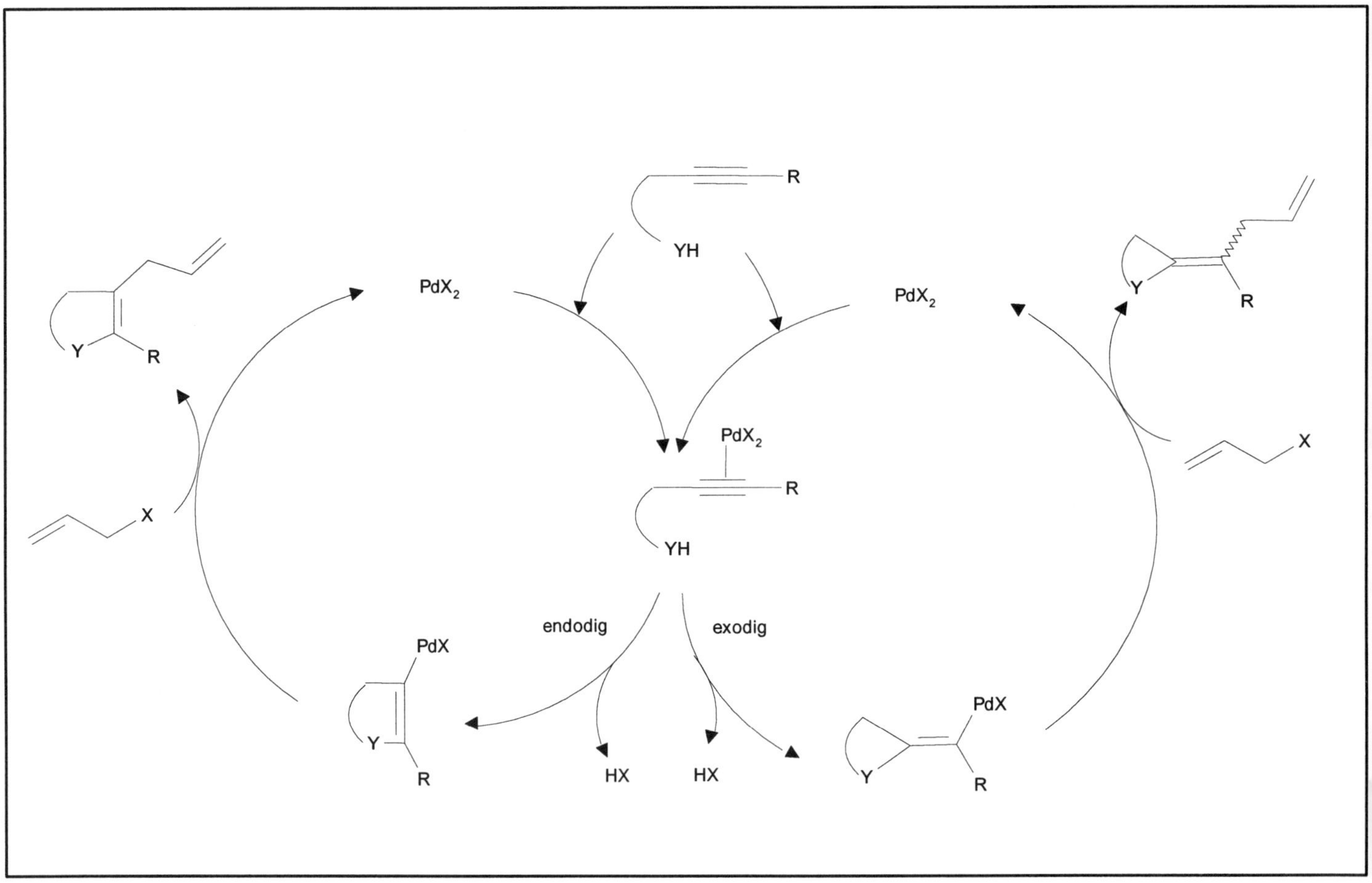

Reagent	YH	Product	Catalyst	References
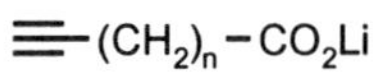	CO_2Li	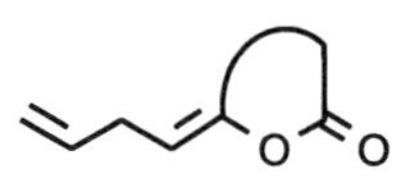	$PdCl_2L_2$	Utimoto K., *J. Am. Chem. Soc.*, 1986; 2753 Murahashi S-I., *Heterocycles*, 1992; 33, 1079
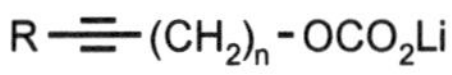	OCO_2Li	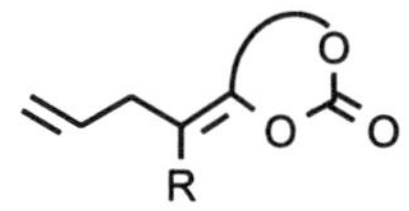	$PdCl_2L_2$	Utimoto K., *J. Org. Chem.*, 1986; 5499
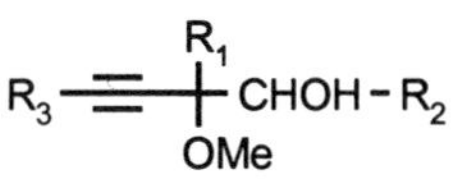	OH	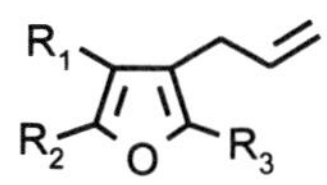	$PdCl_2L_2$	Utimoto K., *Tetrahedron*, 1985; 3655
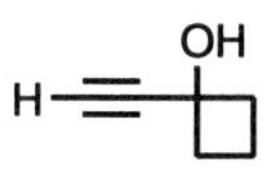	OH	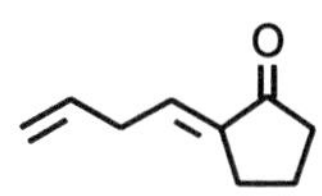	$PdCl_2L_2$ / Hg(II)	Liebeskind L.S., *J. Org. Chem.*, 1994; 1149, (ring extension)
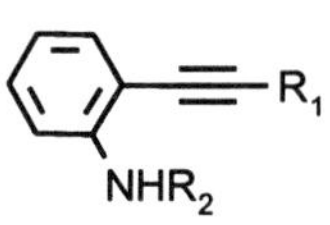	NHAc $NHCO_2Me$	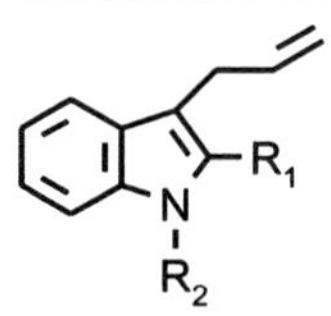	$PdCl_2L_2$	Utimoto K., *Tetrahedron Lett.*, 1988; 1799 Utimoto K., *Tetrahedron Lett.*, 1988; 1799
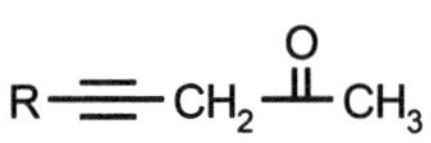	OH	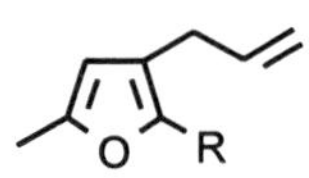	$PdCl_2L_2$	Utimoto K., *J. Org. Chem.*, 1991; 5816

RXN49 Tandem Intramolecular Amination, Alkoxylation or Acyloxylation-Carbonylation of Alkynes

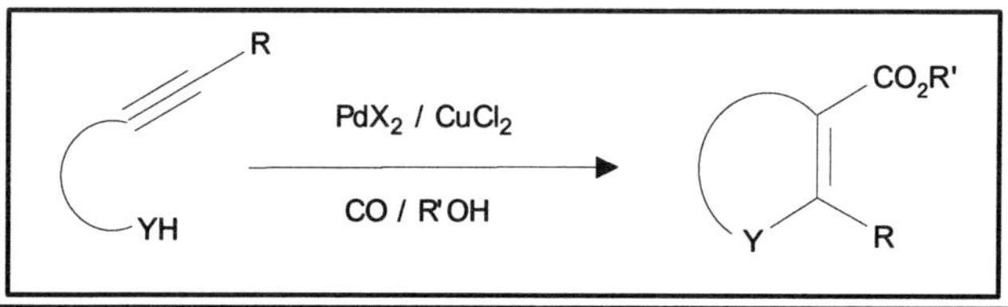

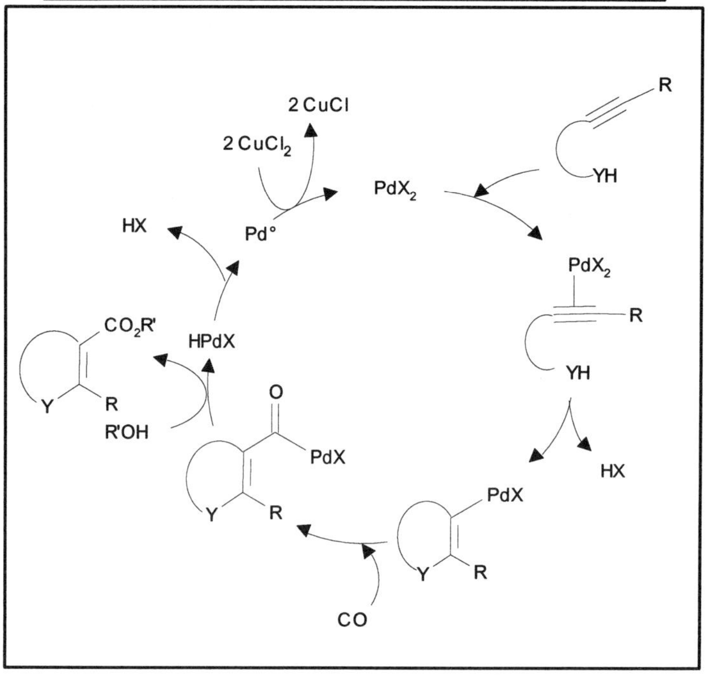

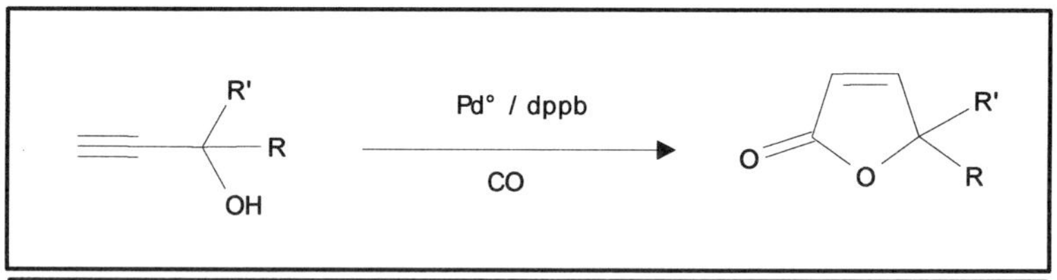

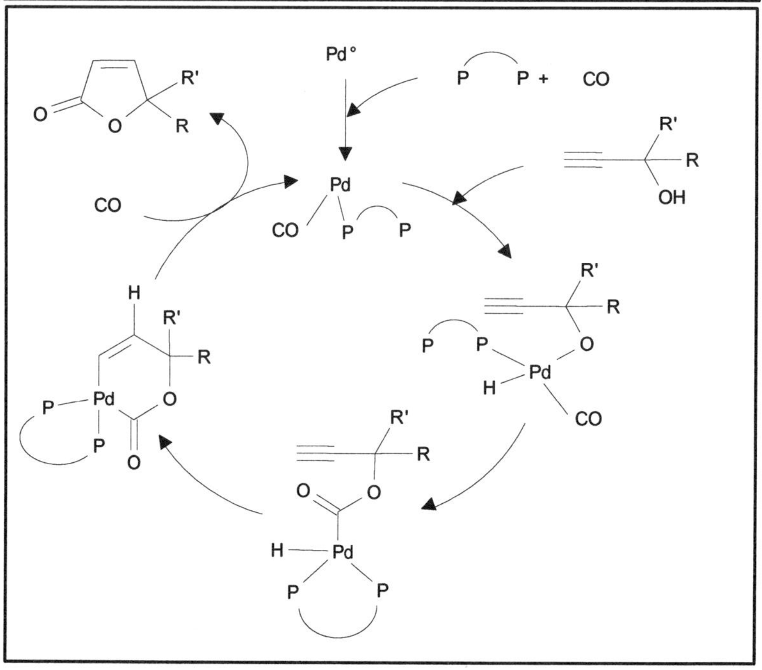

 J.-L. Malleron et al

RXN49 Tandem Intramolecular Amination, Alkoxylation or Acyloxylation-Carbonylation of Alkynes

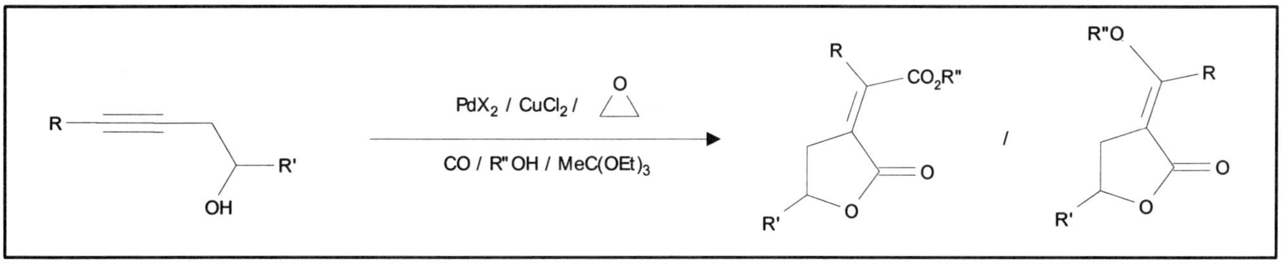

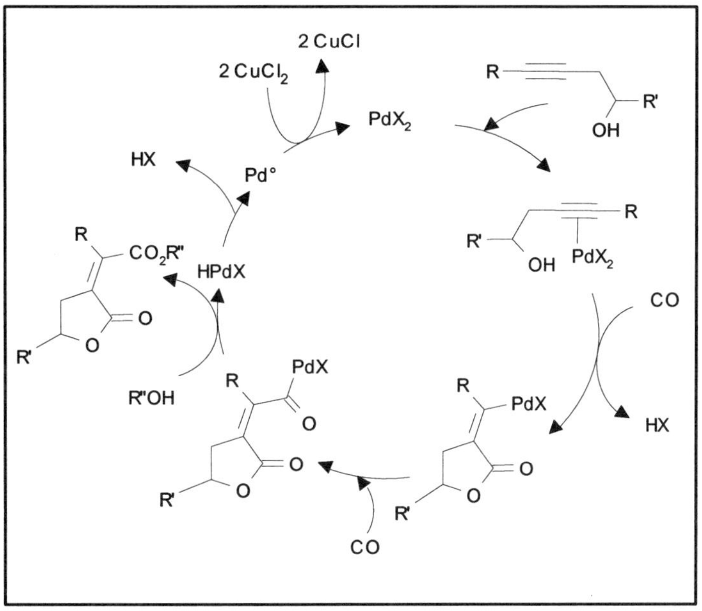

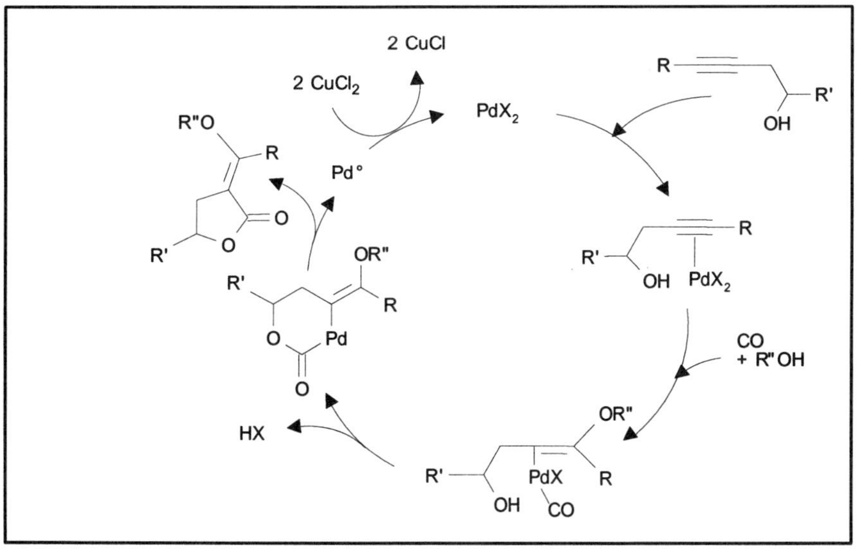

Reagent	YH	Product	Catalyst	References
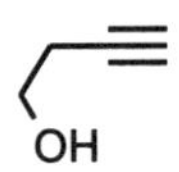	OH	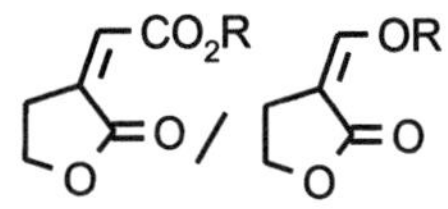	$PdCl_2/CuCl_2$	Tamaru Y., *J. Org. Chem.*, 1991; 1099
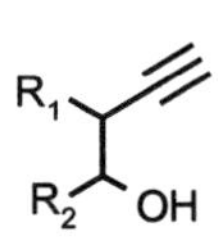	OH	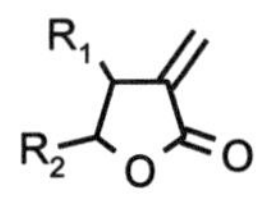	$PdCl_2/SnCl_2$	Norton J.R., *J. Am. Chem. Soc.*, 1991; 7520 Murahashi S.-I., *Heterocycles*, 1992; 33, 1079
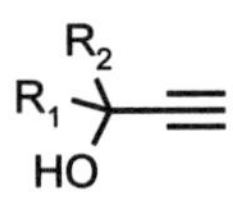	OH	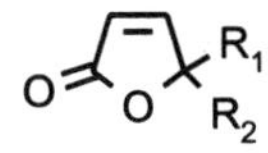	PdL_2	Alper H., *J. Org. Chem.*, 1991; 5357
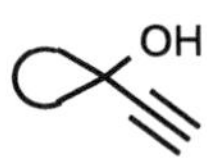	OH	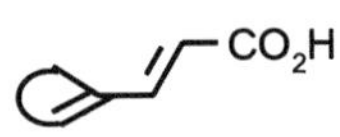	Pd (cationic salt)	Alper H., *J. Org. Chem.*, 1993; 6956
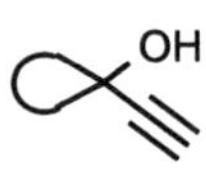	OH	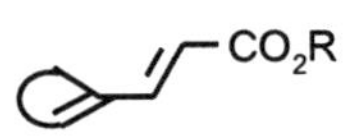	Pd (cationic salt)	Alper H., *J. Org. Chem.*, 1993; 6956
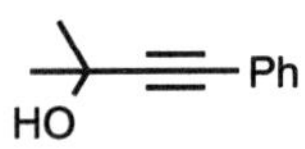	OH	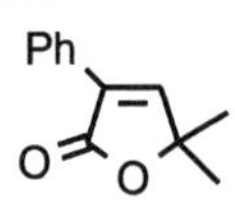	Pd (cationic salt)	Inoue Y., *Tetrahedron Lett.*, 1994; 5889
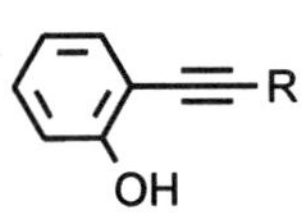	OH	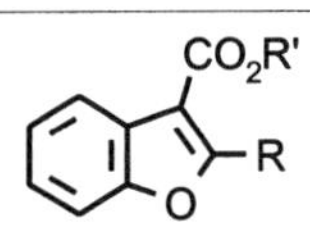	$PdCl_2/CuCl_2$	Yamanaka H., *Heterocycles*, 1989; 29, 1013 Sakamoto T., *Tetrahedron*, 1994; 11803
	NH_2	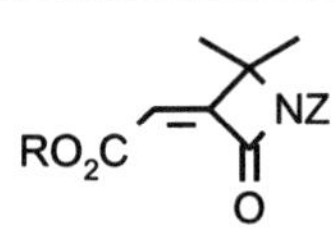	Pd/C / KI-air	Chiusoli G.P., *Tetrahedron Lett.*, 1995; 7495
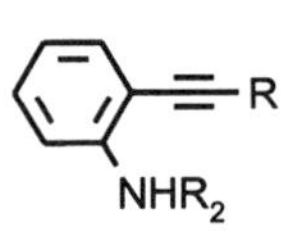	$NHSO_2Me$	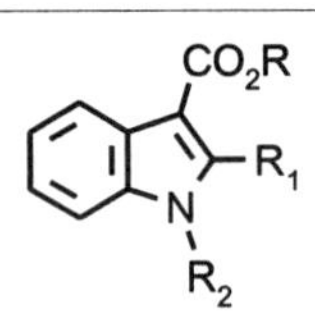	$PdCl_2/CuCl_2$	Yamanaka H., *Heterocycles*, 1989; 29, 1013 Sakamoto T., *Tetrahedron*, 1994; 11803
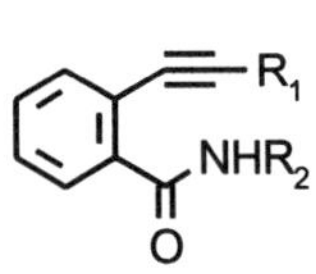	$CONHR_2$	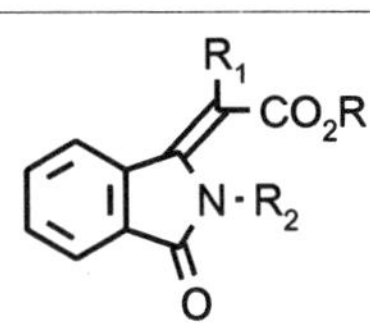	$PdCl_2/CuCl_2$	Sakamoto T., *Tetrahedron*, 1994; 11803

RXN50 Intramolecular Amination or Alkoxylation of Alkenes

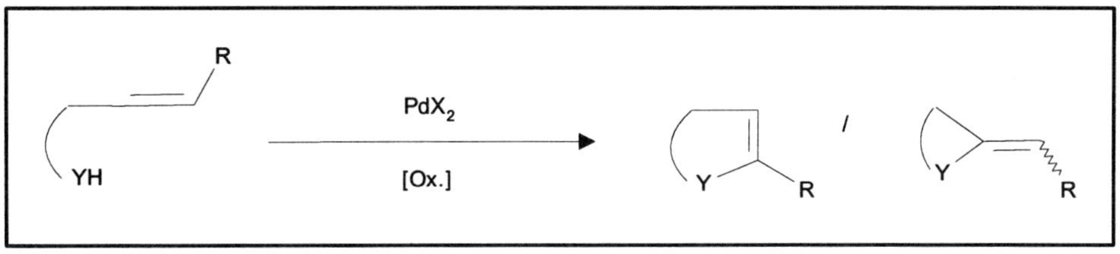

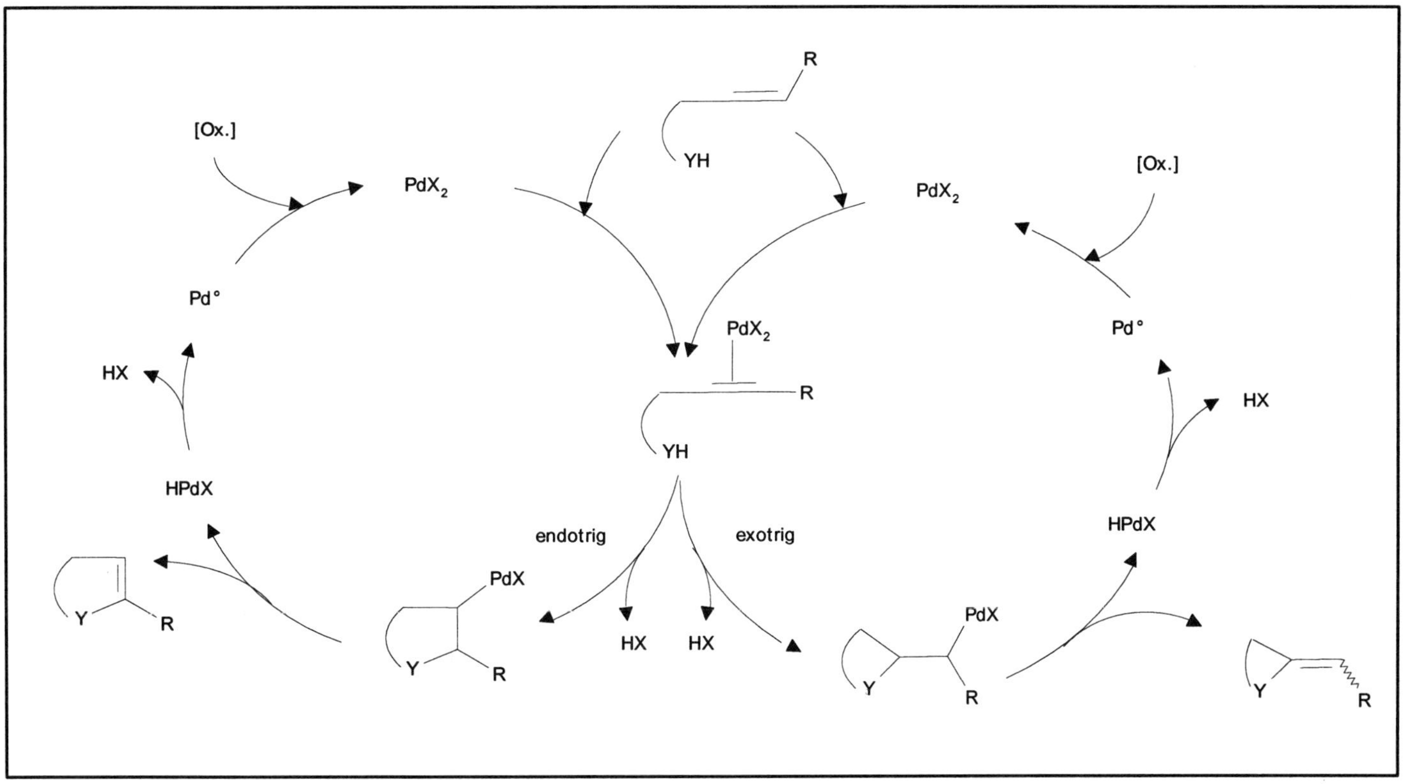

Reagent	YH	Product	Catalyst	References
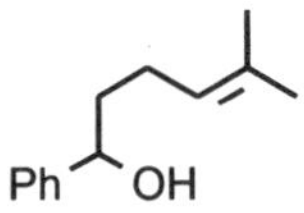	OH	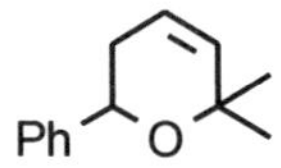	$Pd(OAc)_2$ $Cu(OAc)_2$ / O_2	Hosokawa T., *Tetrahedron Lett.*, 1976; 1821 Murahashi S. I., *Acc. Chem. Res.*, 1990; 49
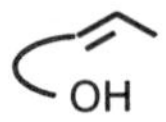	OH	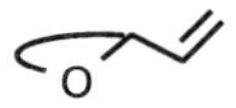	$Pd(OAc)_2$ $Cu(OAc)_2$ / O_2	Hosokawa T., *Tetrahedron Lett.*, 1976; 1821 Murahashi S. I., *Acc. Chem. Res.*, 1990; 49
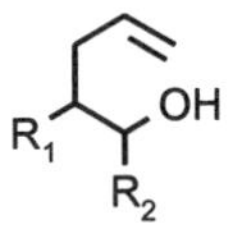	OH	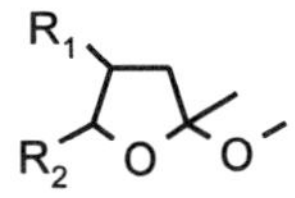	$PdCl_2L_2$ $CuCl/O_2$	Murahashi S-I., *Chem. Lett.*, 1990; 1387 Murahashi S-I., *Heterocycles*, 1992; 33, 1079
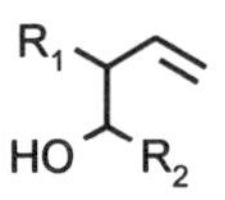	OH	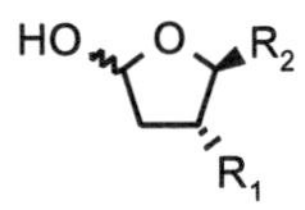	$PdCl_2$ BQ	Nokami J., *Tetrahedron Lett.*, 1988; 5181
	OH		$Pd(OAc)_2$ DMSO / O_2	Andersson P.G., *Tetrahedron Lett.*, 1995; 7749
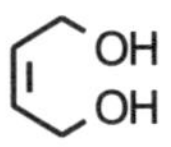	OH		$Pd(OAc)_2$ $Cu(OAc)_2$ O_2	Couturier D., *Synth. Commun.*, 1982; 647
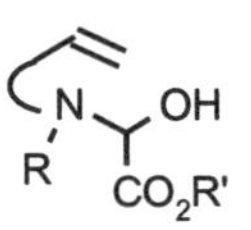	OH	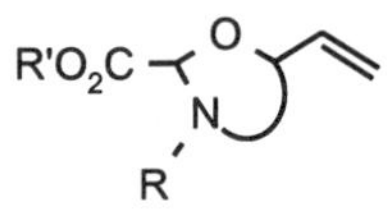	$Pd(OAc)_2$ $Cu(OAc)_2$	Hiemstra H., *J. Org. Chem.*, 1992; 6083
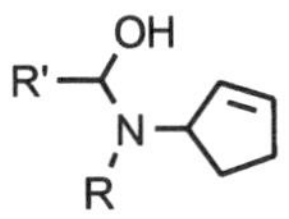	OH	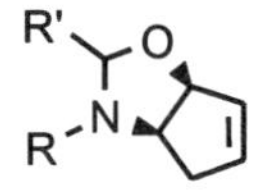	$Pd(OAc)_2$ / O_2	Hiemstra H., *J. Chem. Soc., Chem. Commun.*, 1994; 357 Speckamp W. N., *Angew. Chem. Int. Ed. Engl.*, 1995; 457
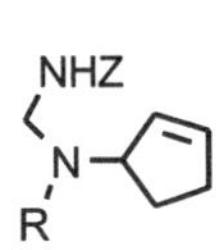	NHZ	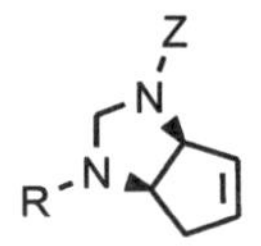	$Pd(OAc)_2$ O_2	Hiemstra H., *J. Tetrahedron Lett.*, 1994; 9281
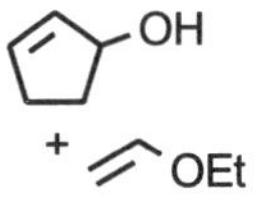	OH	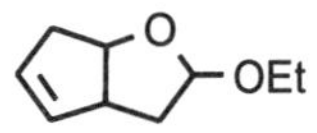	$Pd(OAc)_2$ $Cu(OAc)_2$	Larock R.C., *Tetrahedron Lett.*, 1989; 2767
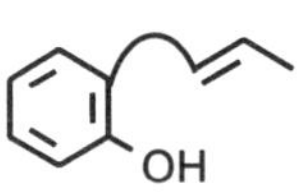	OH	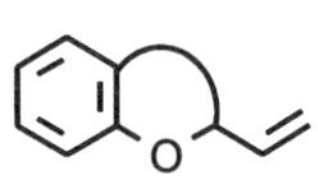	$Pd(OAc)_2$ $Cu(OAc)_2$ / O_2	Hosokawa T., *Bull. Chem. Soc. Jpn.*, 1986; 2191
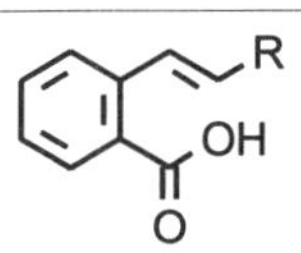	COOH	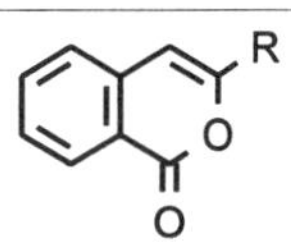	$PdCl_2L_2$ BQ	Hanaoka M., *Chem. Pharm. Bull.*, 1994; 1700
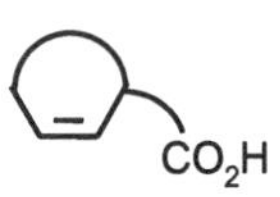	COOH	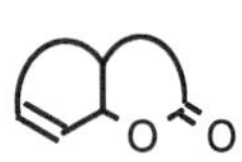	$Pd(OAc)_2$ O_2	Larock R.C., *J. Org. Chem.*, 1993; 5298
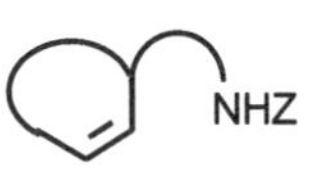	NHZ	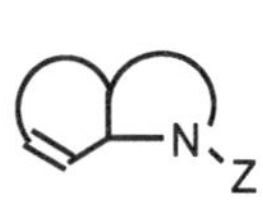	$Pd(OAc)_2$ DMSO / O_2	Andersson P.G., *Tetrahedron Lett.*, 1995; 7749

Reagent	YH	Product	Catalyst	References
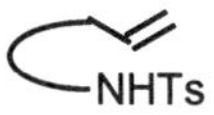	NHTs		$PdCl_2L_2$ BQ / LiCl	Hegedus L.S., *J. Am. Chem. Soc.*, 1982; 2444 Hegedus L.S., *Org. Synth.*, 1984; 62, 48
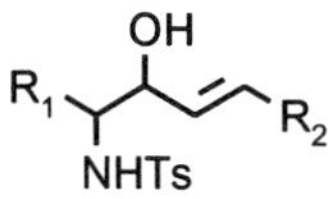	NHTs	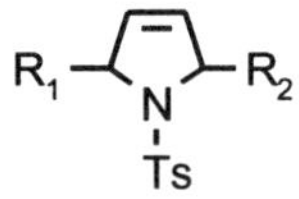	$PdCl_2L_2$	Tamaru Y., *J. Chem. Soc., Chem. Commun.*, 1994; 2531 (-ClPdOH)
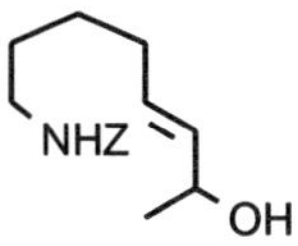	NHZ	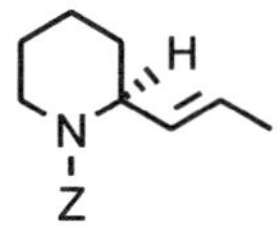	$PdCl_2L_2$	Hirai Y., *Chem. Lett.*, 1994; 21 (-ClPdOH)
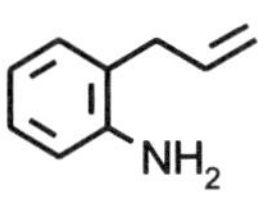	NH_2	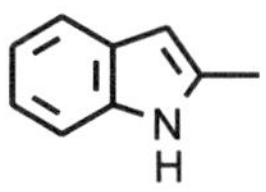	$PdCl_2L_2$ BQ / LiCl or $Cu(OAc)_2$	Hegedus L.S., *J. Am. Chem. Soc.*, 1978; 5800
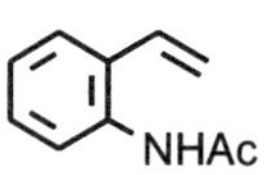	NHAc	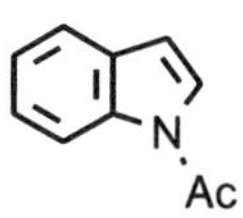	$PdCl_2$ $CuCl_2$ O_2	Kasahara A., *J. Heterocycl. Chem.*, 1989; 1405

RXN51 Tandem Intramolecular Amination or Alkoxylation-Carbonylation of Alkenes and Allenes

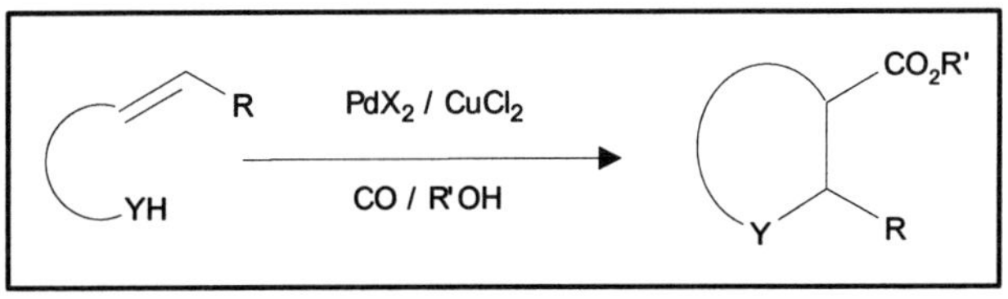

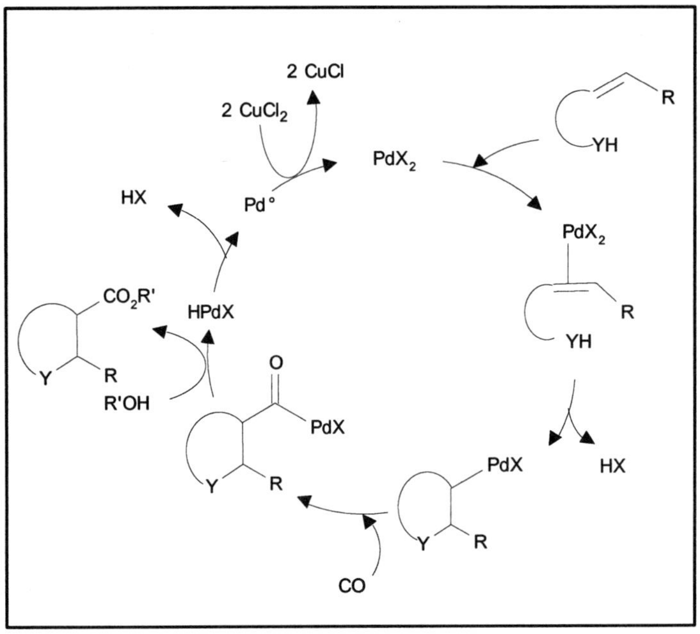

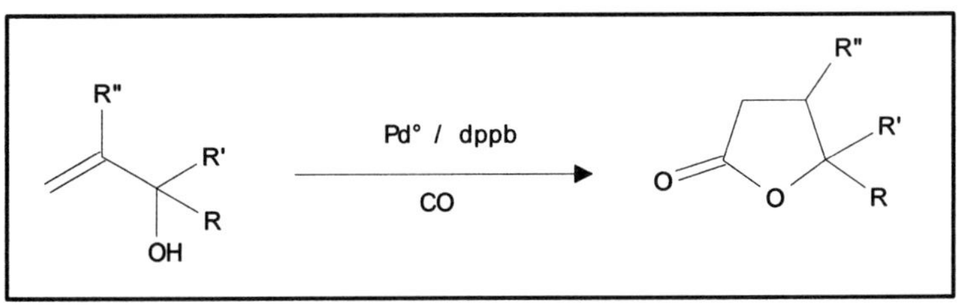

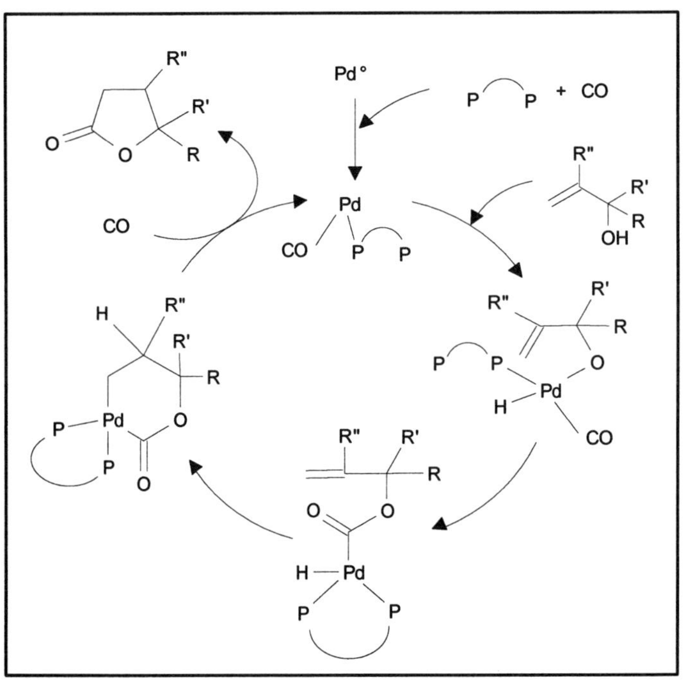

RXN51 Tandem Intramolecular Amination or Alkoxylation-Carbonylation of Alkenes and Allenes

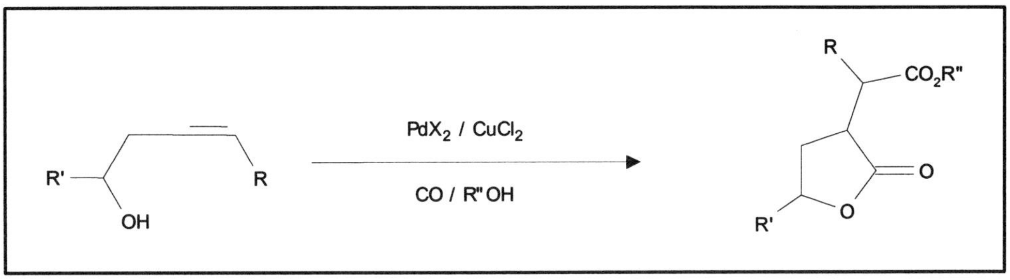

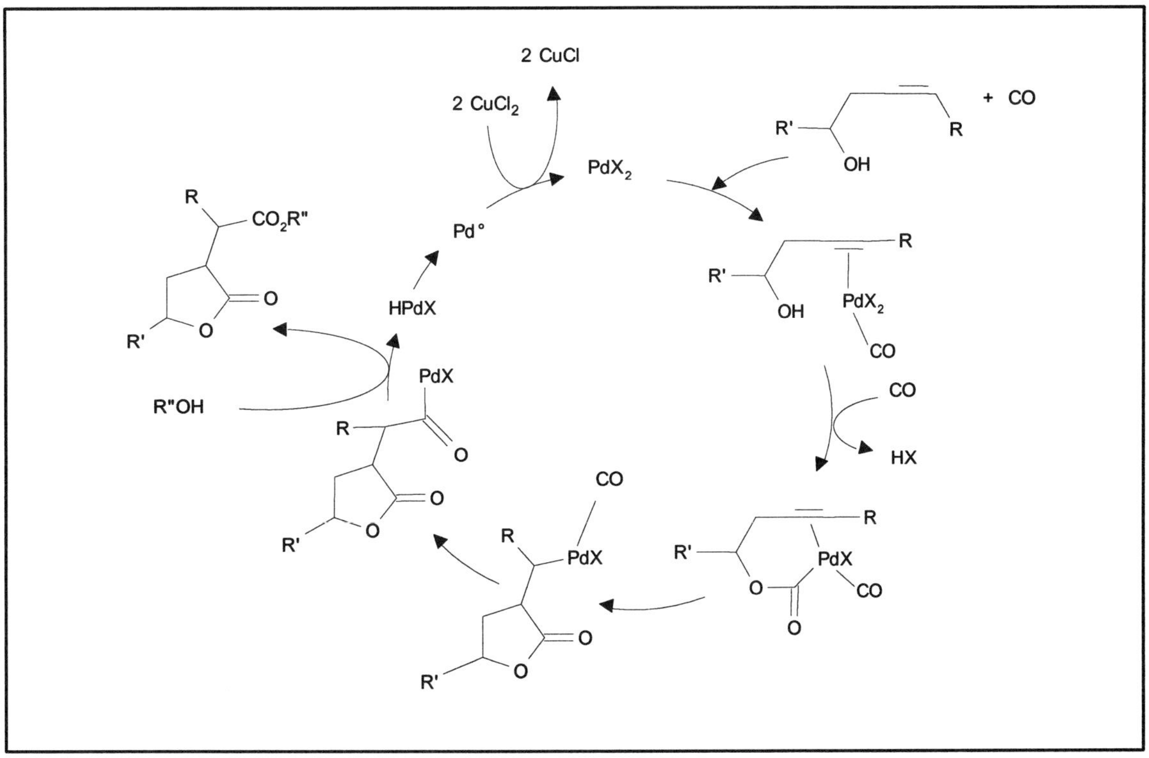

Reagent	YH	Product	Catalyst	References
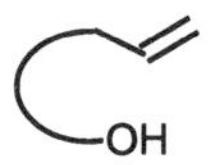	OH	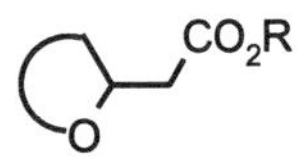	$PdCl_2$ $CuCl_2$	Semmelhack M.F., *J. Am. Chem. Soc.*, 1984; 1496 Semmelhack M.F., *J. Org. Chem.*, 1989; 4483 Semmelhack M.F., *Tetrahedron Lett.*, 1989; 4925 Murahashi S-I., *Acc. Chem. Res.*, 1990; 49
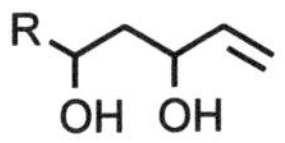	OH	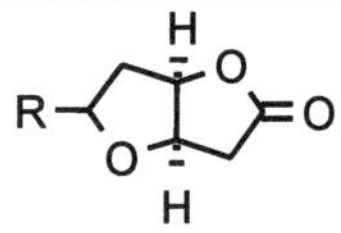	$PdCl_2$ $CuCl_2$	Yoshida Z-I., *Tetrahedron Lett.*, 1985; 3207 Murahashi S- I., *Heterocycles*, 1992; 33, 1079
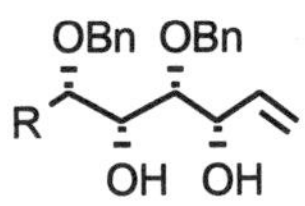	OH	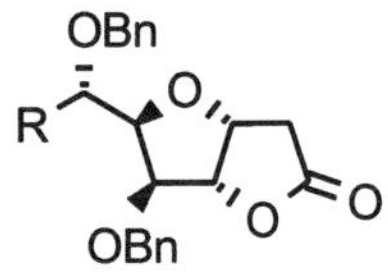	$PdCl_2$ $CuCl_2$	Jäger V., *Synthesis*, 1994; 1359
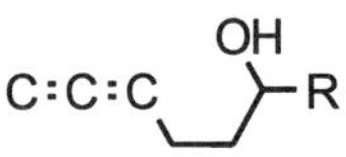	OH	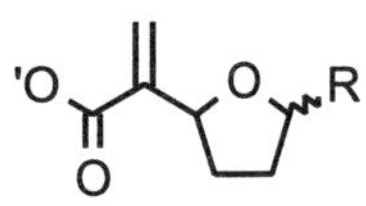	$PdCl_2$ $CuCl_2$	Walkup R.D., *Tetrahedron Lett.*, 1987; 1023
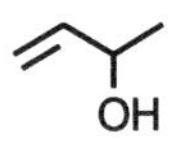	OH	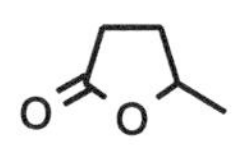	Pd°	Alper H., *J. Org. Chem.*, 1991; 5357
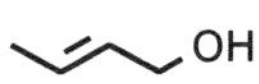	OH	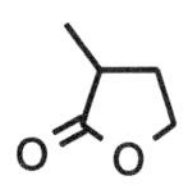	$PdCl_2$ $CuCl_2$ O_2	Alper H., *J. Chem. Soc., Chem. Commun.*, 1990; 135
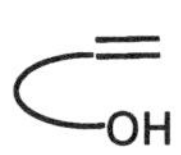	OH	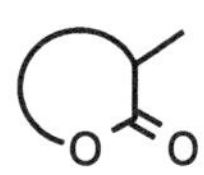	$PdCl_2$ $CuCl_2$ O_2	Alper H., *J. Chem. Soc., Chem. Commun.*, 1985; 511
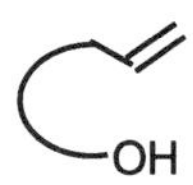	OH	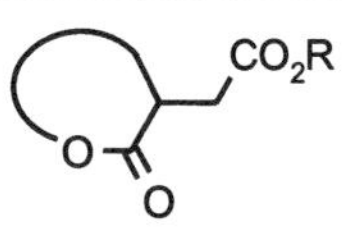	$PdCl_2$ $CuCl_2$	Tamaru Y., *J. Org. Chem.*, 1991; 1099
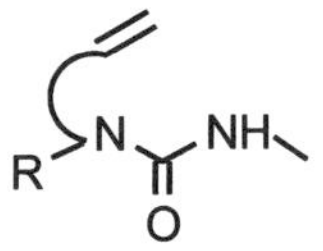	NH-CO-NH	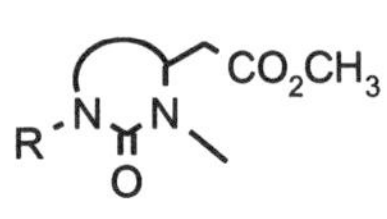	$PdCl_2$ $CuCl_2$	Tamaru Y., *J. Am. Chem. Soc.*, 1988; 3994 Murahashi S-I., *Acc. Chem. Res.*, 1990; 49
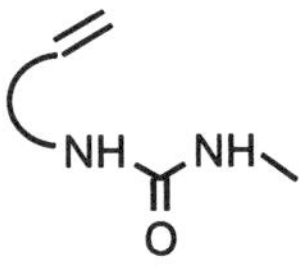	NH-CO-NH	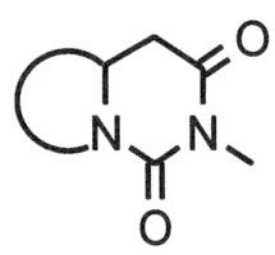	$PdCl_2$ $CuCl_2$	Tamaru Y., *J. Am. Chem. Soc.*, 1988; 3994
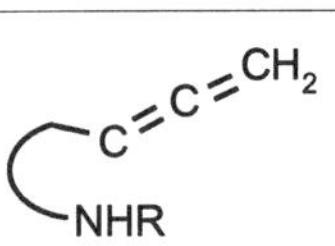	NHR	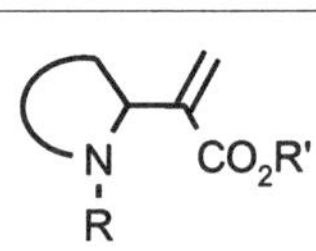	$PdCl_2$ $CuCl_2$ or CuCl	Gallagher T., *Tetrahedron Lett.*, 1986; 6009 Gallagher T., *J. Chem. Soc.*, *Perkin Trans. 1*, 1992; 433
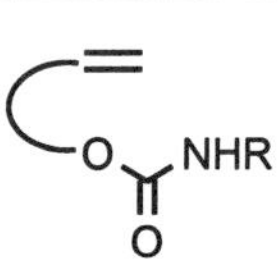	NHR	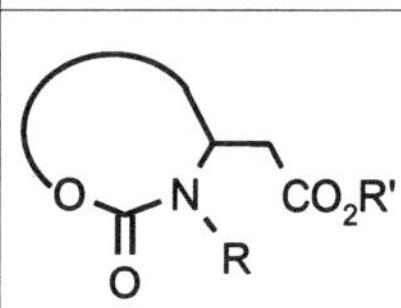	$PdCl_2$ $CuCl_2$	Tamaru Y., *Tetrahedron Lett.*, 1992; 631
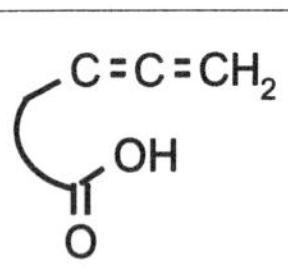	OH	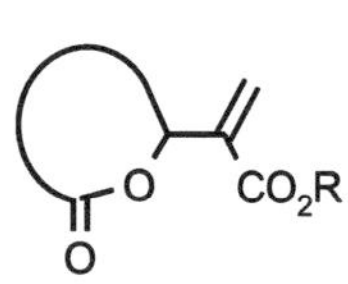	$PdCl_2$ $CuCl_2$	Walkup R.D., *Tetrahedron*, 1993; 9285

Reagent	YH	Product	Catalyst	References
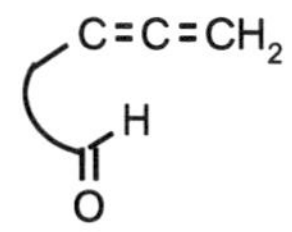	OH	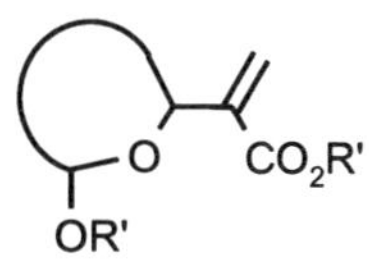	$PdCl_2$ $CuCl_2$	Walkup R. D., *Tetrahedron Lett.*, 1993; 9285 Walkup R. D., *Tetrahedron Lett.*, 1994; 8545
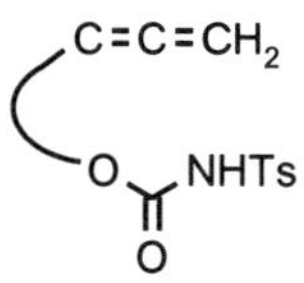	NHTs	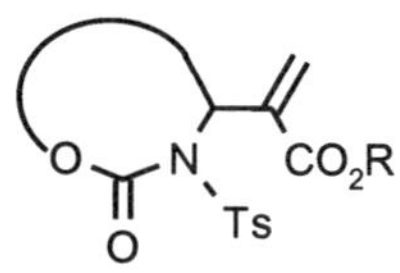	$PdCl_2$ $CuCl_2$	Tamaru Y., *Tetrahedron Lett.*, 1993; 7611

RXN52 Reductive Cyclization with Diynes and Enynes

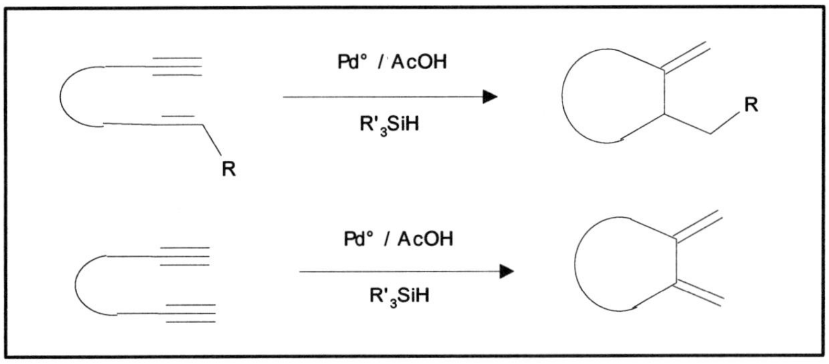

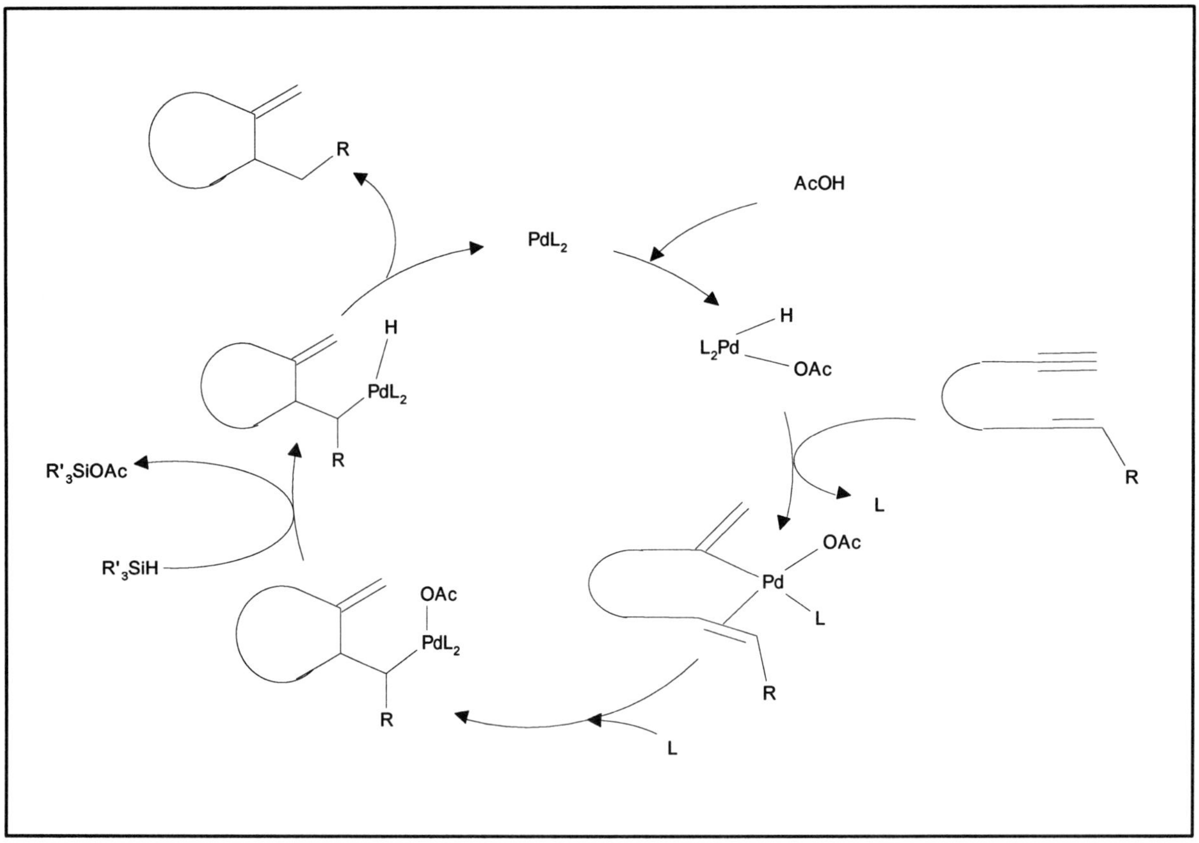

 J.-L. Malleron et al

A ---> B	References
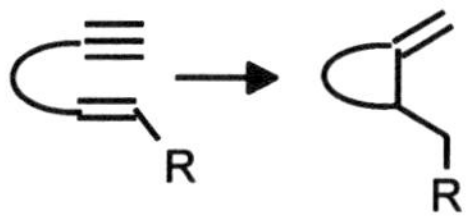	Trost B.M., *J. Am. Chem. Soc.*, 1987; 3161
	Trost B.M., *J. Am. Chem. Soc.*, 1988; 7255

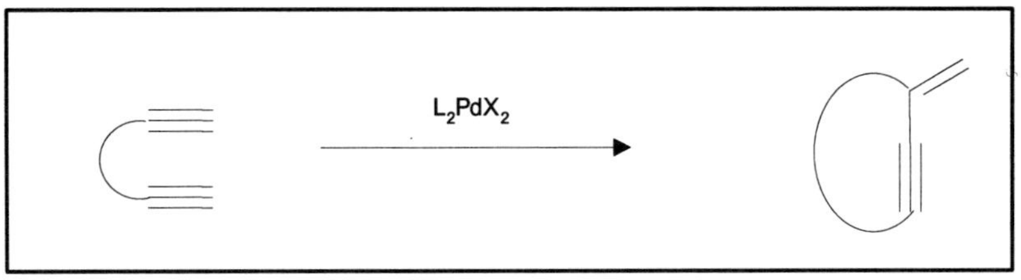
L_2PdX_2

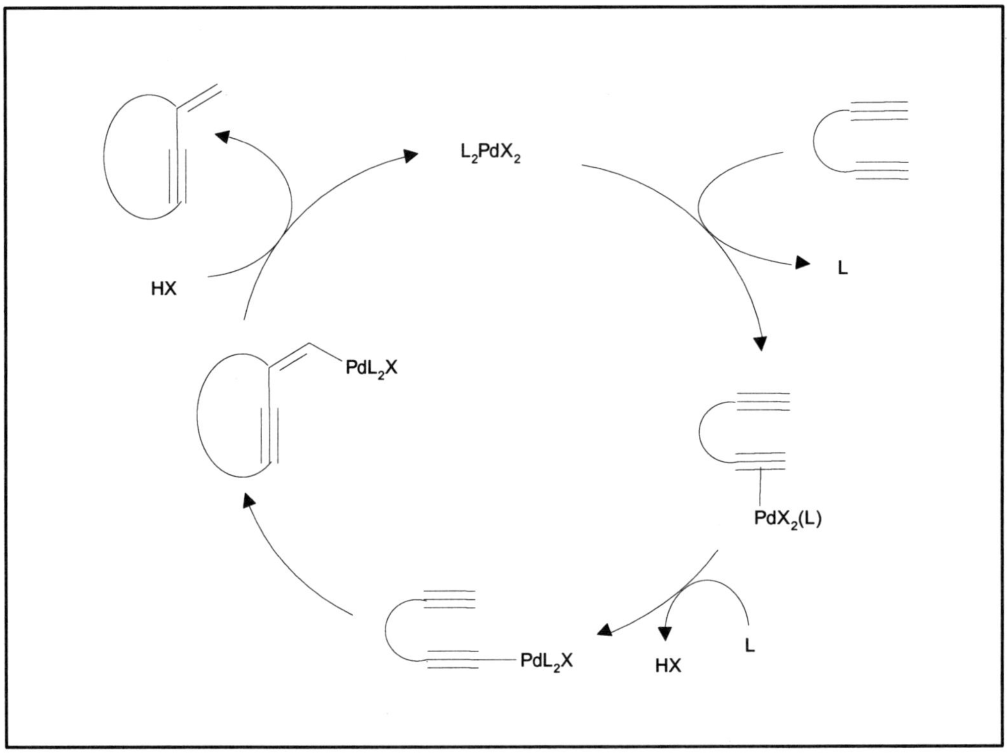
L_2PdX_2
L
$PdX_2(L)$
HX
L
PdL_2X
PdL_2X
HX

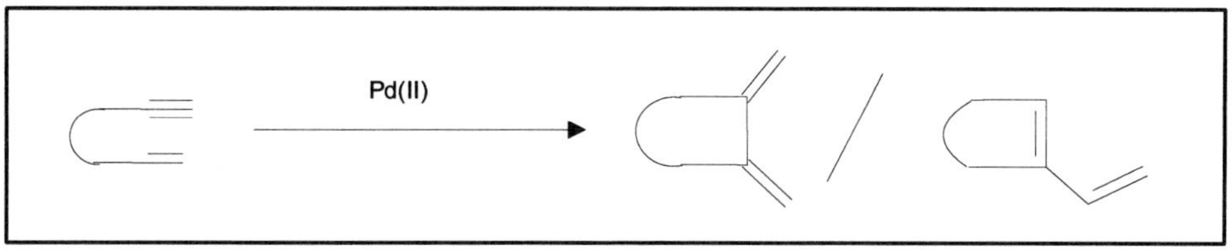
Pd(II)

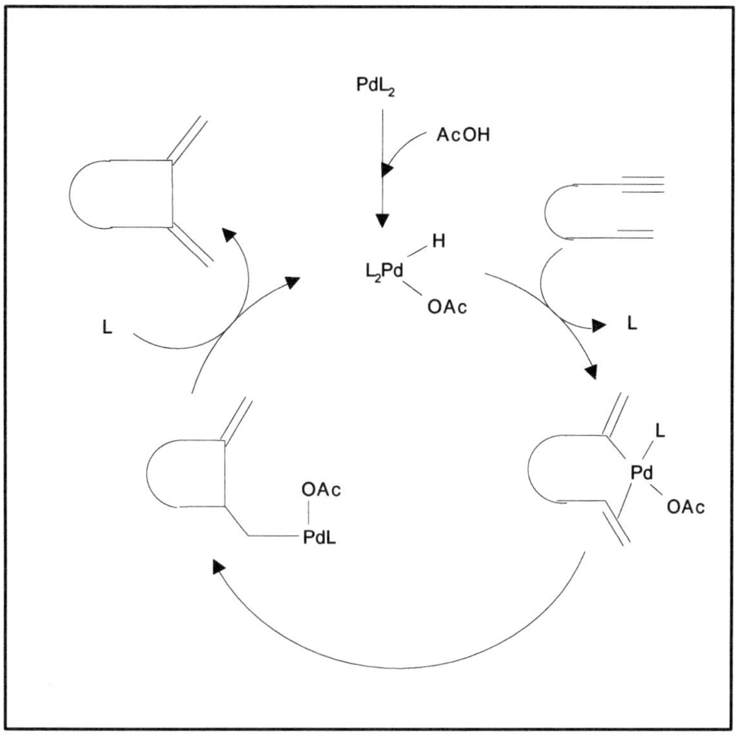
PdL2
AcOH
L2Pd
H
OAc
L
L
L
Pd
OAc
OAc
PdL

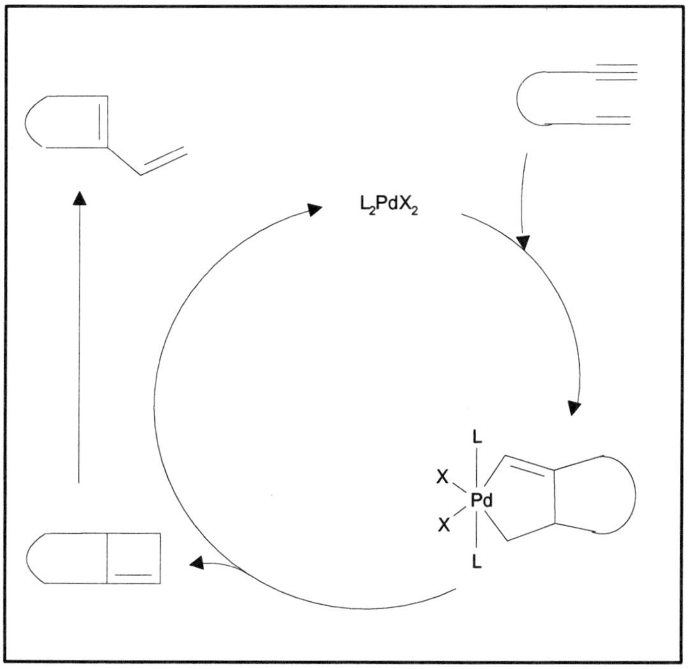
L2PdX2
L
X
Pd
X
L

A ---> B	References
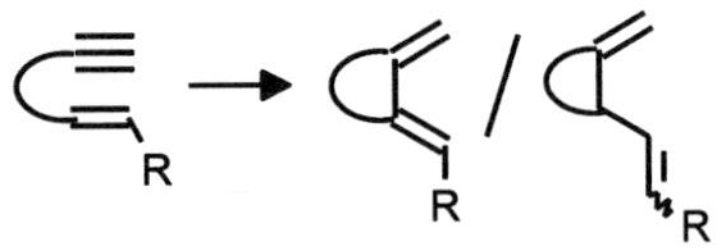	Trost B.M., *Tetrahedron Lett.*, 1989; 651 Trost B.M., *J. Org. Chem.*, 1989; 2271 Trost B.M., *Acc. Chem. Res.*, 1990; 34 Trost B.M., *J. Am. Chem. Soc.*, 1991; 636 Trost B.M., *Tetrahedron Lett.*, 1994; 1361 Trost B.M., *J. Am. Chem. Soc.*, 1994; 4255 Trost B.M., *J. Am. Chem. Soc.*, 1994; 4268 De Meijere A., *Synlett.*, 1994; 191
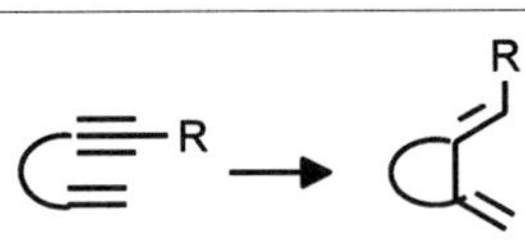	Trost B.M., *J. Am. Chem. Soc.*, 1987; 3483 Trost B.M., *Angew. Chem. Int. Ed. Engl.*, 1989; 1502
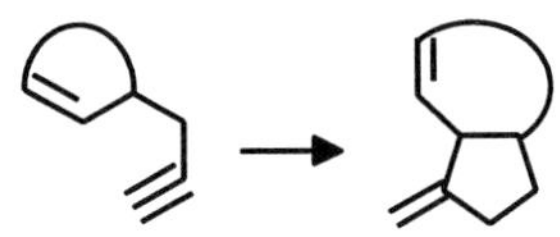	Trost B.M., *J. Am. Chem. Soc.*, 1985; 1781 Fukumoto K., *Synlett.*, 1995; 761
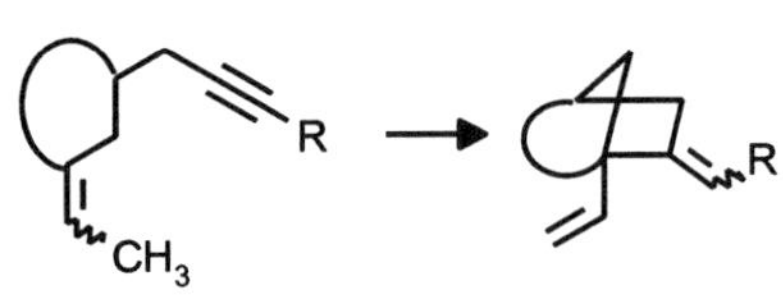	Fukumoto F., *Tetrahedron Lett.*, 1995; 5379
	Trost B.M., *J. Am. Chem. Soc.*, 1993; 5294
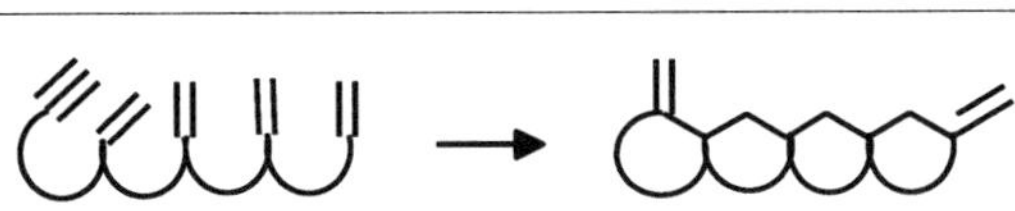	Trost B.M., *J. Am. Chem. Soc.*, 1993; 9421
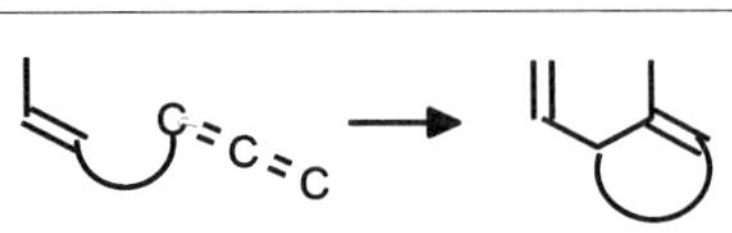	Trost B.M., *J. Am. Chem. Soc.*, 1988; 5233
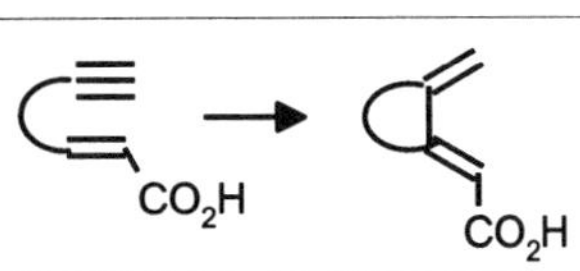	Trost B.M., *Tetrahedron Lett.*, 1993; 8233
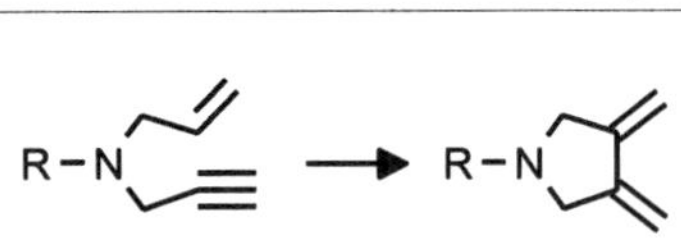	Trost B.M., *J. Am. Chem. Soc.*, 1986; 6053 Blechert S., *Tetrahedron Lett.*, 1994; 9537
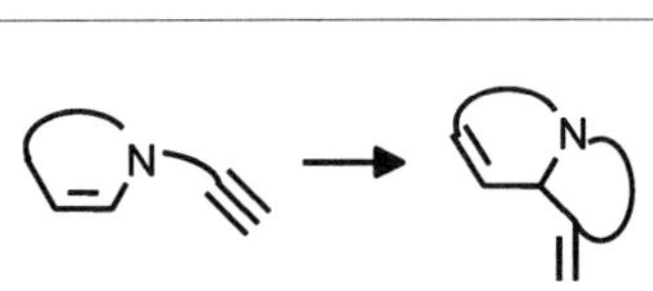	Trost B.M., *J. Am. Chem. Soc.*, 1992; 7292
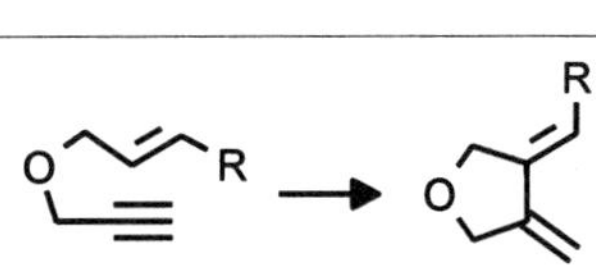	Trost B.M., *J. Org. Chem.*, 1989; 4489

A ---> B	References
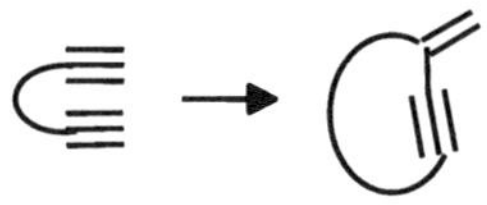	Trost B.M., *J. Am. Chem. Soc.*, 1989; 8745
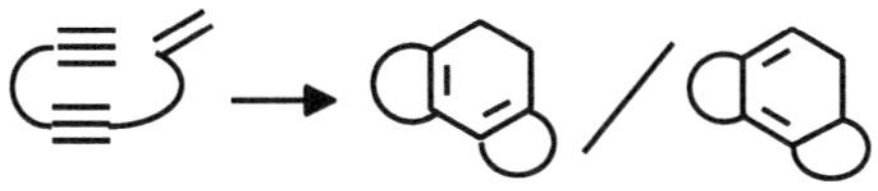	Trost B.M., *J. Am. Chem. Soc.*, 1993; 12491

RXN54 Cycloaddition of Aziridines with Carbodiimides

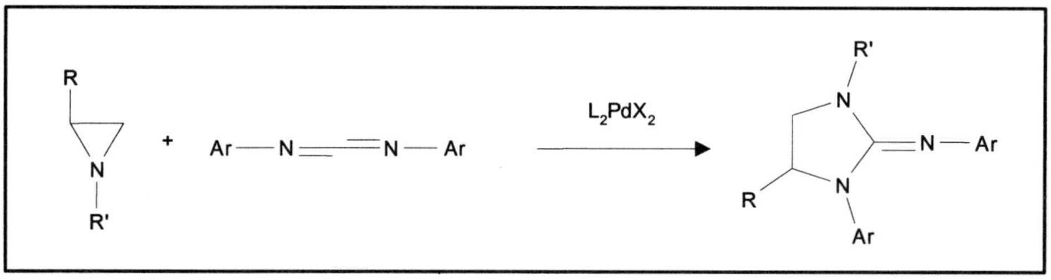

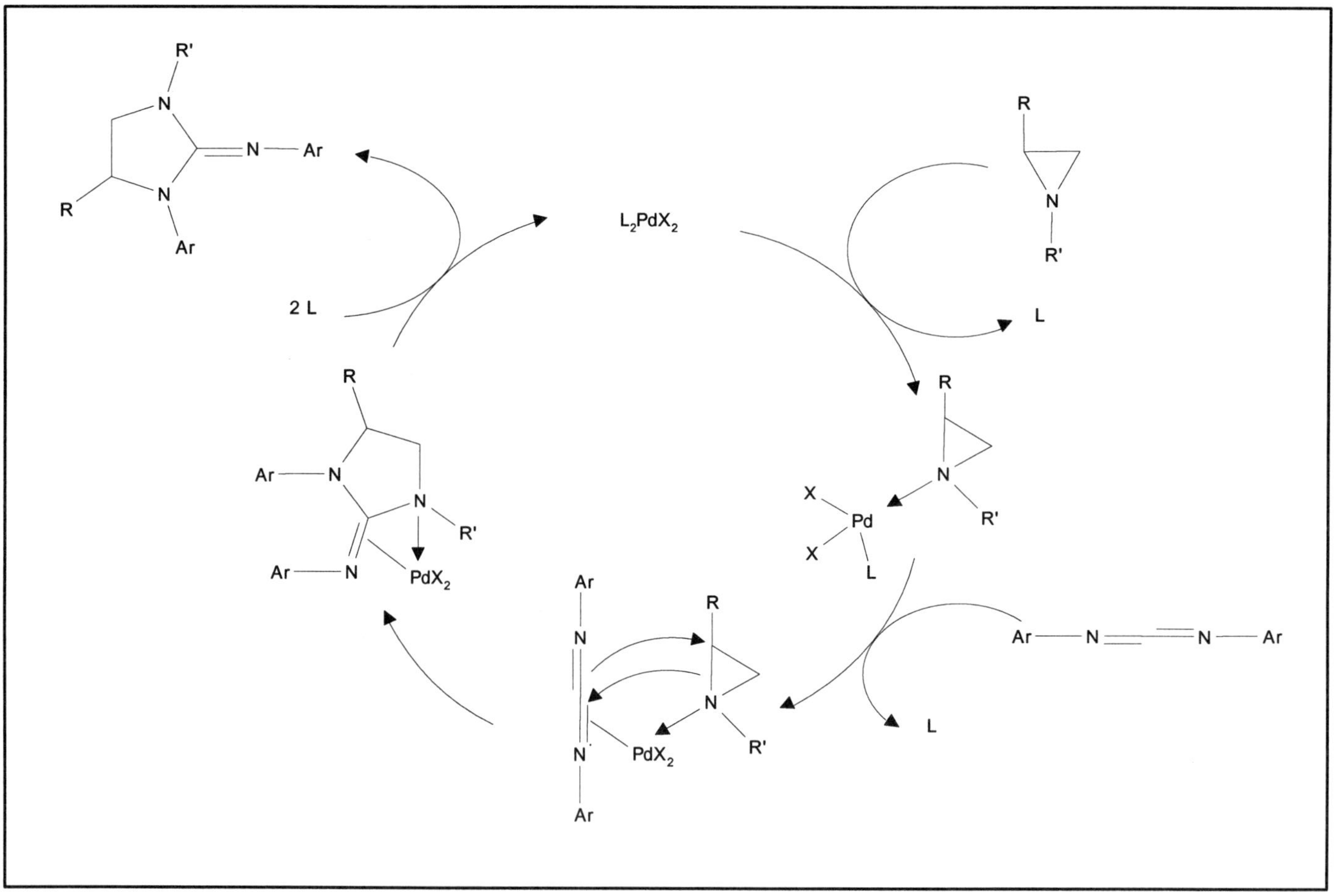

A + B ---> C	References
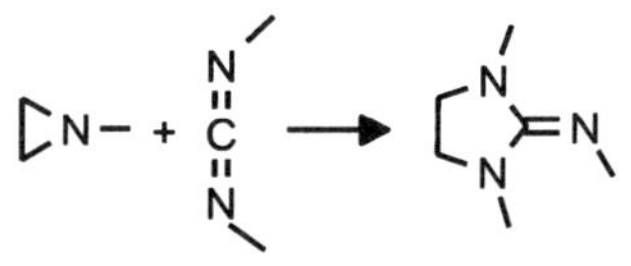	Alper H., *J. Org. Chem.*, 1992; 157 Alper H., *J. Am. Chem. Soc.*, 1995; 4700
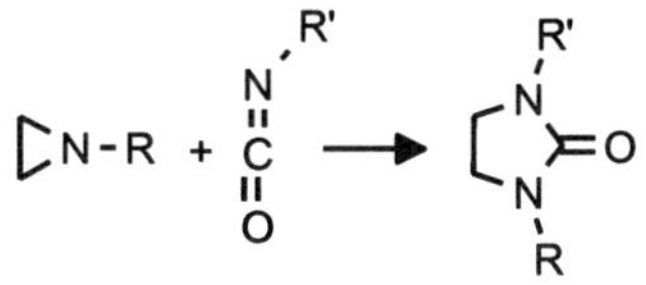	Alper H., *J. Am. Chem. Soc.*, 1995; 4700
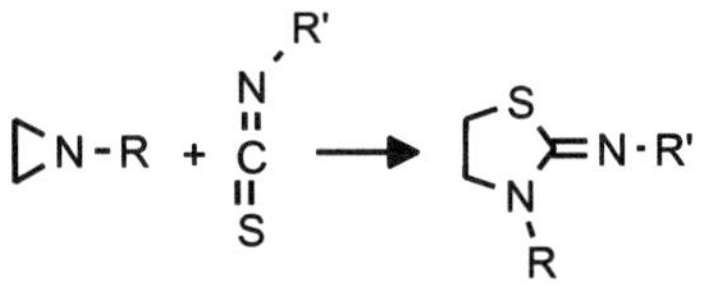	Alper H., *J. Am. Chem. Soc.*, 1995; 4700
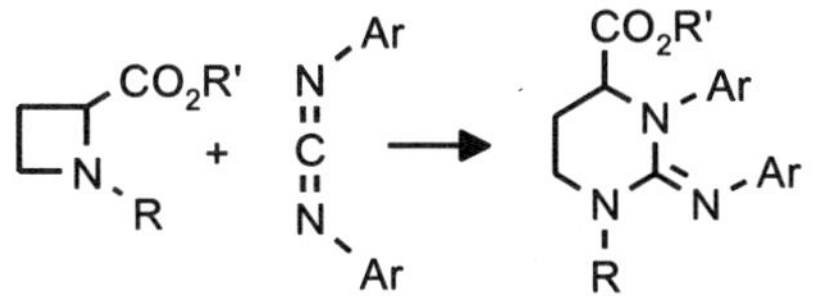	Alper H., *J. Org. Chem.*, 1995; 253
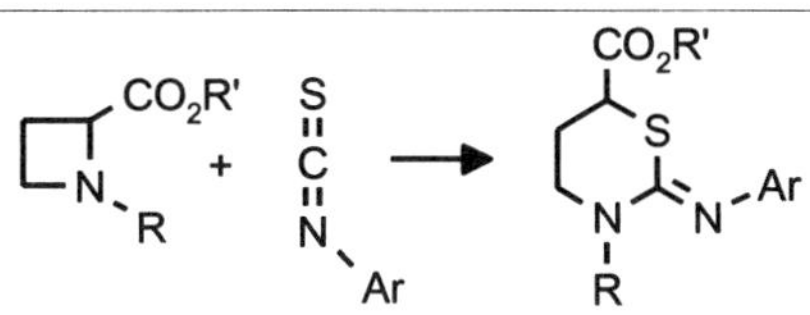	Alper H., *J. Org. Chem.*, 1995; 3092

RXN55 Telomerization of 1,3-Dienes with Nucleophiles

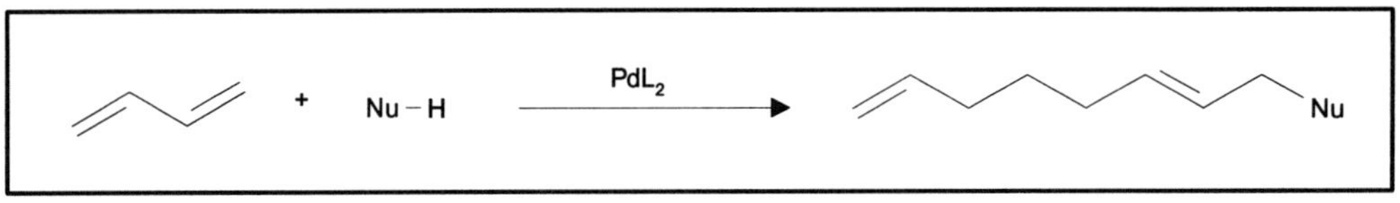

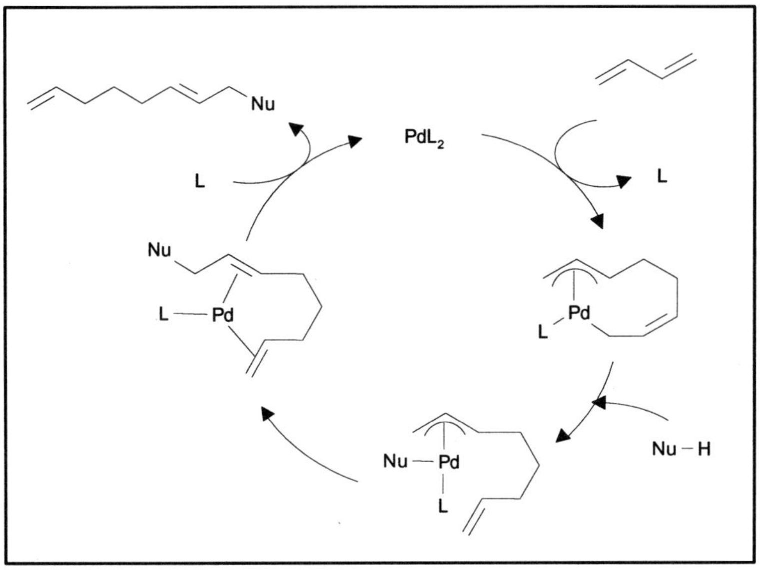

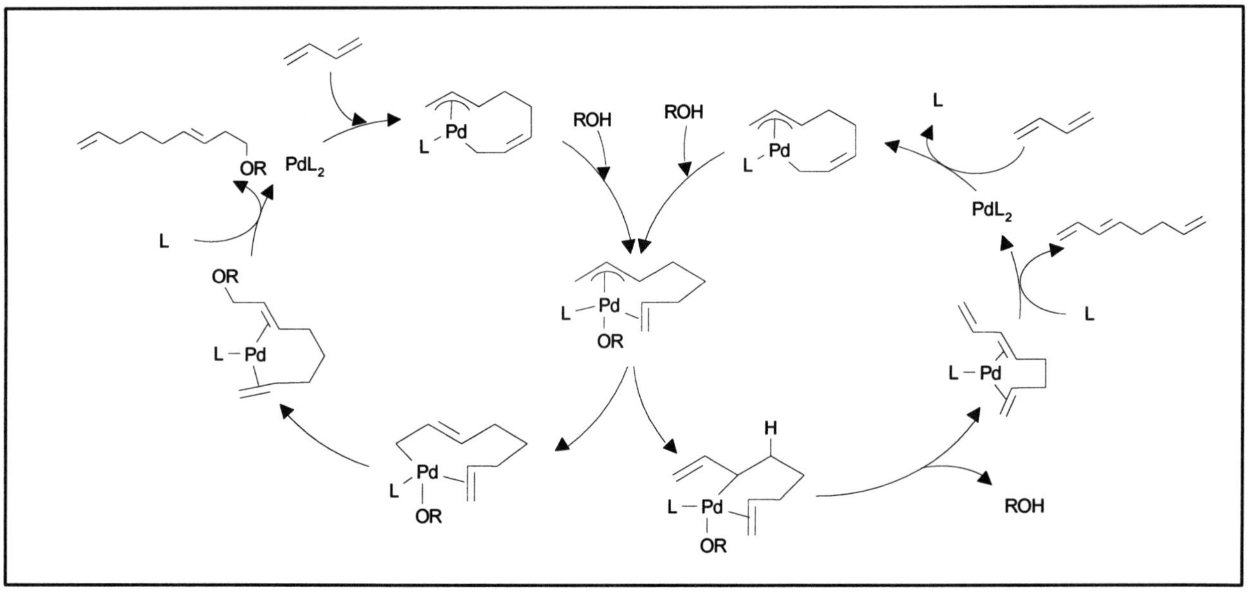

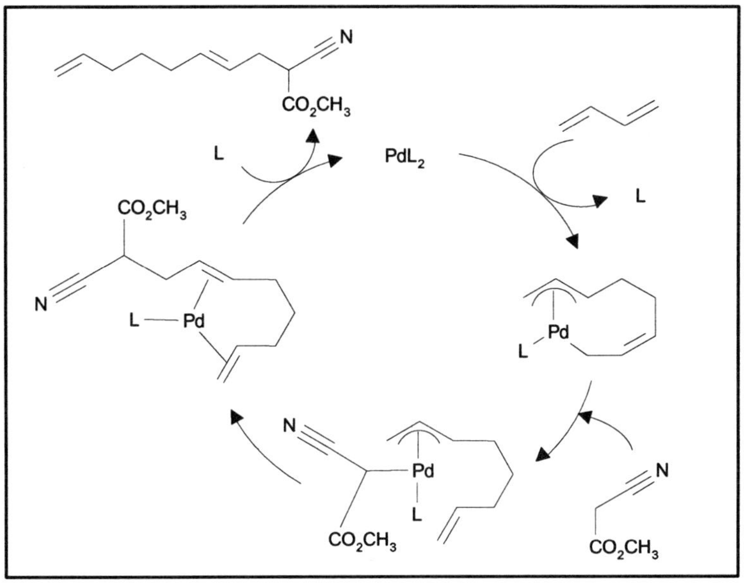
N
CO_2CH_3
L
PdL_2
L
CO_2CH_3
N
L — Pd
Pd
L
N
Pd
L
N
CO_2CH_3
CO_2CH_3

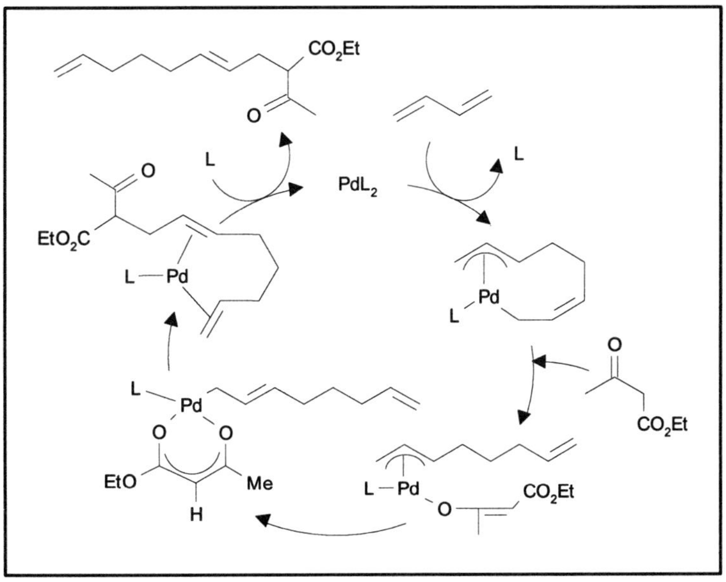
CO_2Et
O
L
PdL_2
L
O
EtO_2C
L — Pd
Pd
L
L
Pd
O
O
EtO
Me
H
O
CO_2Et
L — Pd
O
CO_2Et

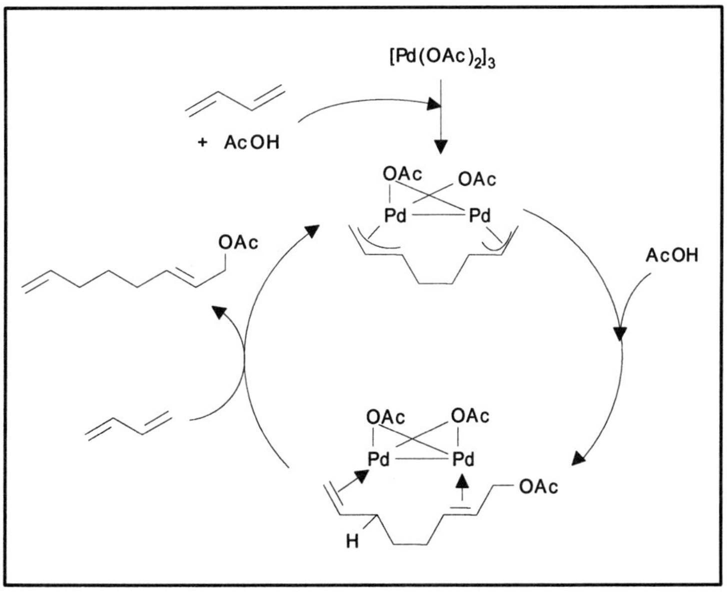
$[Pd(OAc)_2]_3$
+ AcOH
OAc
OAc
Pd
Pd
OAc
AcOH
OAc
OAc
Pd
Pd
OAc
H

Diene - Product	References
	Tsuji J., *Acc. Chem. Res.*, 1973; 8 Jolly P.W., *Angew. Chem. Int. Ed. Engl.*, 1985; 283 Jolly P.W., *Organometallics*, 1986; 5, 473 Keim W., *Organometallics*, 1986; 5, 514 Hata G., *J. Org. Chem.*, 1971; 2116
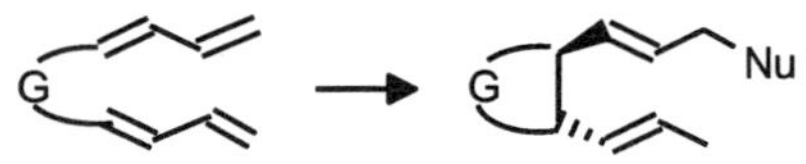	Takacs J.M., *J. Org. Chem.*, 1989; 5193 Takacs J.M., *Tetrahedron Lett.*, 1990; 1117
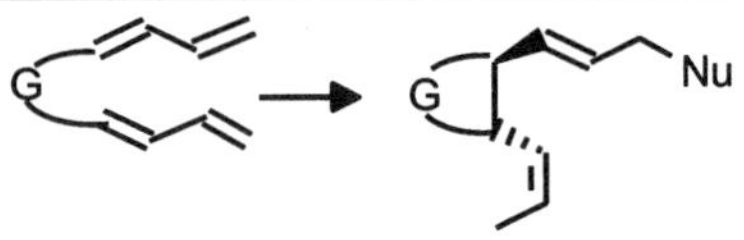	Takacs J.M., *Organometallics*, 1990; 9, 2877
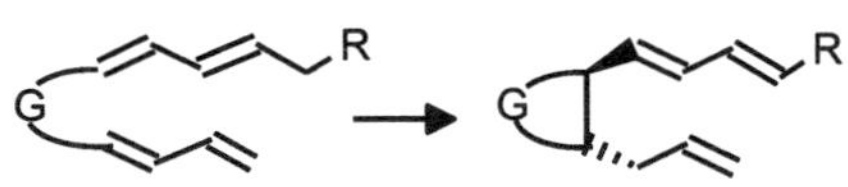	Takacs J.M., *J. Am. Chem. Soc.*, 1992; 773
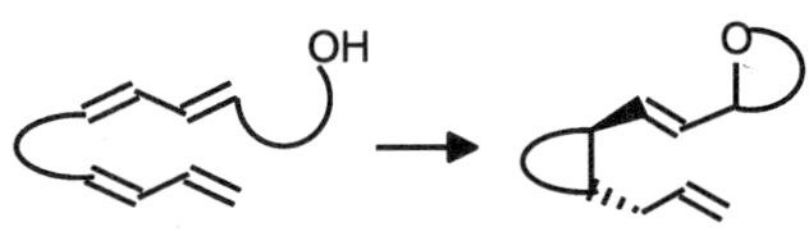	Takacs J.M., *J. Org. Chem.*, 1993; 7315 Takacs J.M., *Tetrahedron Lett.*, 1994; 9161 Takacs J.M., *Tetrahedron Lett.*, 1994; 9165

Nucleophile	References
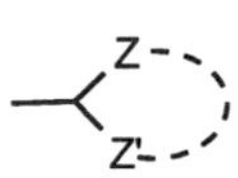	Hata G., *J. Org. Chem.*, 1971; 2116 Tsuji J., *Tetrahedron Lett.*, 1977; 2267
$RCHNO_2$	Tsuji J., *J. Chem. Soc., Chem. Commun.*, 1971; 345
CO/ROH	Tsuji J., *Tetrahedron*, 1972; 3721
NH_2	Tsuji J., *Acc. Chem. Res.*, 1973; 8
RR'N	Takahashi S., *Bull. Chem. Soc. Jpn.*, 1968; 454 Tsuji J., *Acc. Chem. Res.*, 1973; 8 Moberg C., *Organometallics*, 1985; 4, 1083 Moberg C., *Tetrahedron Lett.*, 1981; 4827 Takacs J.M., *Organometallics*, 1994; 4787
OH	Monflier E., *J. Mol. Catal.*, 1995; 97, 29 Inoue Y., *Bull. Chem. Soc. Jpn.*, 1984; 3345
RO	Takahashi S., *Tetrahedron Lett.*, 1967; 2451 Inoue Y., *Bull. Chem. Soc. Jpn.*, 1984; 3345

Nucleophile	References
R-CO-O	Takahashi S., *Tetrahedron Lett.*, 1967; 2451
Me_3Si	Takahashi S., *J. Chem. Soc., Chem. Commun.*, 1969; 161 Tsuji J., *J. Chem. Soc., Chem. Commun.*, 1971; 247
$PhSO_2$	Inoue Y., *Bull. Chem. Soc. Jpn.*, 1986; 3705
H	Gardner S., *Tetrahedron Lett.*, 1972; 163 Roffia P., *J. Organomet. Chem.*, 1973; 55, 405

RXN56 WACKER Process

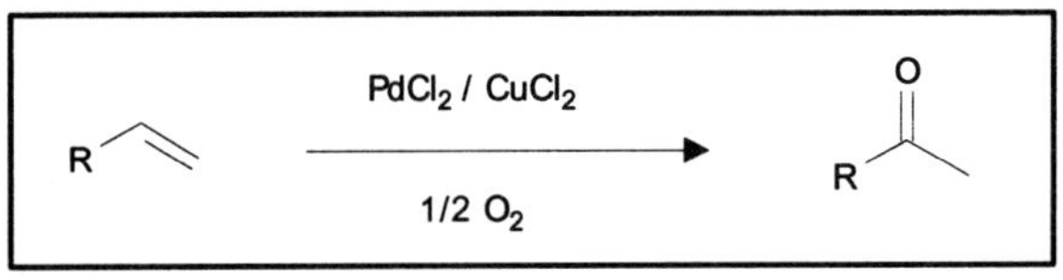

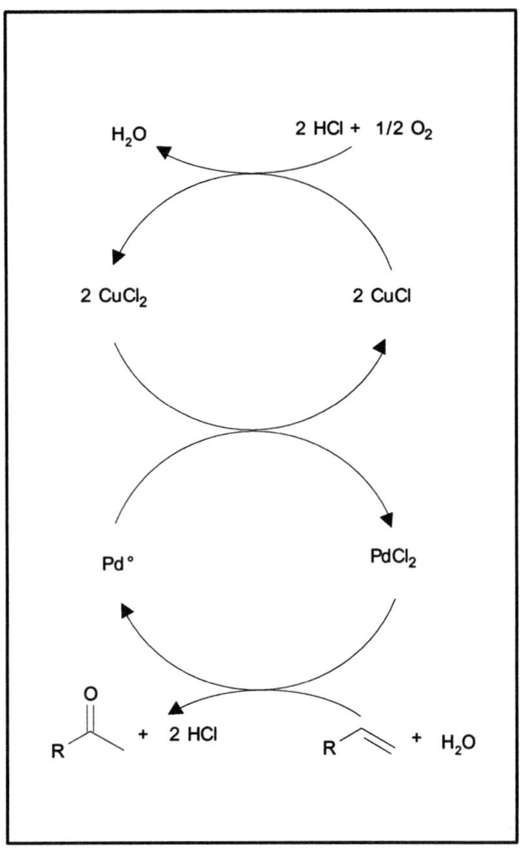

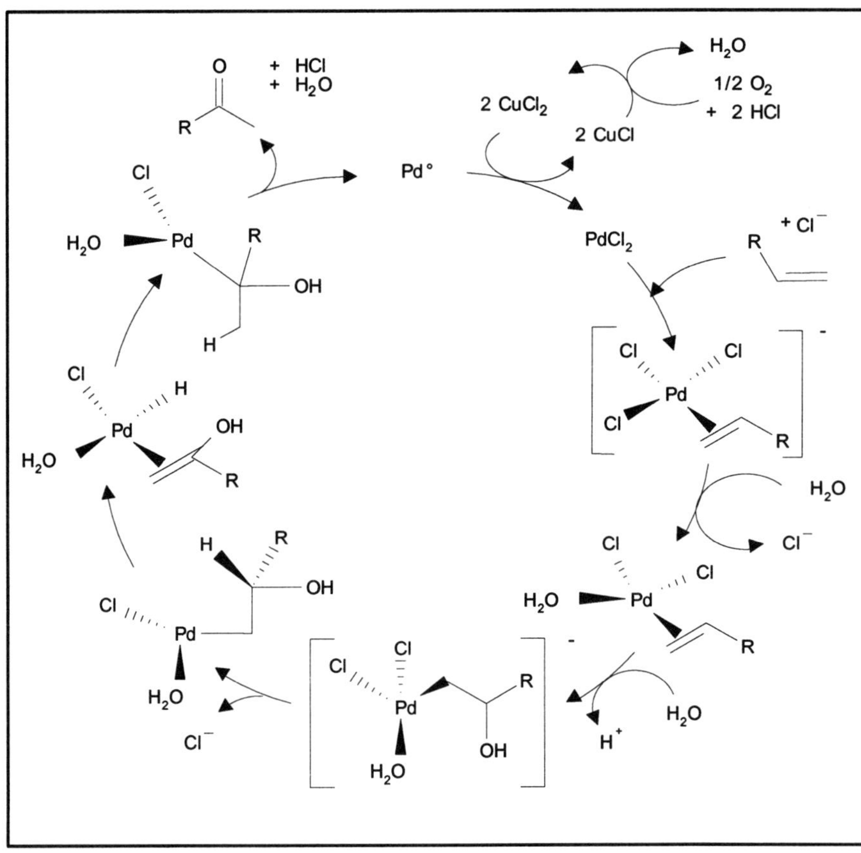

RXN56 WACKER Process

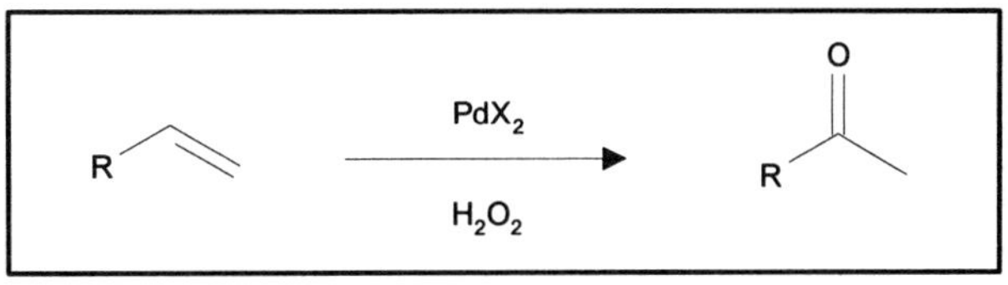

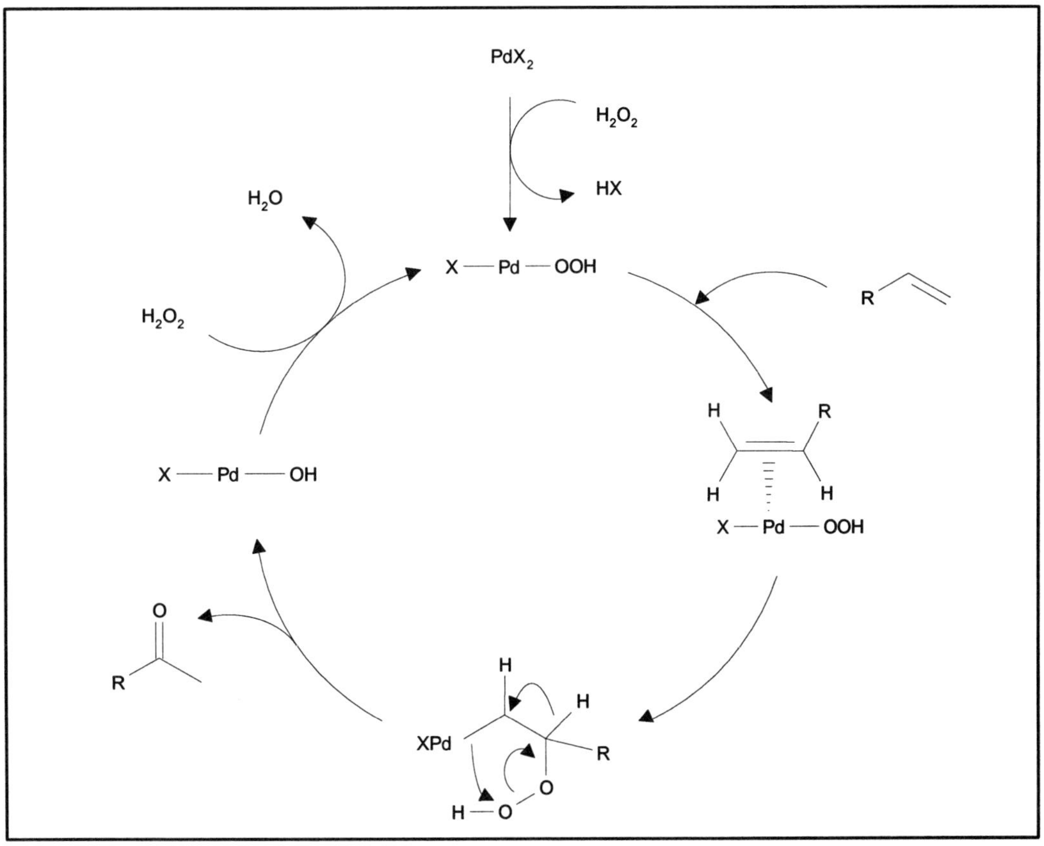

Alkene	Oxidizer	References
C=C	O_2 / $CuCl_2$	Tsuji J., *Org. Synth. By Means of Transition Metal Complexes*, 1975; 113 Alper H., *Tetrahedron Lett.*, 1985; 2263 (ethylene glycol) Takehira K., *J. Mol. Catal.*, 1987; 42, 247 (EtOH) Feringa B.L., *Tetrahedron Lett.*, 1992; 2403
C=C	O_2 / CuCl	Tsuji J., *Tetrahedron Lett.*, 1977; 2267 Tsuji J., *Chem. Lett.*, 1982; 859 (internal olefin) Fukumoto F., *Tetrahedron Lett.*, 1995; 5379 Kang S.K., *J. Org. Chem.*, 1995; 4678 (anti-Markovnikov hydration)
CH_2=CH_2	O_2 / $CuCl_2$	Bäckvall J.E., *J. Am. Chem. Soc.*, 1979; 2411
RO-C=C	O_2 / CuCl	Mereyala H.B., *Tetrahedron Lett.*, 1993; 6929
C=C	O_2 / Fe(phthalocyanine) / Hydroquinone	Bäckvall J.E., *Tetrahedron Lett.*, 1988; 2885 Bäckvall J.E., *J. Am. Chem. Soc.*, 1990; 5160
C=C	O_2 / Coenzyme PPQ	Ohshiro Y., *Chem. Lett.*, 1989; 785
C=C	O_2 / $CuSO_4$ Cyclodextrin	Monflier E., *Tetrahedron Lett.*, 1995; 387
C=C	H_2O_2	Mimoun H., *J. Org. Chem.*, 1980; 5387
C=C	^{t}BuOOH	Tsuji J., *Chem. Lett.*, 1980; 257
C=C-CO_2R	^{t}BuOOH	Bhat S.V., *J. Chem. Soc., Chem. Commun.*, 1994; 903

RXN57 Preparation of Ketones from Alcohols or Derivatives via a β-Hydride Elimination

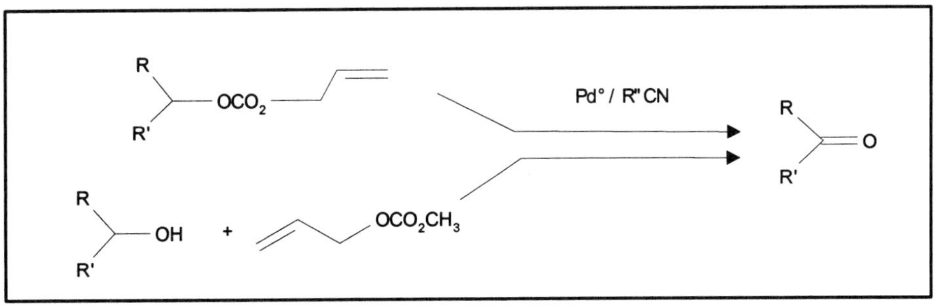

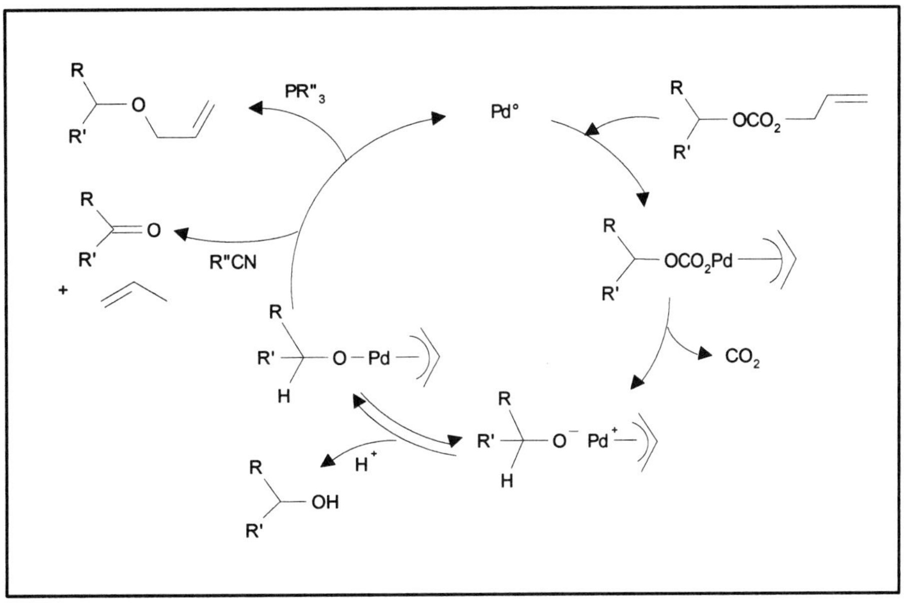

RXN57 Preparation of Ketones from Alcohols or Derivatives via a β-Hydride Elimination

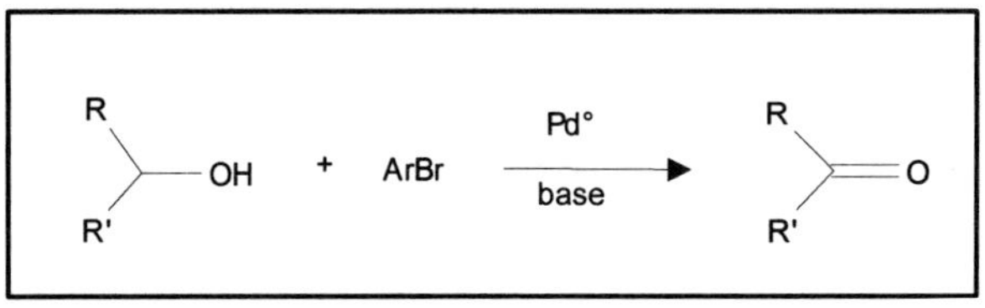

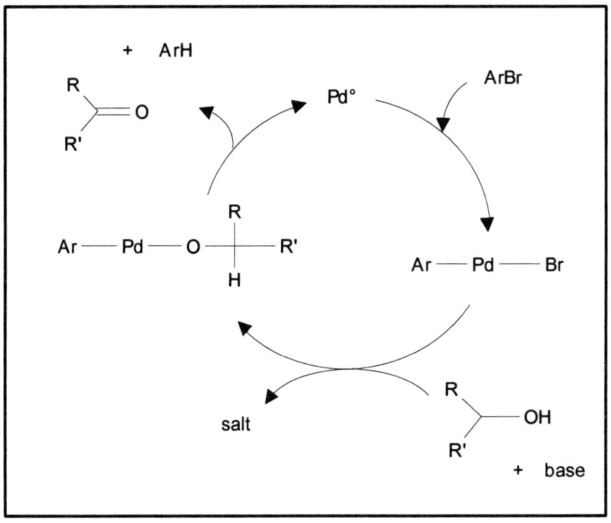

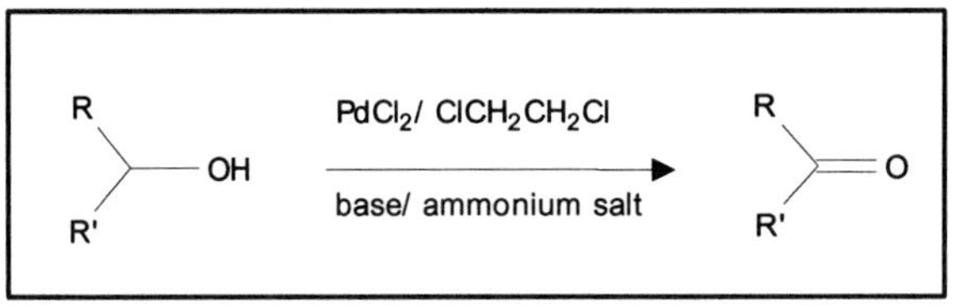

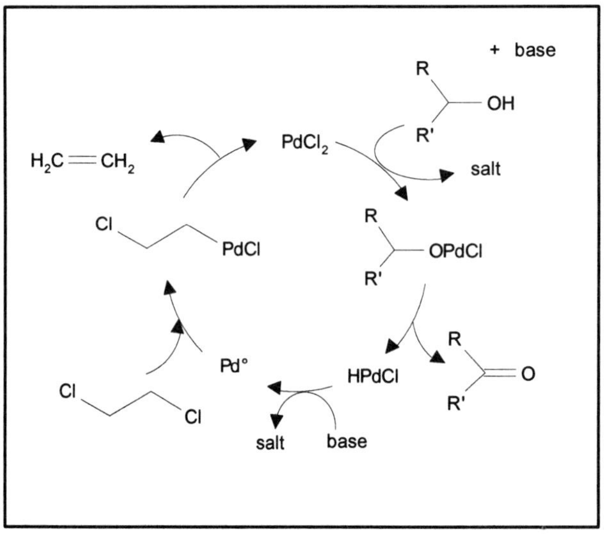

A ---> B	References
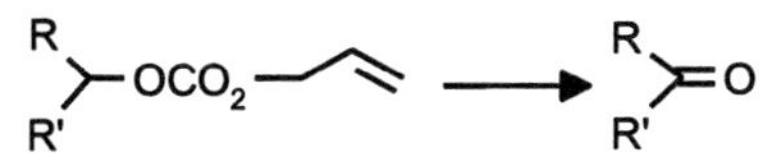	Tsuji J., *Tetrahedron*, 1987; 3903
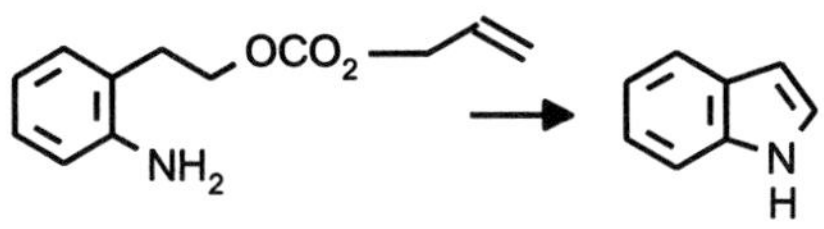	Izumi T., *J. Het. Chem.*, 1992; 1085
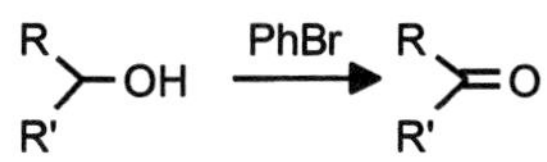	Yoshida Z.I., *Tetrahedron Lett.*, 1979; 1401
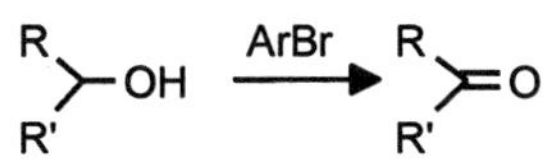	Yoshida Z., *J. Org. Chem.*, 1983; 1286
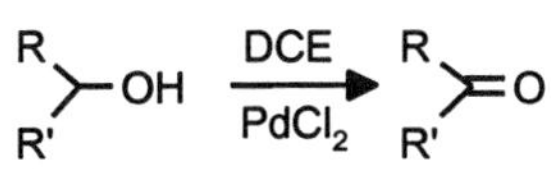	Muzart J., *Tetrahedron Lett.*, 1995; 2473
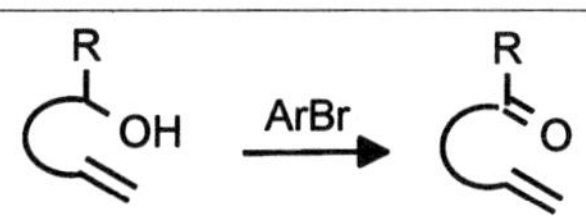	Yoshida Z., *Tetrahedron Lett.*, 1981; 1801
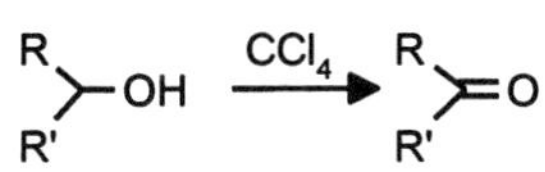	Tsuji J., *Chem. Lett.*, 1981; 1171
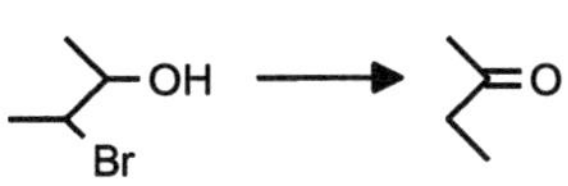	Tsuji J., *Tetrahedron Lett.*, 1982; 3085
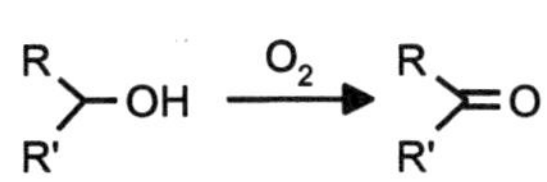	Schwartz J., *J. Chem. Soc., Chem. Commun.*, 1977; 157 Echavarren A.M., *Tetrahedron Lett.*, 1994; 7097
C=C-CH-OH ⟶ C=C-C=O	Echavarren A.M., *Tetrahedron Lett.*, 1994; 7097

RXN58 Preparation of α,β-Unsaturated Carbonyl Derivatives via a β-Hydride Elimination

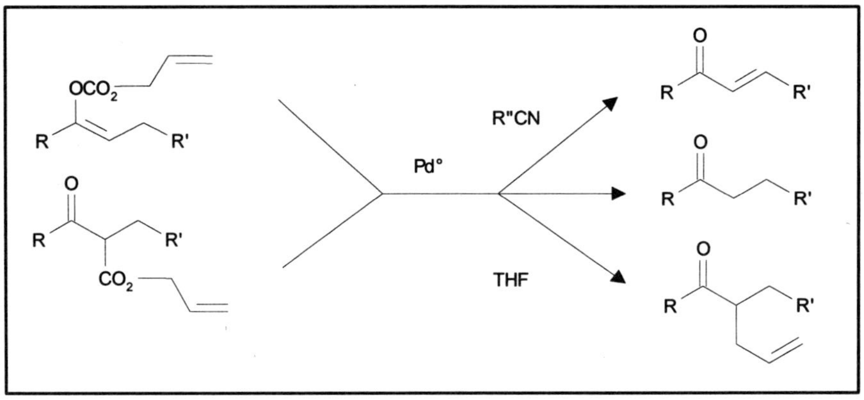

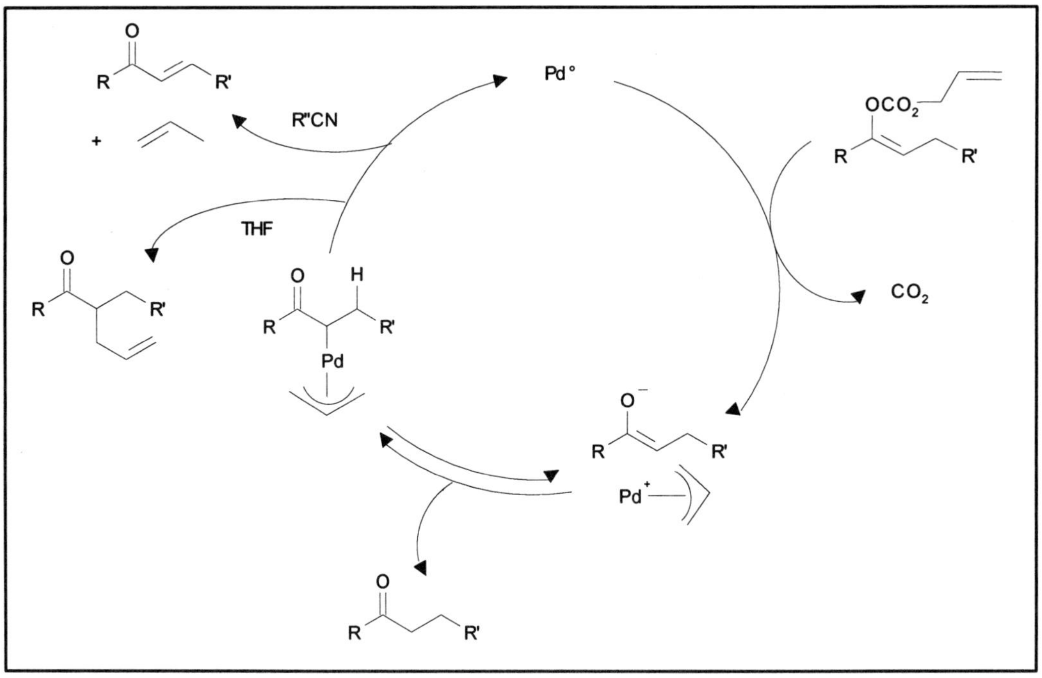

RXN58 Preparation of α,β-Unsaturated Carbonyl Derivatives via a β-Hydride Elimination

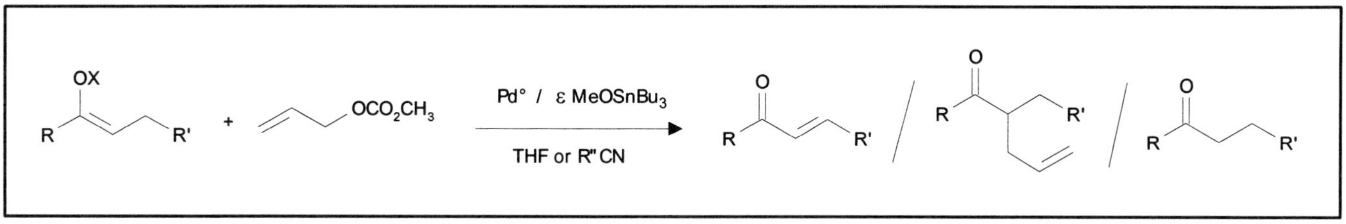

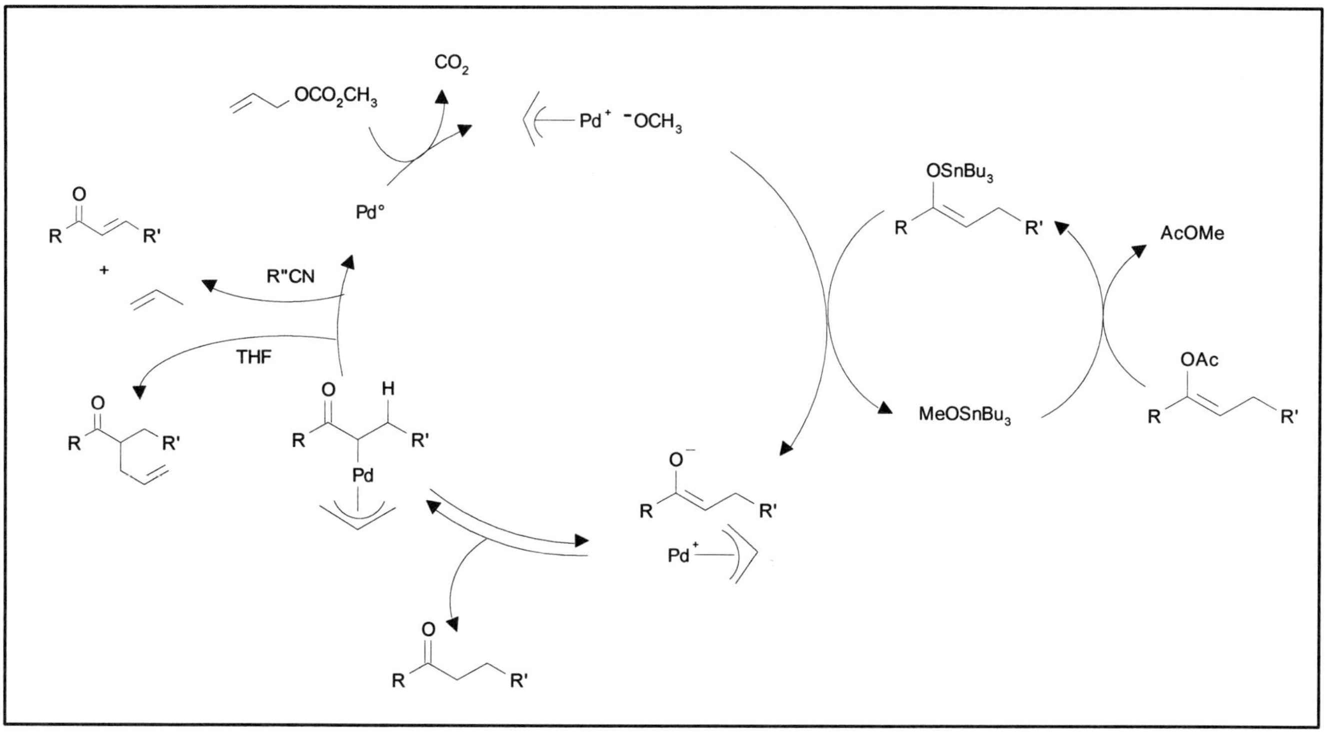

A ---> B	References
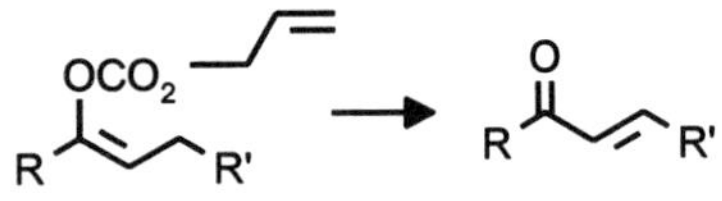	Tsuji J., *Tetrahedron*, 1986; 4361 Tsuji J., *Chem. Lett.*, 1984; 1133
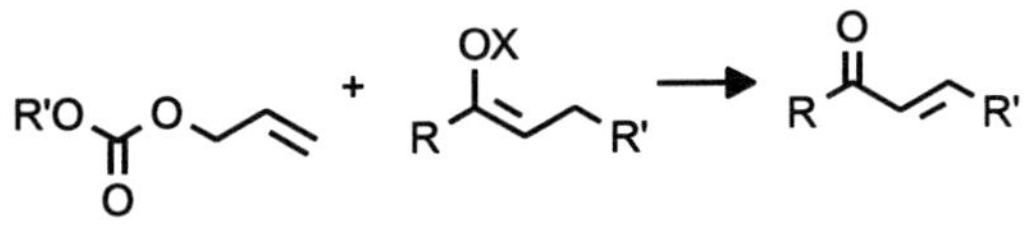	Tsuji J., *Tetrahedron*, 1986; 2971 Tsuji J., *Tetrahedron*, 1986; 4361 Tsuji J., *Chem. Lett.*, 1984; 1133 Tsuji J., *Tetrahedron Lett.*, 1984; 4783
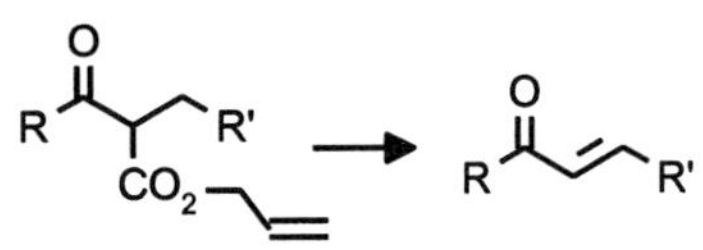	Tsuji J., *J. Am. Chem. Soc.*, 1982; 5844
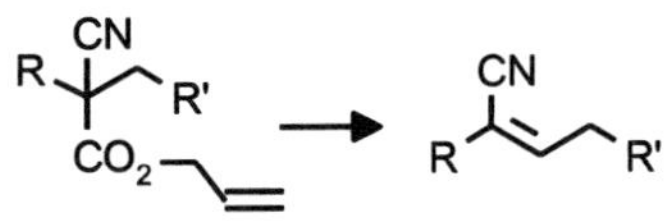	Tsuji J., *J. Chem. Soc., Chem. Commun.*, 1986; 118

A ---> B	References
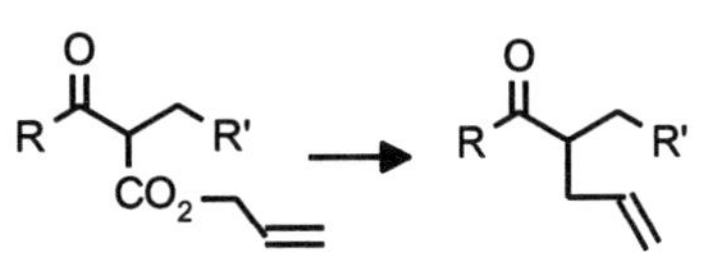	Shimizu I., *Chem. Lett.*, 1989; 577 Tsuji J., *Tetrahedron*, 1986; 4361 Tsuji J., *J. Org. Chem.*, 1987; 2988 Saegusa T., *J. Am. Chem. Soc.*, 1980; 6381 Tsuji J., *Tetrahedron Lett.*, 1989; 4829
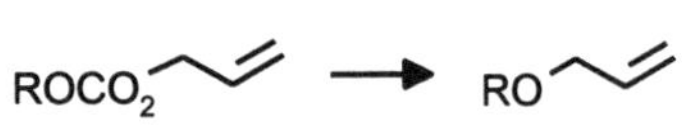	Tsuji J., *Tetrahedron*, 1987; 3903
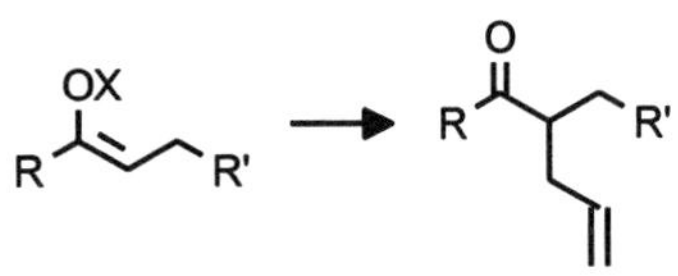	Tsuji J., *Tetrahedron Lett.*, 1984; 4783 Tsuji J., *Tetrahedron*, 1986; 4361 Watson S. P., *Tetrahedron Lett.*, 1994; 9763
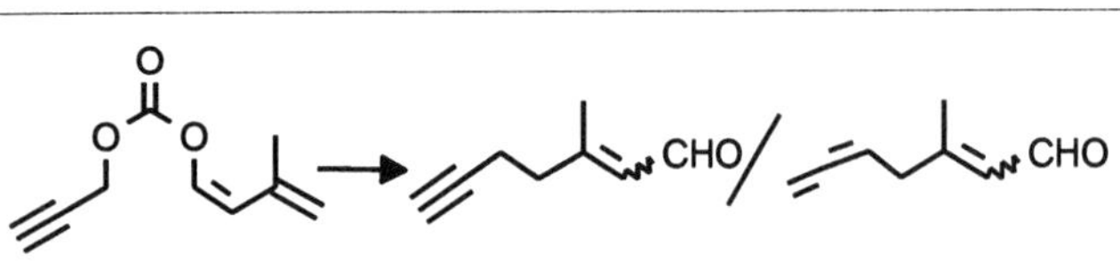	Bienayme H., *Tetrahedron Lett.*, 1994; 7387

RXN59 Preparation of α-Diketones Derivatives via an Oxidative Rearrangement of a Propargyl Acetate

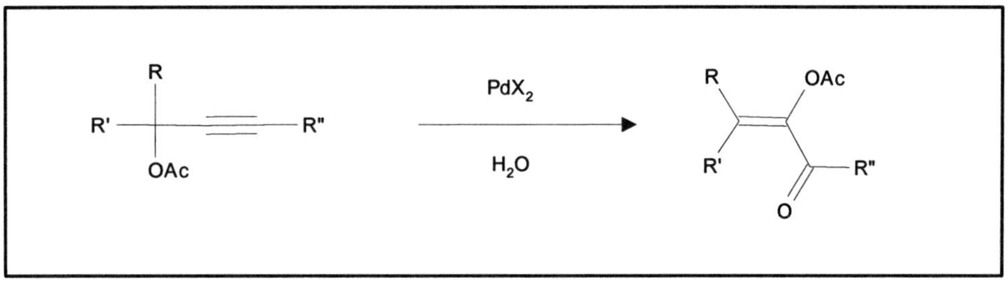

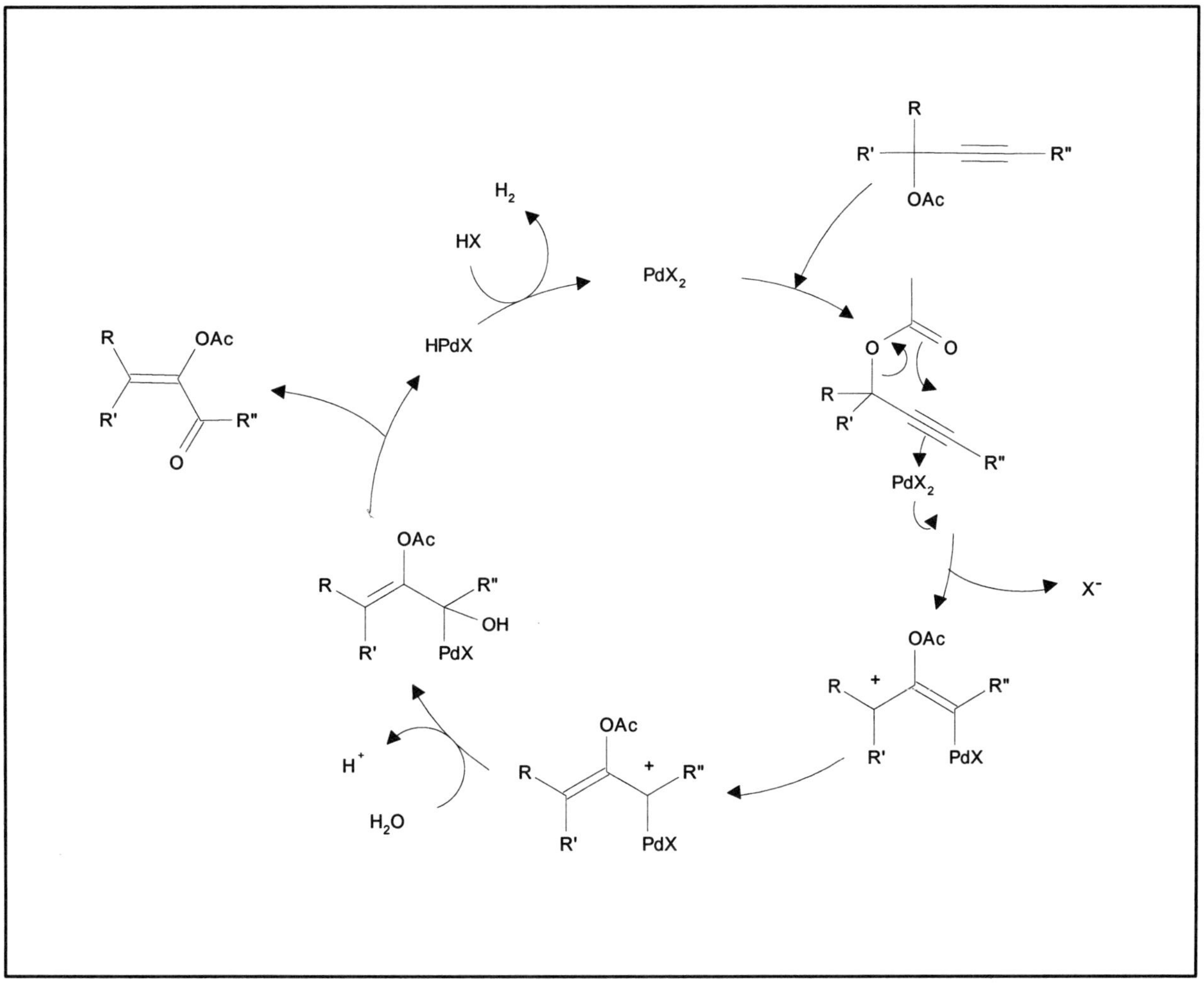

RXN59

R"	Catalyst	References
C	PdX_2 / H_2O	Mahrwald R., *Angew. Chem. Int. Ed. Engl.*, 1991; 593
H	PdX_2 / O_2	Kataoka H., *Tetrahedron Lett.*, 1990; 4181

RXN60 Carbonylation

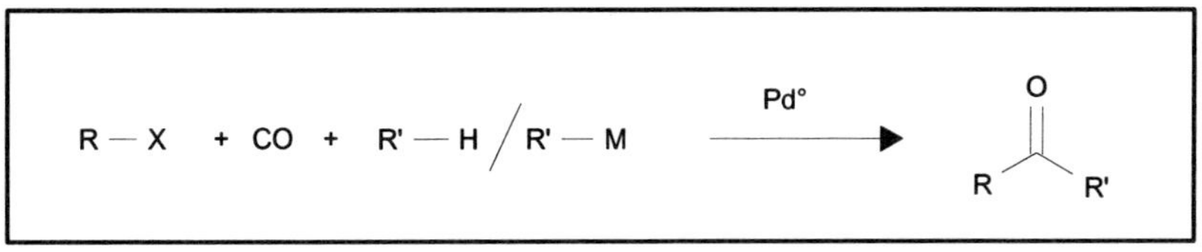

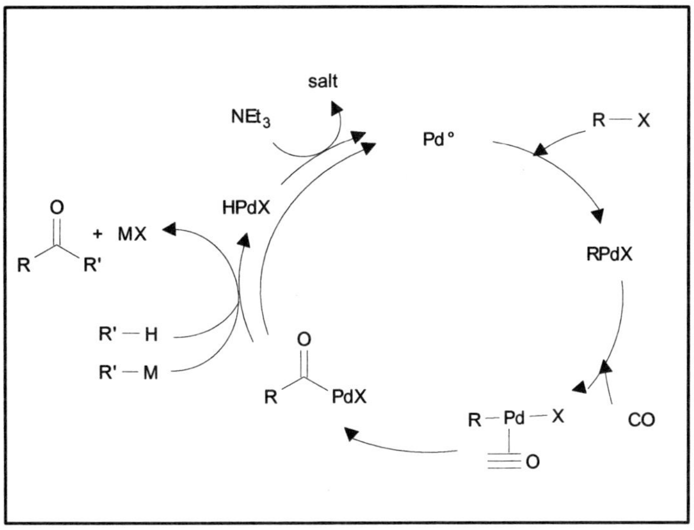

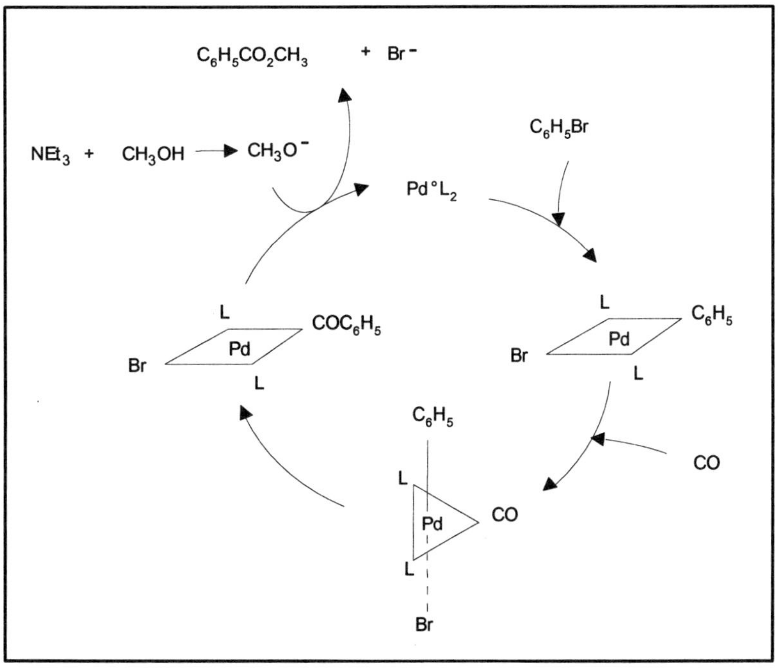

RXN60 Carbonylation

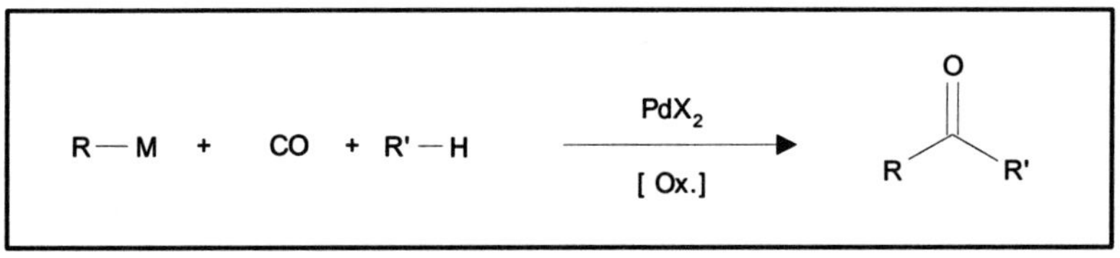

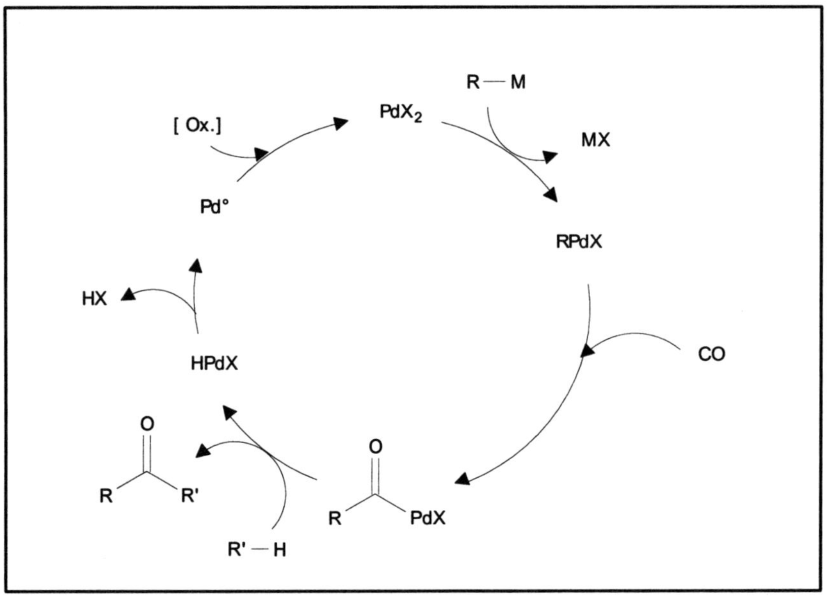

 J.-L. Malleron et al

R	References
Ar	Cacchi S., *Tetrahedron Lett.*, 1986; 3931 Matsuda T., *J. Org. Chem.*, 1980; 2365 Moser W.R., *J. Am. Chem. Soc.*, 1988; 2816 Huang X., *Chem. Ind.*, 1990; 548 Suzuki A., *Tetrahedron Lett.*, 1993; 7595 Fuchikami T., *J. Org. Chem.*, 1991; 4320 Beletskaya I.P., *J. Organomet. Chem.*, 1995; 486, 297 Chen L., *Tetrahedron Lett.*, 1995; 8715
Het	Spurr P.R., *Tetrahedron Lett.*, 1995; 2745 Hibino S., *J. Org. Chem.*, 1995; 5899
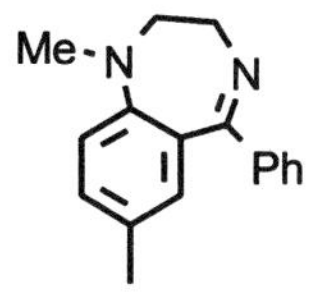	Kraus G.A., *Tetrahedron Lett.*, 1994; 9189
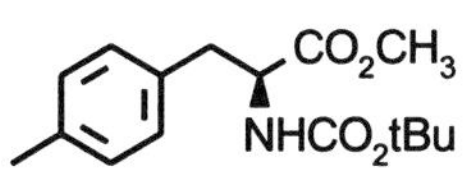	Weinstock J., *Org. Prep. Proced. Int.*, 1994; 533
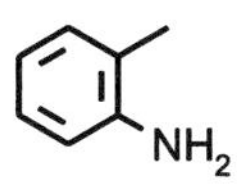	Kalinin V.N., *Tetrahedron Lett.*, 1992; 373
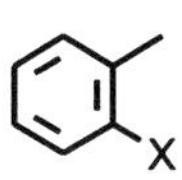	Perry R.J., *J. Org. Chem.*, 1991; 6573
$ArCH_2$	Stille J.K., *J. Am. Chem. Soc.*, 1980; 4193
Alkyl	Fuchikami T., *J. Org. Chem.*, 1991; 4320
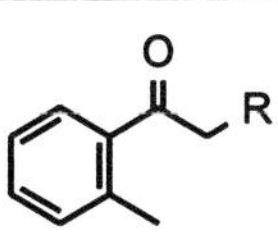	Ortar G., *Tetrahedron Lett.*, 1993; 3763 (cyclisation)
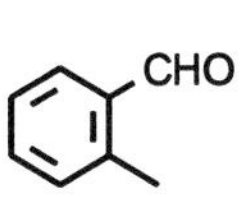	Shim S.C., *J. Heterocycl. Chem.*, 1995; 363
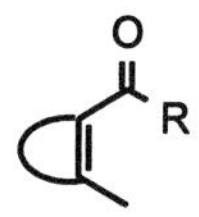	Mori M., *Heterocycles*, 1994; 37, 167 (cyclisation)
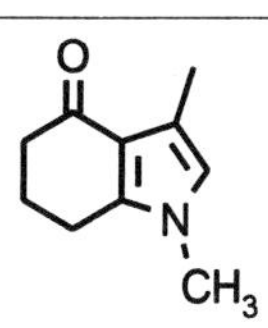	Edstrom E.D., *Synlett.*, 1995; 49
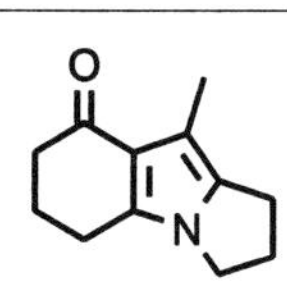	Edstrom E.D., *Tetrahedron Lett.*, 1995; 7035

R	References
HOH_2C-C=C-	Stille J.K., *J. Am. Chem. Soc.*, 1980; 4193 (cyclisation) Hoye T.R., *Tetrahedron Lett.*, 1994; 7517 (cyclisation)
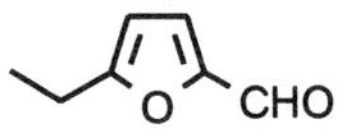	Sheldon R.A., *J. Chem. Soc., Chem. Commun.*, 1994; 2659
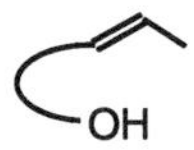	Negishi E.I., *Tetrahedron Lett.*, 1994; 695 (cyclisation)
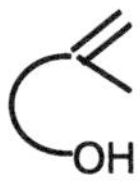	Adam W., *Synthesis*, 1994; 567 (cyclisation)
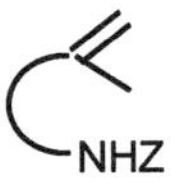	Negishi E.I., *Tetrahedron Lett.*, 1995; 1771 (cyclisation)
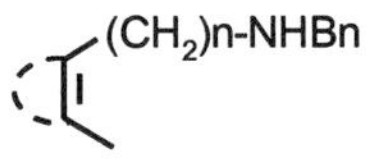	Crisp G.T., *Tetrahedron*, 1995; 5585 (cyclisation)
C=C	Ortar G., *Tetrahedron Lett.*, 1985; 1109 Freskos J.N., *Tetrahedron Lett.*, 1994; 835 Stille J.K., *Org. Synth.*, 1990; 68, 116
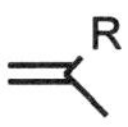	Crisp G.T., *Tetrahedron*, 1995; 5831
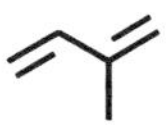	Snider B.B., *J. Org. Chem.*, 1994; 5419
C=C-C	Kiji J., *Chem. Lett.*, 1988; 957 Stille J.K., *J. Am. Chem. Soc.*, 1984; 4833 Dent W.T., *J. Chem. Soc.*, 1964; 1588 Liao S., *React. Polym., Ion Exch., Sorbents*, 1994; 23 (2-3), 113
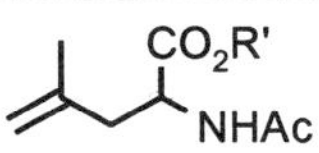	Crisp G.T., *Tetrahedron*, 1994; 2623
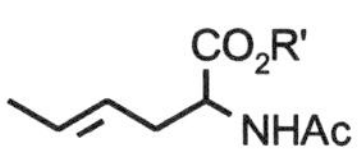	Crisp G.T., *Tetrahedron*, 1994; 2623
	Liebeskind L.S., *Tetrahedron*, 1993; 5461
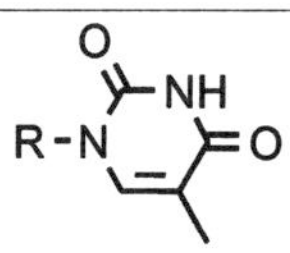	Matsuda A., *Bioorg. Med. Chem. Lett.*, 1994; 361 Eaton B.E., *J. Am. Chem. Soc.*, 1995; 8474
Me_2N-C=C(Z)-	Torii S., *Synlett.*, 1991; 695
C=C=C	Nokami J., *Tetrahedron Lett.*, 1990; 5629 Alper H., *J. Org. Chem.*, 1994; 1956

R	References
C=C=C-C	Elsevier C.J., *J. Organomet. Chem.*, 1987; 325, C23
C≡C	Kitamura T., *J. Chem. Soc., Chem. Commun.*, 1990; 614
C≡C—C	Tsuji J., *Tetrahedron Lett.*, 1986; 731 Tsuji J., *Synlett.*, 1991; 697 Tsuji J., *Tetrahedron Lett.*, 1991; 7683 Tsuji J., *Tetrahedron Lett.*, 1991; 7687 Arzoumanian H., *J. Mol. Catal.*, 1993; 85, 287 Elsevier C.J., *J. Organomet. Chem.*, 1987; 325, C23
	Tsuji J., *J. Organomet. Chem.*, 1995; 488, 127 (cyclisation)
R'-N=(R)C-	Watanabe H., *Tetrahedron Lett.*, 1992; 4333
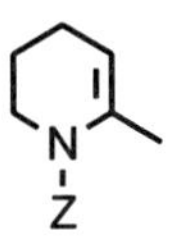	Comins D.L., *J. Org. Chem.*, 1995; 2656
$(RCH_2)_2N$-	Saegusa T., *J. Org. Chem.*, 1971; 858

X	References
H	Fujiwara Y., *Chem. Lett.*, 1995; 345 (Pd II as catalyst)
Halogen	Collin J., *Bull. Soc. Chim. Fr.*, 1988; 976 Moser W.R., *J. Am. Chem. Soc.*, 1988; 2816 Kiji J., *Chem. Lett.*, 1988; 957
Br	Sugi Y., *Synlett.*, 1994; 515 Alper H., *Organometallics*, 1993; 1890 Okano T., *Bull. Chem. Soc. Jpn.*, 1994; 2329 Koga H., *Tetrahedron Lett.*, 1995; 87
Cl	Yamamoto A., *Bull. Chem. Soc. Jpn.*, 1995; 433 (double carboxylation)
I	Beletskaya I.P., *J. Organomet. Chem.*, 1995; 486, 297
OH	Alper H., *J. Org. Chem.*, 1994; 1956 Sheldon R.A., *J. Chem. Soc., Chem. Commun.*, 1994; 2659
OR	Tkatchenko I., *J. Organomet. Chem.*, 1989; 366, C9

X	References
OCHO	Yamamoto A., *Bull. Chem. Soc. Jpn.*, 1995; 433
$OCOC_6H_5$	Tamaru Y., *Angew. Chem. Int. Ed. Engl.*, 1992; 645
OAc	Murahashi S.I., *J. Org. Chem.*, 1993; 1538 Sato F., *Tetrahedron Lett.*, 1994; 4389
OCO_2R	Tsuji J., *J. Org. Chem.*, 1984; 1341 Tsuji J., *Tetrahedron Lett.*, 1982; 5189
OCO_2CH_3	Tsuji J., *Synlett.*, 1991; 697 Tsuji J., *Tetrahedron Lett.*, 1991; 7683 Tsuji J., *Tetrahedron Lett.*, 1991; 7687 Mandai T., *Tetrahedron Lett.*, 1994; 5697 Mandai T., *Tetrahedron Lett.*, 1994; 5701
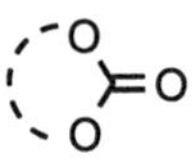	Tamaru Y., *Bull. Chem. Soc. Jpn.*, 1992; 97
OCONRR'	Yamamoto K., *Synlett.*, 1992; 323 Tamaru Y., *Bull. Chem. Soc. Jpn.*, 1992; 97
$OPO(OR)_2$	Murahashi S.I., *J. Org. Chem.*, 1993; 1538 Murahashi S.I., *Chem. Lett.*, 1985; 1477 Tamaru Y., *J. Org. Chem.*, 1995; 1365
OSO_2CF_3	Cacchi S., *Tetrahedron Lett.*, 1986; 3931 Ortar G., *Tetrahedron Lett.*, 1985; 1109 Stille J.K., *Org. Synth.*, 1990; 68, 116 Rizzo C.J., *Tetrahedron Lett.*, 1988; 2793 Crisp G.T., *J. Org. Chem.*, 1992; 6972
OSO_2F	Roth G.P., *Tetrahedron Lett.*, 1992; 1959 Mc Guire M.A., *J. Org. Chem.*, 1994; 6683
NEt_2	Murahashi S.I., *Tetrahedron*, 1994; 453
NRR'	Murahashi S.I., *J. Chem. Soc., Chem. Commun.*, 1988; 1578
N_2^+	Matsuda T., *J. Org. Chem.*, 1980; 2365 Kikukawa K., *Chem. Lett.*, 1982; 35 Kikukawa K., *J. Chem. Soc., Perkin Trans. 1,* 1987; 1511
$C_6H_5\text{-}I^+$	Kitamura T., *J. Chem. Soc., Chem. Commun.*, 1990; 614
IO_2	Alper H., *J. Org. Chem.*, 1993; 4794
RS	Alper H., *J. Org. Chem.*, 1995; 5579 (CO insertion)
SO_2Cl	Miura M., *J. Mol. Catal.*, 1990; 59, 325 Miura M., *Chem. Lett.*, 1989; 77

X	References
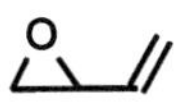	Shimizu I., *Tetrahedron Lett.*, 1993; 2135
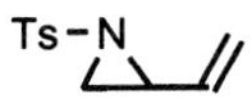	Tanner D., *Bioorg. Med. Chem. Lett.*, 1993; 2415
$B(OH)_2$	Uemura S., *J. Organomet. Chem.*, 1995; 496, 221 (Pd° as catalyst)
$B(OiPr)_2$	Suzuki A., *Tetrahedron Lett.*, 1989; 6555 (Pd II as catalyst)

R'H	References
HCO_2Na	Okano T., *Bull. Chem. Soc. Jpn.*, 1994; 2329
H-$SnBu_3$	Stille J.K., *J. Am. Chem. Soc.*, 1983; 7175
H-$SiEt_3$	Hidai M., *Organometallics*, 1995; 1770 (Pd-Co as catalysts)
HCO_2R	Petit F., *Tetrahedron Lett.*, 1991; 4705 Lin I.J.B., *J. Mol. Catal.*, 1992; 73, 167
H_2	Heck R.F., *J. Am. Chem. Soc.*, 1974; 7761 Hidai M., *J. Mol. Catal.*, 1993; 78, 1
H_2O	Matsuda T., *J. Org. Chem.*, 1980; 2365 Alper H., *J. Organomet. Chem.*, 1983; 2, 801 Kalck Ph., *J. Organomet. Chem.*, 1994; 482, 45 Beletskaya I.P., *Izv. Akad. Nauk. SSSR, Ser. Khim.*, 1988; 1450 (CA110, 212282)
KOH	Alper H., *Organometallics*, 1993; 1890
R'OH	Cacchi S., *Tetrahedron Lett.*, 1986; 3931 Ortar G., *Tetrahedron Lett.*, 1985; 1109 Moser W.R., *J. Am. Chem. Soc.*, 1988; 2816 Smith A.B., *Tetrahedron Lett.*, 1988; 2793 Miura M., *J. Mol. Catal.*, 1992; 75, 117
ArOH	Sugi Y., *Synlett.*, 1994; 515
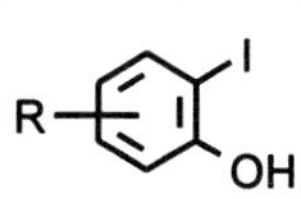	Marinelli F., *Synthesis*, 1995; 831
C=C-CH_2OH	Crisp G.T., *J. Org. Chem.*, 1992; 6972

R'H	References
HNRR'	Cacchi S., *Tetrahedron Lett.*, 1986; 3931 Ortar G., *Tetrahedron Lett.*, 1985; 1109 Collin J., *Bull. Soc. Chim. Fr.*, 1988; 976 (double carbonylation) Perry R.J., *J. Org. Chem.*, 1992; 6351 Yamamoto A., *Bull. Chem. Soc. Jpn.*, 1988; 1251 Yamamoto A., *Pure Appl. Chem.*, 1991; 687 Haffner C., *Tetrahedron Lett.*, 1994; 1349
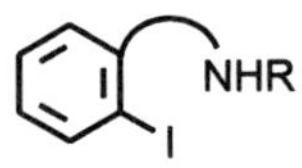	Grigg R., *Tetrahedron*, 1995; 295 (cyclisation)
$ArNH_2$	Hartstock F.W., *Tetrahedron Lett.*, 1994; 8761 (Pd II as catalyst)
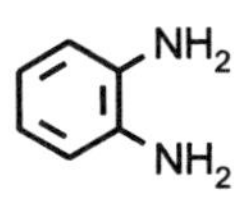	Perry R.J., *J. Org. Chem.*, 1993; 7016
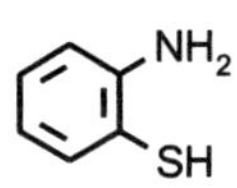	Perry R.J., *Organometallics.*, 1994; 3346
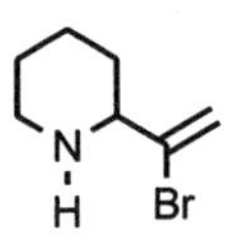	Mori M., *Heterocycles*, 1989; 29, 853
$HN-CH_2-OH$	Meyers A.I., *Tetrahedron Lett.*, 1992; 1181
C=C-OH	Negishi E.I., *Tetrahedron Lett.*, 1990; 2841
N=C(R)-OH	Grigg R., *Tetrahedron Lett.*, 1994; 3197
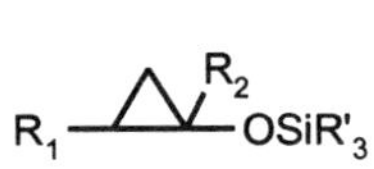	Nakamura E., *Synlett.*, 1990; 741
Z-CH-COR	Tanaka M., *Tetrahedron Lett.*, 1986; 4745 Negishi E.I., *Tetrahedron*, 1994; 425 (cyclisation) Negishi E.I., *Tetrahedron Lett.*, 1990; 2841 (cyclisation) Negishi E.I., *J. Am. Chem. Soc.*, 1989; 8018 (cyclisation)
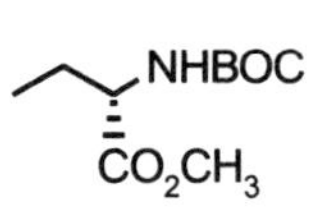	Jackson R.F.W., *J. Chem. Soc., Chem. Commun.*, 1995; 2207

R'H	References
R−C≡CH	Tanaka M., *J. Chem. Soc., Chem. Commun.*, 1981; 333 Torii S., *Tetrahedron Lett.*, 1991; 237 (cyclisation) Kalinin V.N., *Tetrahedron Lett.*, 1990; 4073 (cyclisation) Ortar G., *Tetrahedron Lett.*, 1991; 6449 (cyclisation) Kalinin V.N., *Tetrahedron*, 1993; 6773 (cyclisation) Torii S., *Synlett.*, 1992; 513 (cyclisation) Catellani M., *Tetrahedron Lett.*, 1994; 5923 Cacchi S., *Synlett.*, 1995; 823
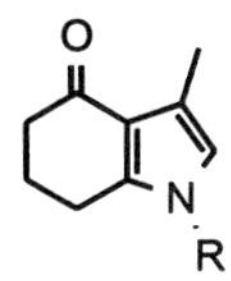	Inoue Y., *Bull. Chem. Soc. Jpn.*, 1989; 3518
$CHCl_3$	Alper H., *Organometallics*, 1993; 3846

R'M	References
$R'CO_2Na$	Matsuda T., *Tetrahedron Lett.*, 1980; 2877 Matsuda T., *J. Org. Chem.*, 1980; 2365
AcOK	Koga H., *Tetrahedron Lett.*, 1995; 87 Edstrom E.D., *Synlett.*, 1995; 49
$R'CO_2K$	Cacchi S., *Tetrahedron Lett.*, 1992; 3939
KCN	Tanaka M., *Bull. Chem. Soc. Jpn.*, 1981; 637
KF	Okano T., *Bull. Chem. Soc. Jpn.*, 1992; 1741
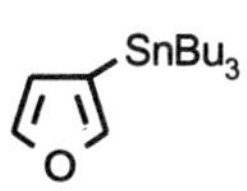	Wong N.C., *Tetrahedron*, 1994; 9583
$C{=}C(OR)\text{-}SnR'_3$	Stille J.K., *J. Org. Chem.*, 1990; 3114
Z-C(MgX)-Z'	Tanaka M., *Tetrahedron Lett.*, 1986; 4745
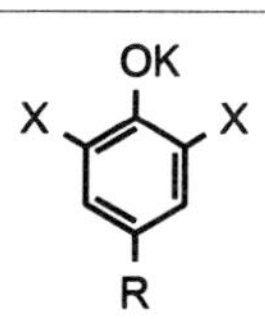	Sugi Y., *J. Chem. Soc., Chem. Commun.*, 1994; 1553
Ar_3Sb	Uemura S., *J. Org. Chem.*, 1995; 274
$R'OSnR''_3$	Bumagin N.A., *J. Organomet. Chem.*, 1985; 285, 415

R'M	References
$RR'NSnR''_3$	Bumagin N.A., *J. Organomet. Chem.*, 1985; 285, 415
$C_6H_5SSnR''_3$	Bumagin N.A., *J. Organomet. Chem.*, 1985; 285, 415
9-alkyl-9-BBN	Suzuki A., *Bull. Chem. Soc. Jpn.*, 1991; 1999 Suzuki A., *Tetrahedron Lett.*, 1992; 4465
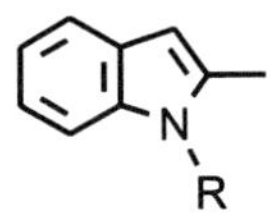	Ishikura M., *J. Org. Chem.*, 1994; 2634 Ishikura M., *Heterocycles*, 1995; 41, 1385
$ArB(OH)_2$	Miyaura M., *Tetrahedron Lett.*, 1993; 7595
RHgI	Beletskaya I.P., *J. Organomet. Chem.*, 1989; 365, 379
$ArSiRF_2$	Hiyama T., *Tetrahedron*, 1992; 2113
$R'SnR''_3$	Tanaka M., *Tetrahedron Lett.*, 1979; 2601 Kikukawa K., *Chem. Lett.*, 1982; 35
$C{=}C\text{-}SnMe_3$	Stille J.K., *Org. Synth.*, 1990; 68, 116
$IZn\text{-}(CH_2)n\text{-}CO_2Et$	Tamaru Y., *Angew. Chem. Int. Ed. Engl.*, 1992; 645 Tamaru Y., *J. Org. Chem.*, 1995; 1365
$IZn\text{-}(CH_2)n\text{-}CN$	Tamaru Y., *J. Org. Chem.*, 1995; 1365
$Ti(OiPr)_4$	Miura M., *J. Mol. Catal.*, 1992; 75, 117
$R-C\equiv CH$	Brandsma L., *Synth. Commun.*, 1994; 85 (Pd II as catalyst) Takano S., *Synthesis*, 1993; 1253 (Pd II as catalyst) Tsuji J., *Tetrahedron Lett.*, 1980; 849 (Pd II as catalyst) Alper H., *Synthesis*, 1994; 1149 (Pd II as catalyst) Temkin O.N., *Mendeleev Commun.*, 1994; 2 (Pd II as catalyst)

RXN61 Isomerization of Alkynes

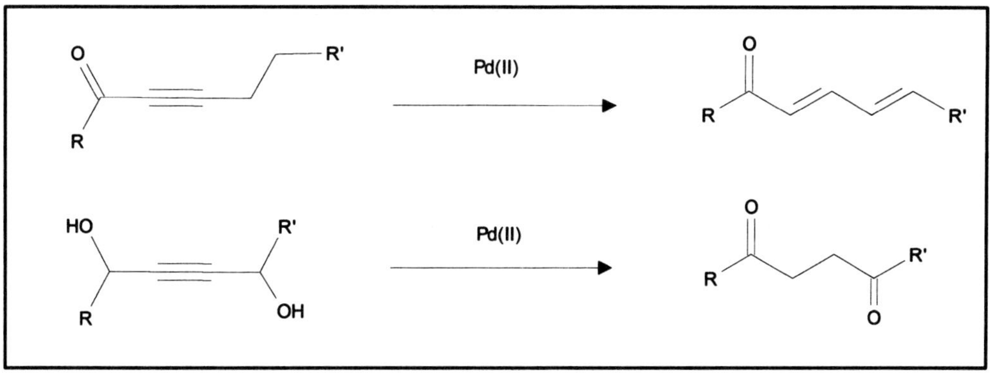

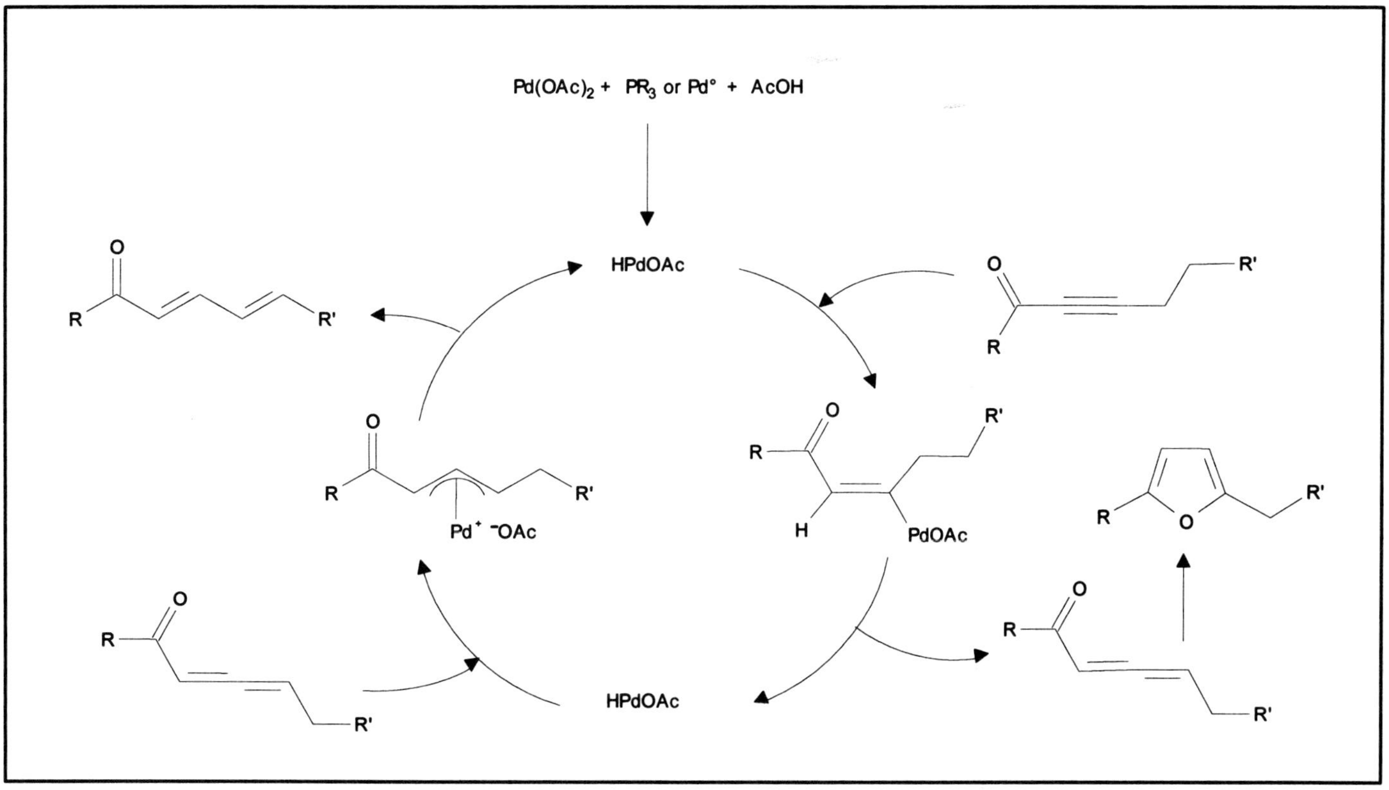

RXN61 Isomerization of Alkynes

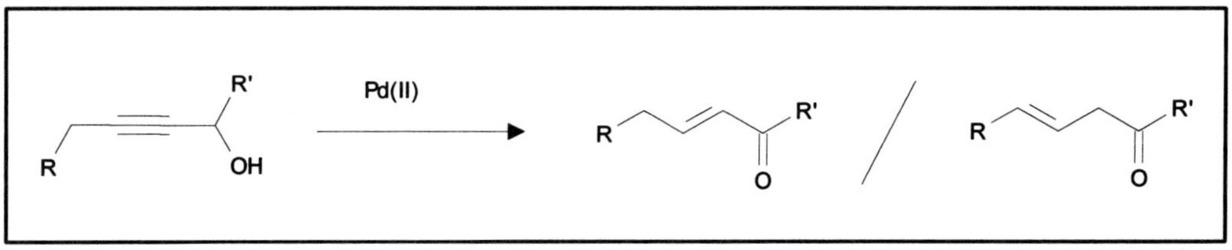

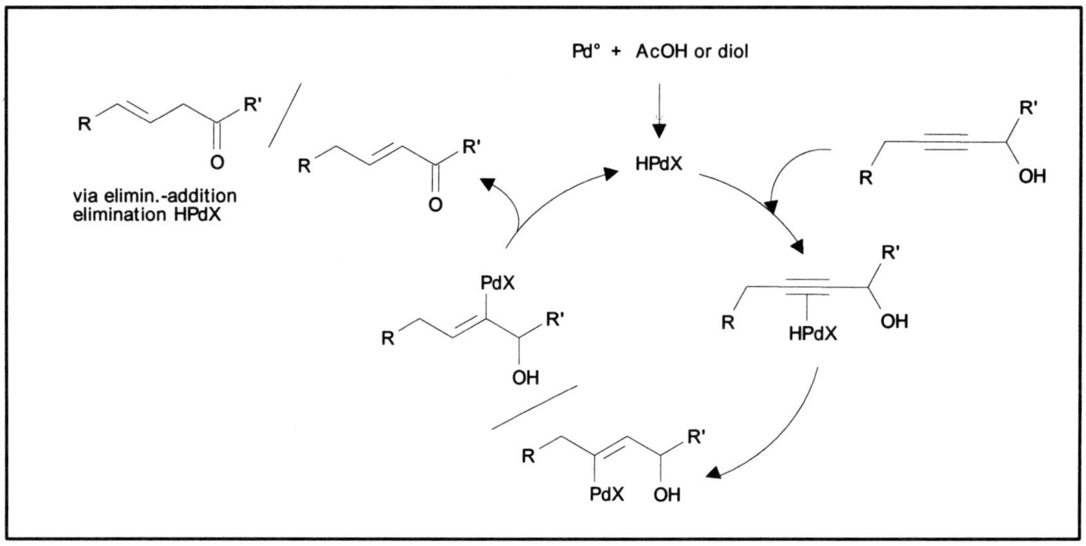

A ---> B	References
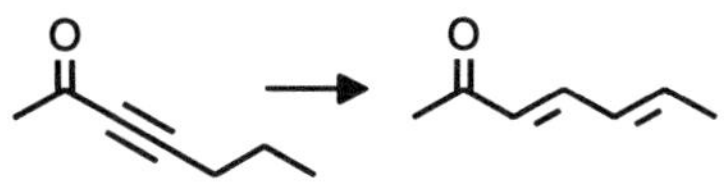	Trost B.M., *J. Am. Chem. Soc.*, 1988; 2301
	Lu X., *Tetrahedron*, 1995; 11765
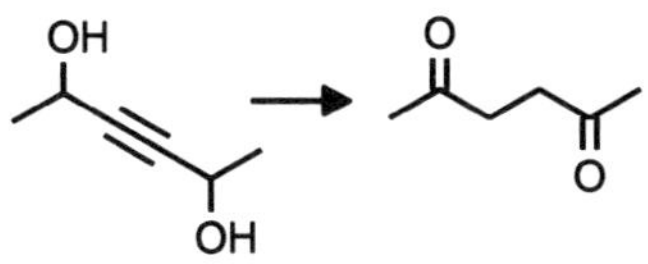	Lu X., *J. Org. Chem.*, 1991; 5774
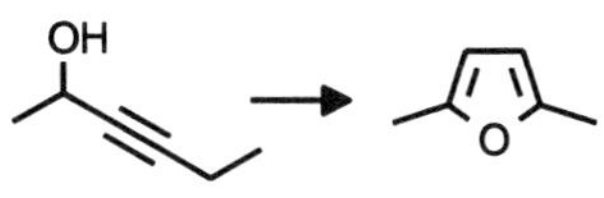	Huang Y.Z., *Tetrahedron Lett.*, 1986; 4893
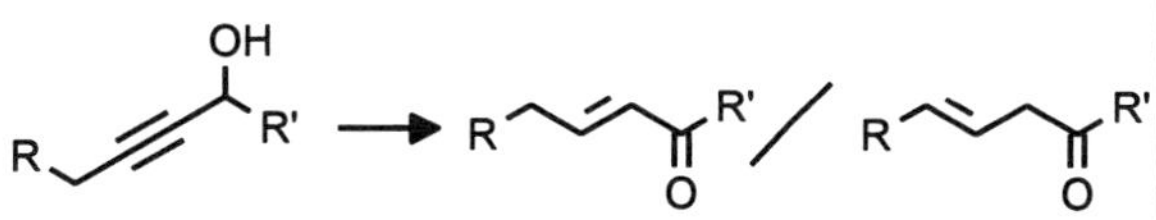	Lu X., *J. Organomet. Chem.*, 1992; 428, 259

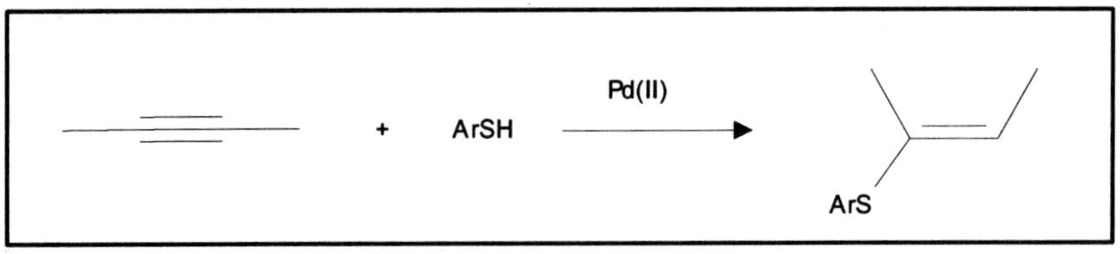
+
ArSH
Pd(II)
ArS

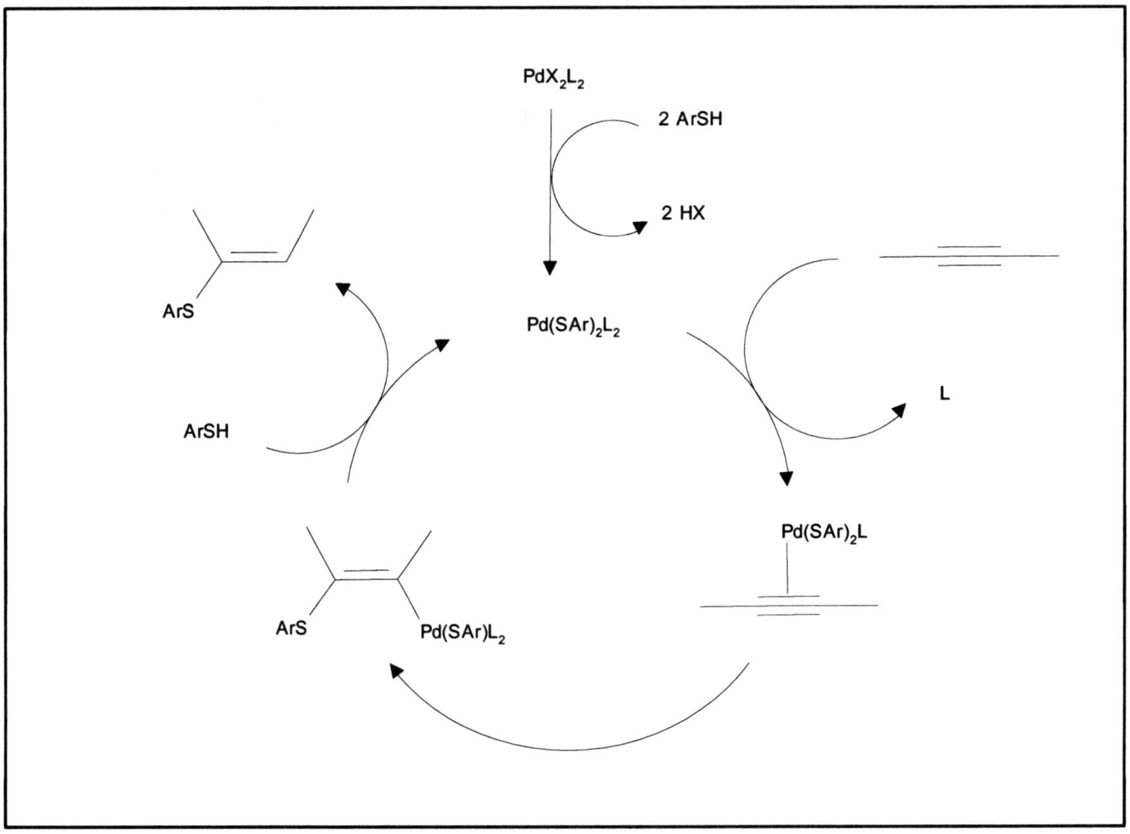
PdX2L2
2 ArSH
2 HX
ArS
Pd(SAr)2L2
L
ArSH
Pd(SAr)2L
ArS
Pd(SAr)L2

A ---> B	References
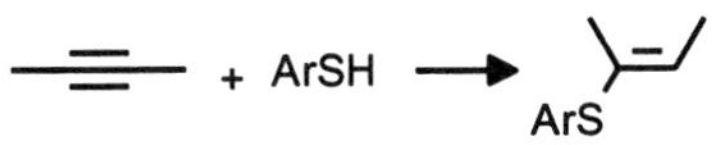	Sonoda N., *J. Am. Chem. Soc.*, 1992; 5902
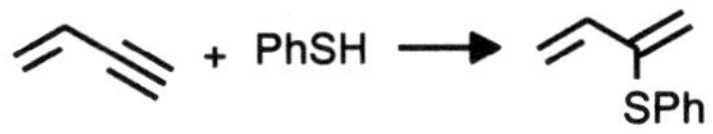	Bäckvall J.E., *J. Org. Chem.*, 1994; 5850

RXN62

RXN63 Preparation of Allylic Acetates from Alkynes by Tandem Redox-Addition

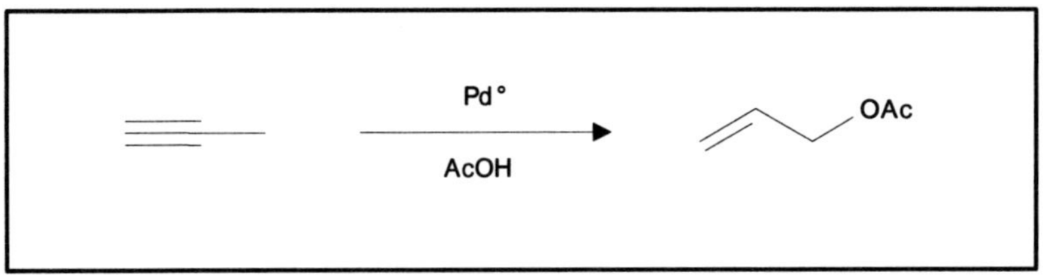

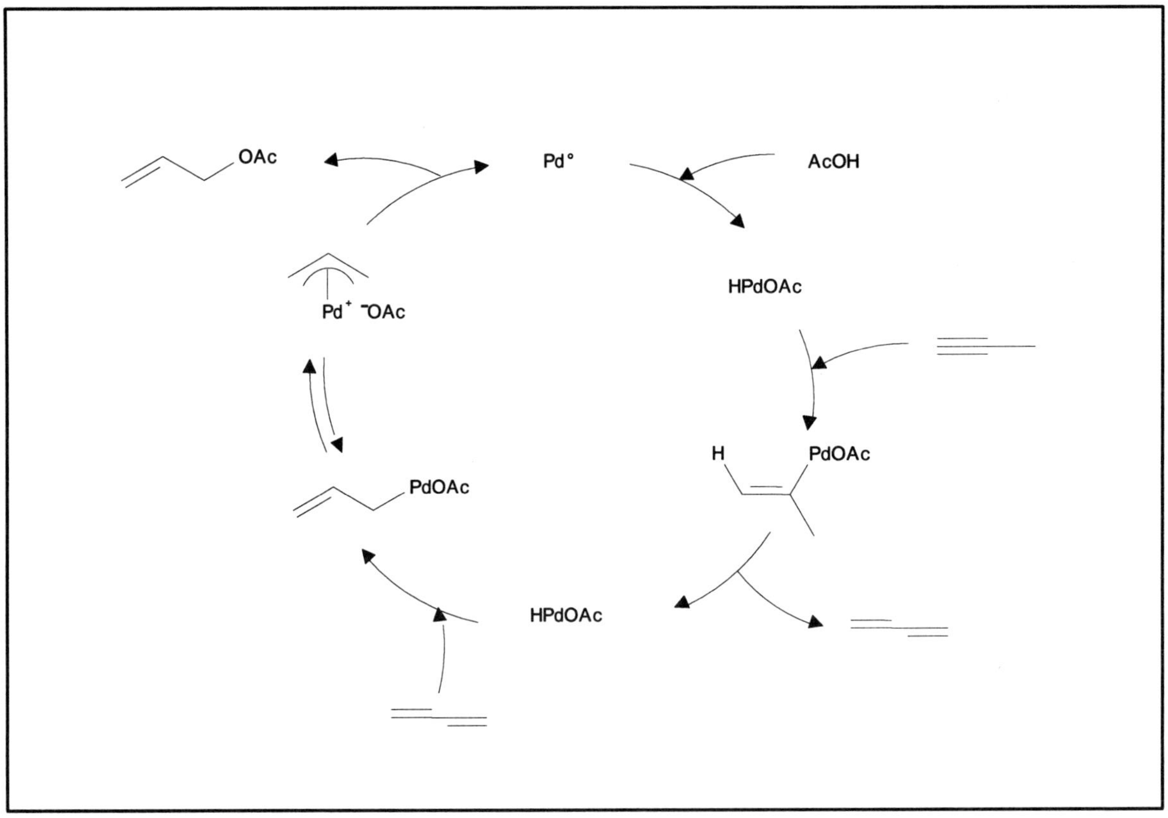

A ---> B	References
	Trost B.M., *Angew. Chem. Int. Ed. Engl.*, 1992; 1335

RXN63

RXN64 Tandem Cyclization-Capture Process of Enynes

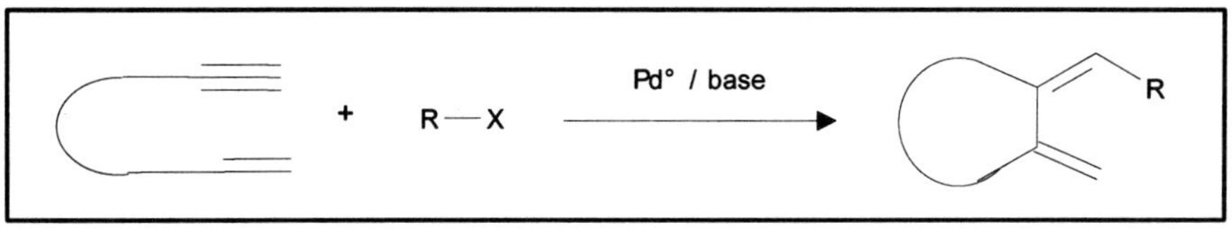

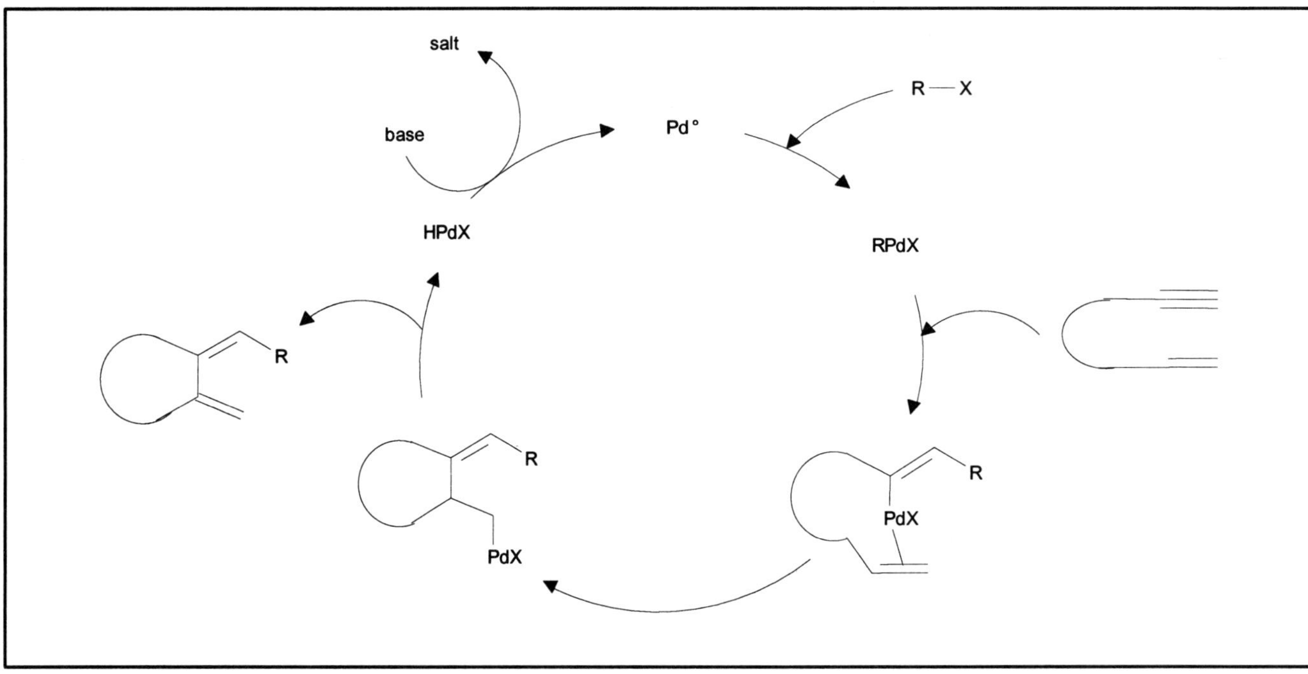

A ---> B	References
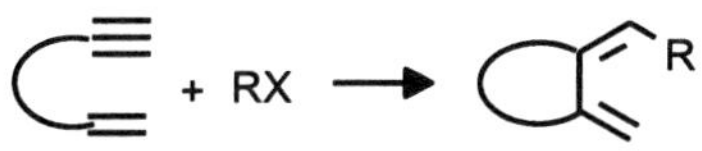	Negishi E.I., *J. Am. Chem. Soc.*, 1992; 10091 Trost B.M., *J. Am. Chem. Soc.*, 1992; 9836 Trost B.M., *Tetrahedron Lett.*, 1993; 19 Tanaka T., *Tetrahedron*, 1995; 5543
	Grigg R., *Tetrahedron Lett.*, 1993; 157

RXN64

RXN65 Aldol-Like Condensation of Enol Esters with Aldehydes

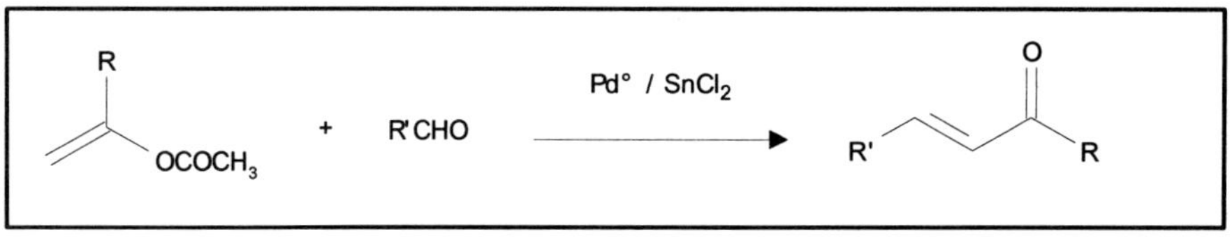

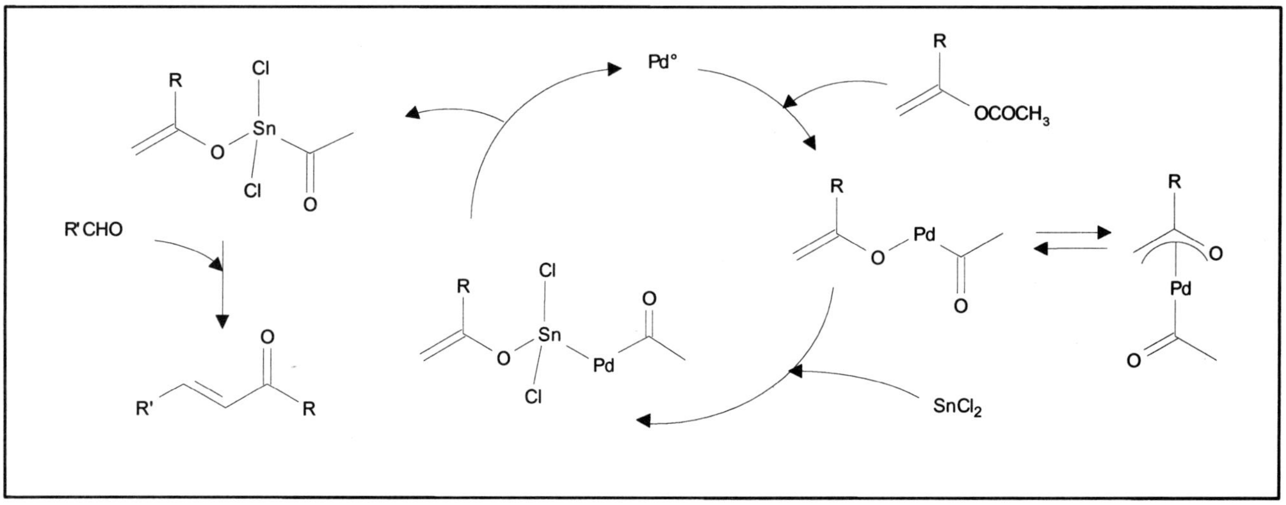

A ---> B	References
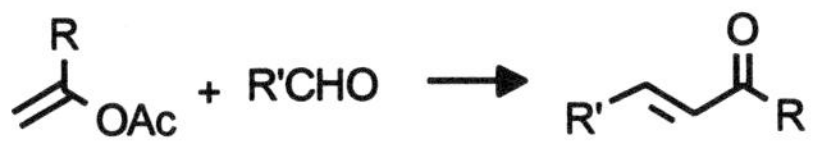	Masuyama Y., *Tetrahedron Lett.*, 1993; 653

RXN65

RXN66 Addition of Fluoroalkyl Iodides to Alkenes or Alkynes via a Pd(I) Species

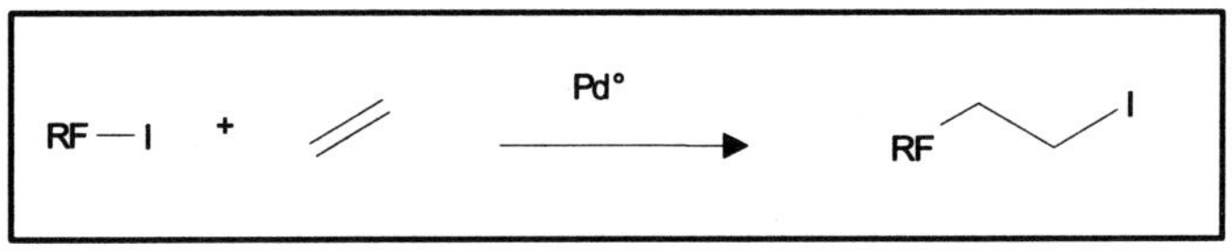

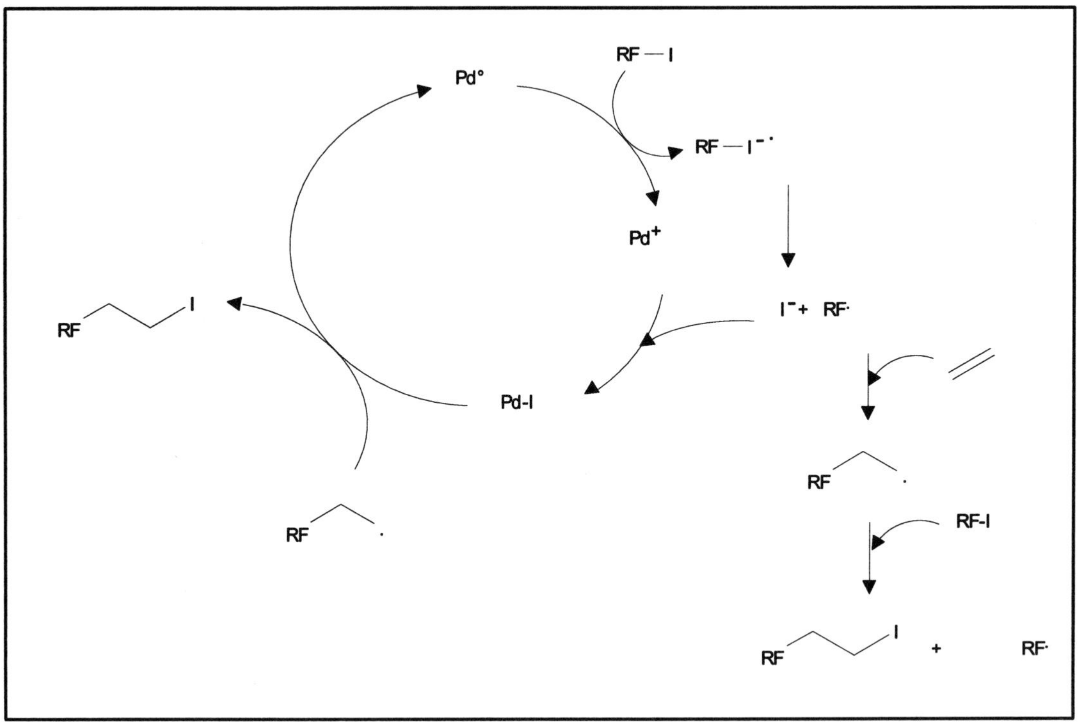

A ---> B	References
	Chen Q.Y., *J. Chem. Soc., Perkin Trans. 1*, 1988; 563 Ishihara T., *Chem. Lett.*, 1986; 1895 Burton D., *J. Org. Chem.*, 1993; 419
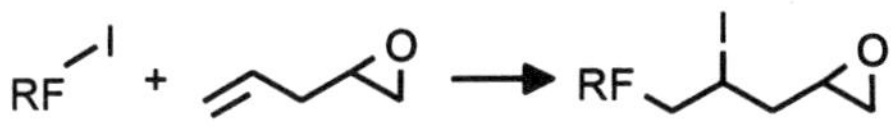	Burton D., *J. Org. Chem.*, 1993; 419
	Ishihara T., *Chem. Lett.*, 1986; 1895

RXN67 Intermolecular Tandem Carbonylation-Coupling-Cyclization Process of Aryl Halides with Terminal Alkynes

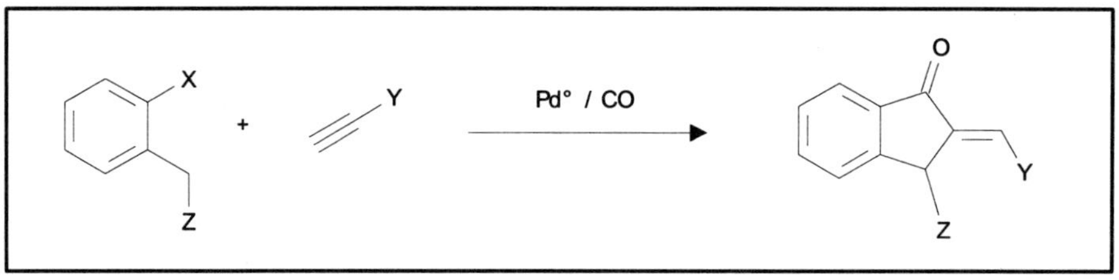

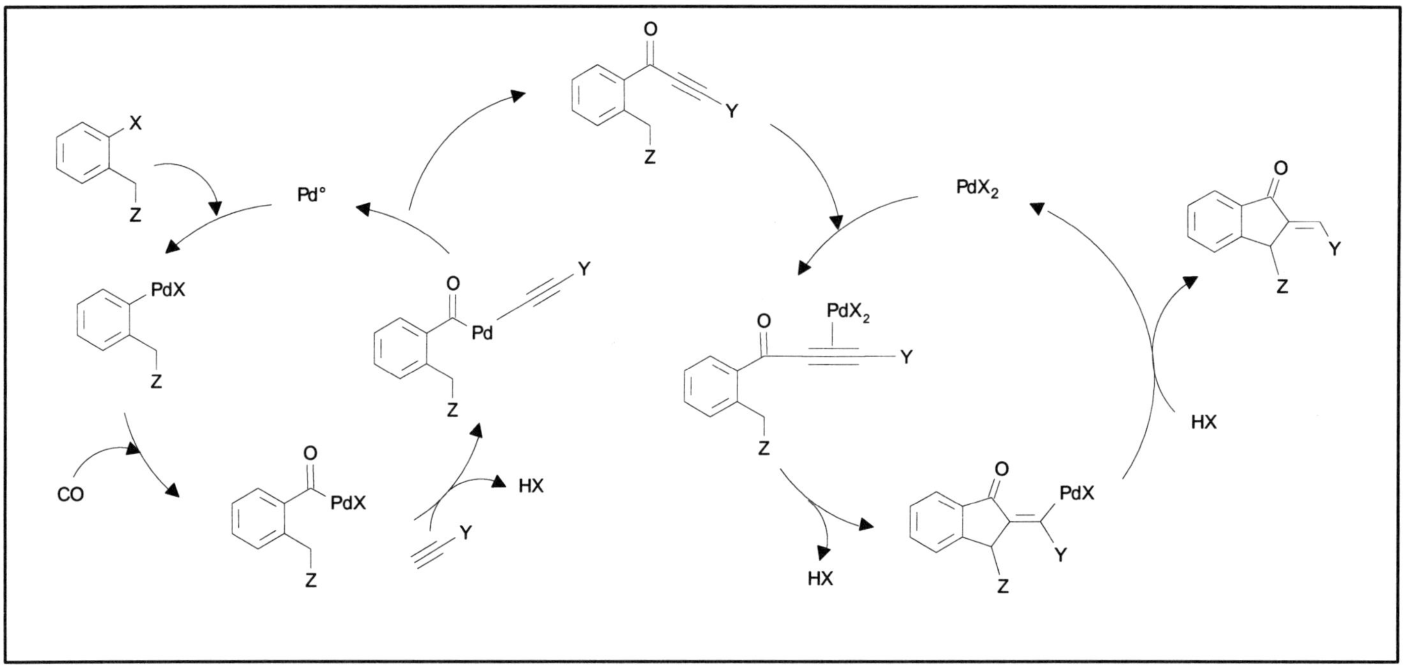

A ---> B	References
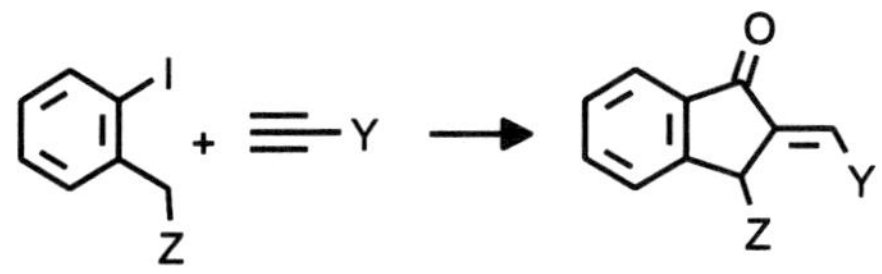	Chiusoli G.P., *Tetrahedron Lett.*, 1992; 7433

RXN67

RXN68 Intramolecular Coupling of Aryl Halides with Arenes

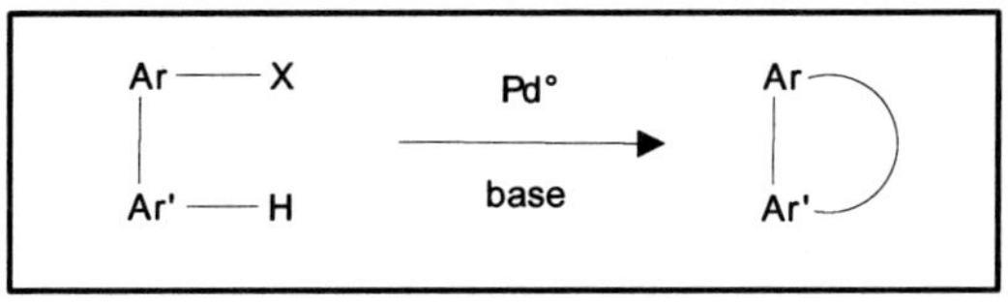

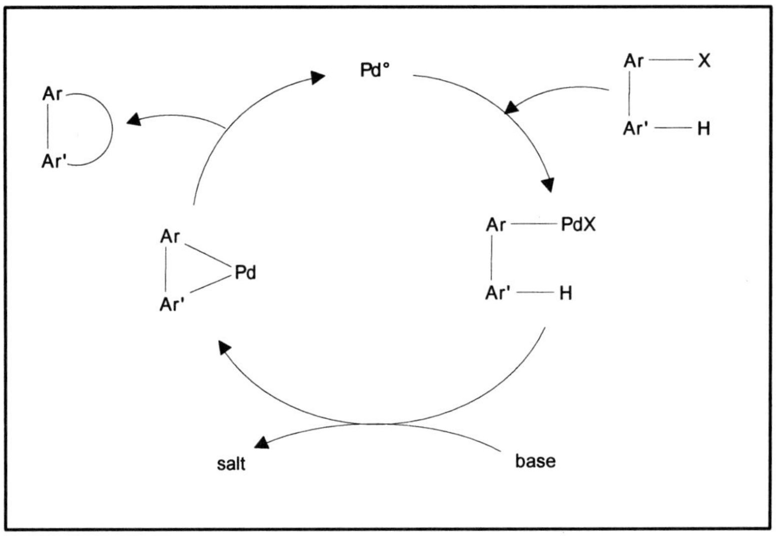

A ---> B	References
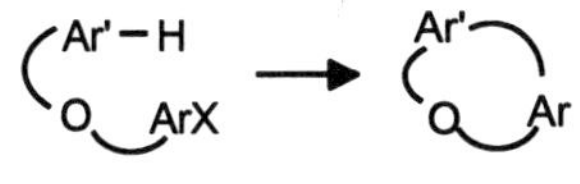	Ames D.E., *Synthesis*, 1983; 234 Ames D.E., *Tetrahedron*, 1984; 1919
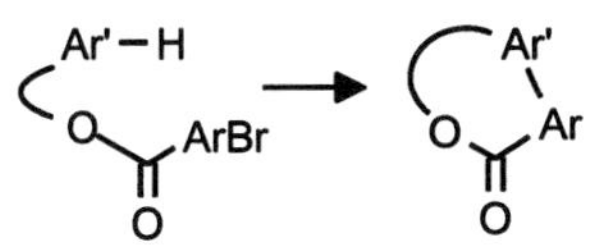	Bringmann G., *Angew. Chem., Int. Ed. Engl.*, 1986; 913 Bringmann G., *Tetrahedron Lett.*, 1989; 5249 Bringmann G., *Liebigs Ann. Chem.*, 1992; 225
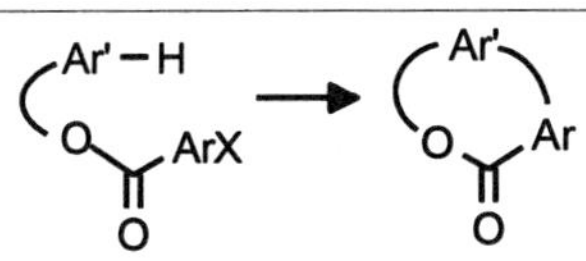	Martin O.R., *Tetrahedron Lett.*, 1990; 6313
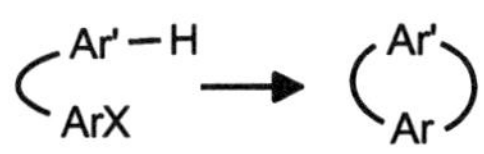	Rice J.E., *J. Org. Chem.*, 1993; 1415 Ames D.E., *Synthesis*, 1983; 234

RXN69 Tandem Intramolecular Alkoxylation-Vinylation of Alkenes

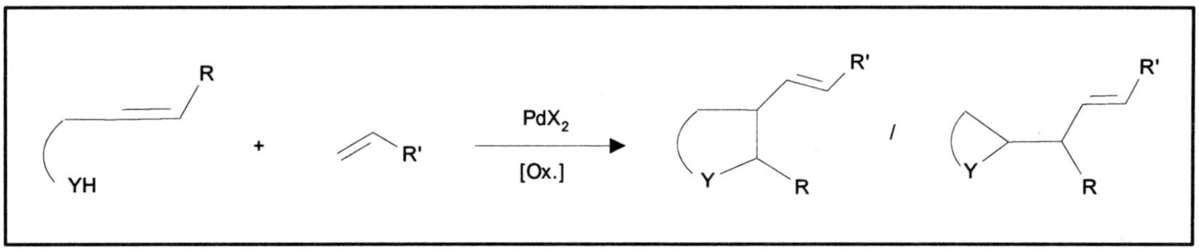

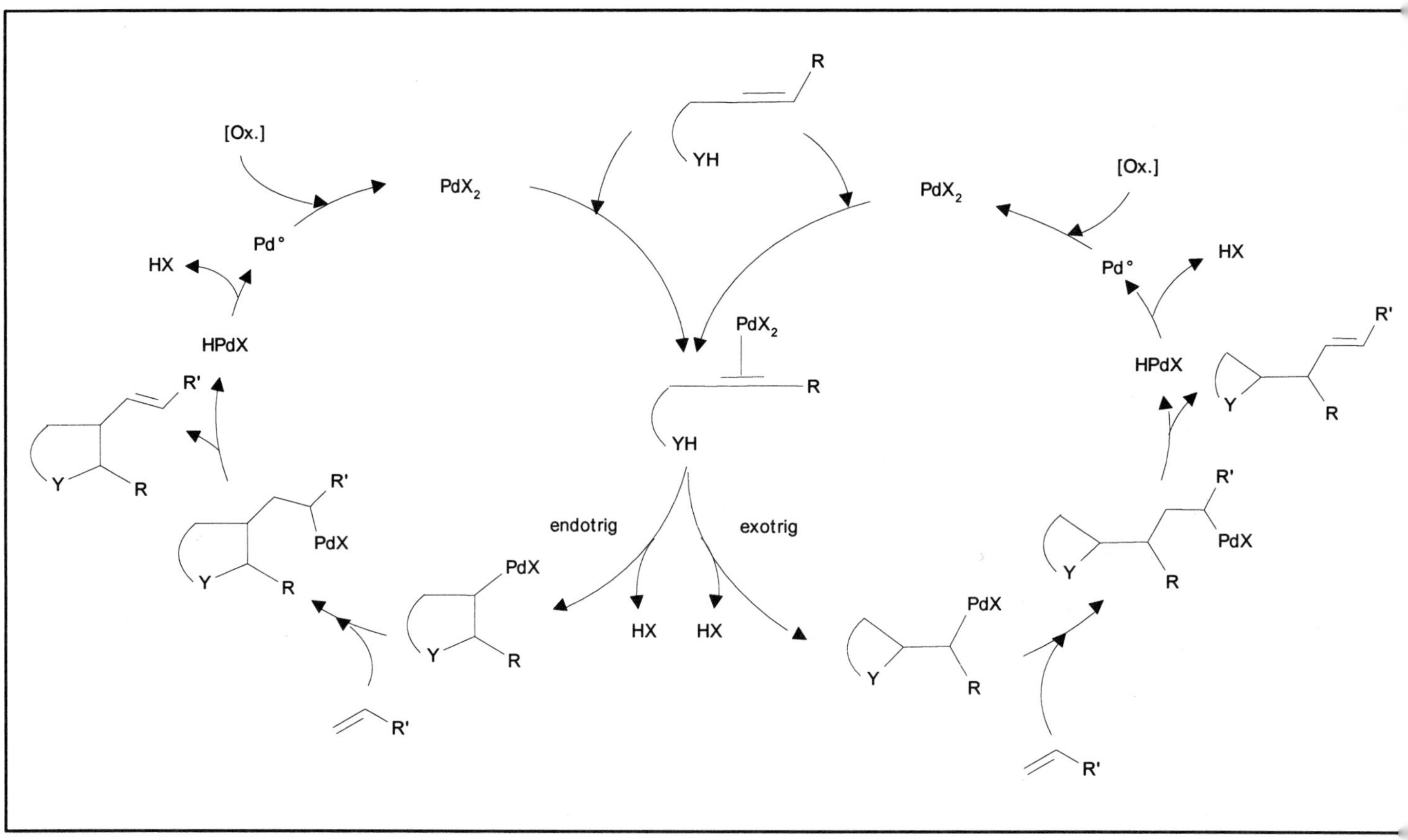

Reagent	YH	Product	Catalyst	References
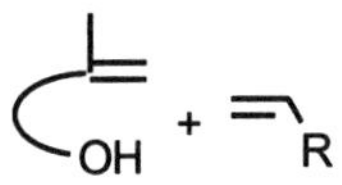	OH	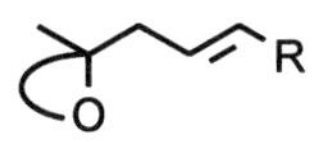	PdX_2 / CuCl / O_2	Semmelhack M.F., *Tetrahedron Lett.*, 1993; 7205

RXN70 Carbonylative [2+2] Cycloaddition

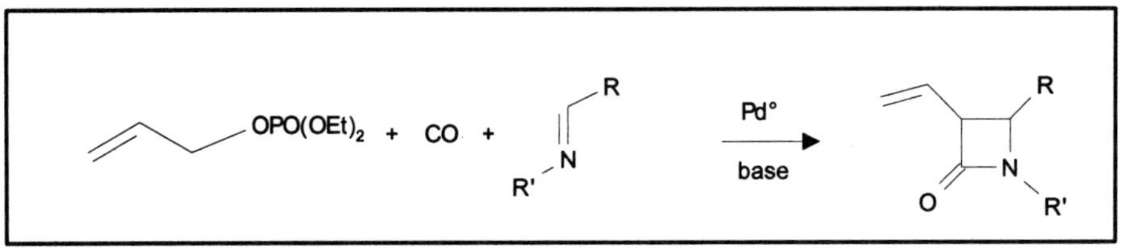

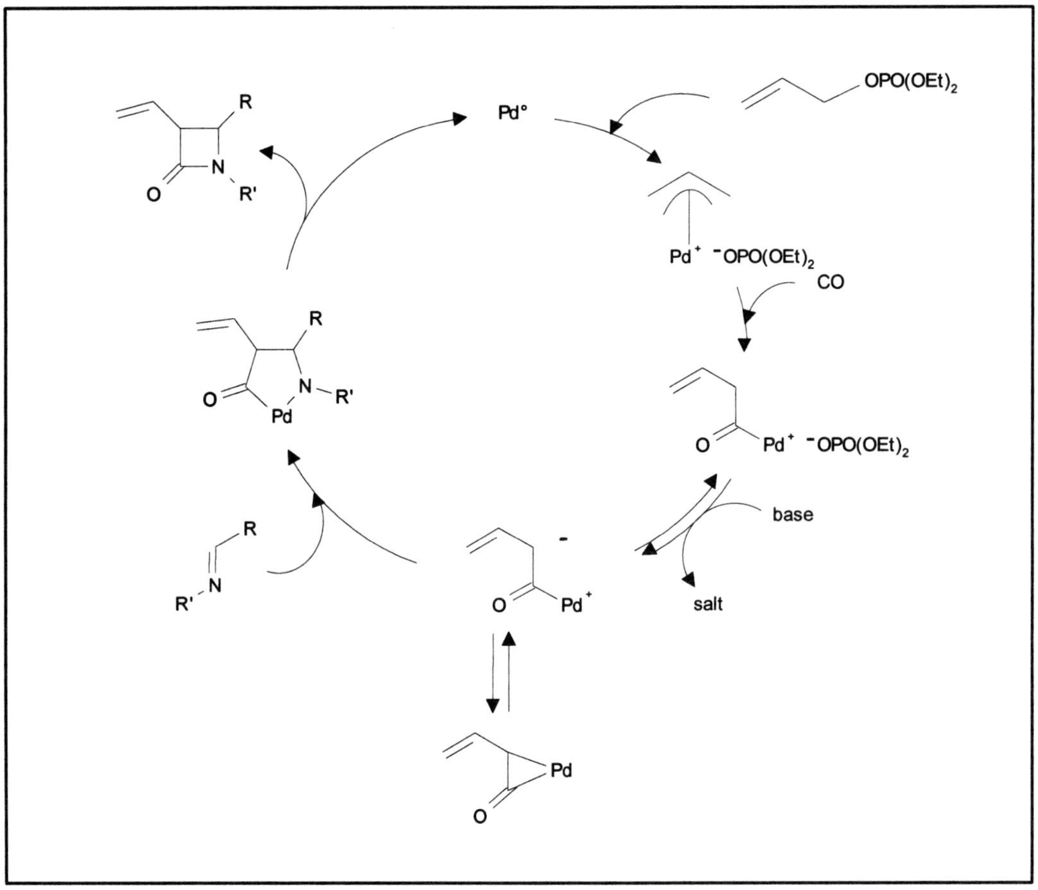

A ---> B	References
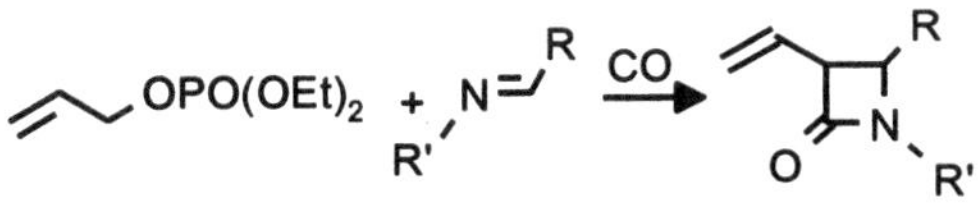	Torii S., *Tetrahedron Lett.*, 1993; 6553 Torii S., *J. Org. Chem.*, 1994; 3040

RXN71 Dicarboalkoxylation of Alkenes

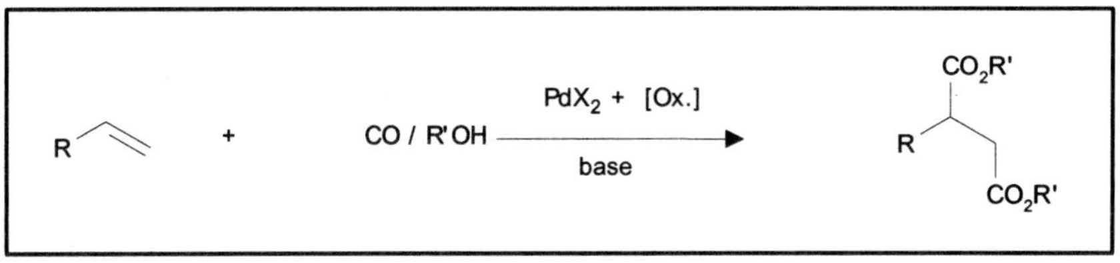

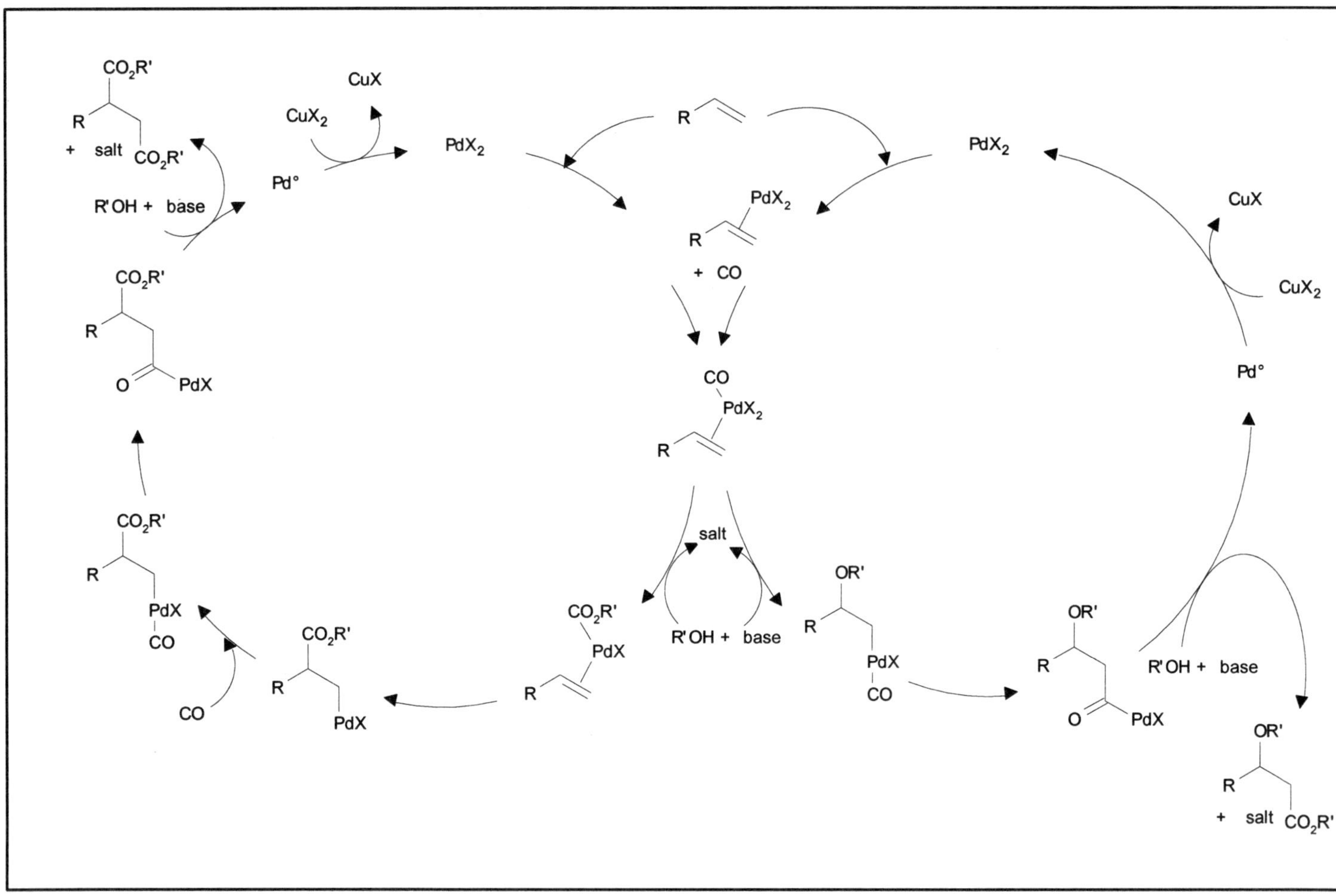

A ---> B	References
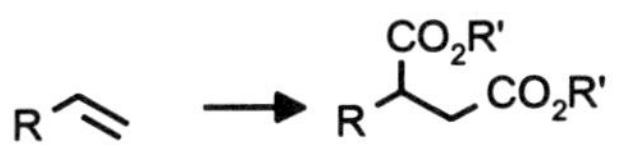	Stille J.K., *J. Am. Chem. Soc.*, 1976; 1810 Stille J.K., *J. Org. Chem.*, 1979; 3474 Consiglio G., *Angew. Chem. Int. Ed. Engl.*, 1993; 1719
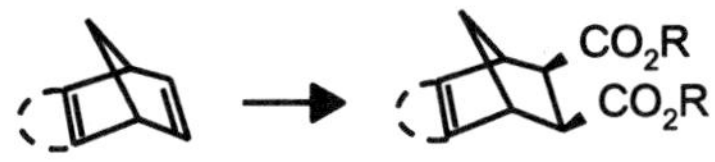	Paddon-Row M.N., *Tetrahedron*, 1995; 2689
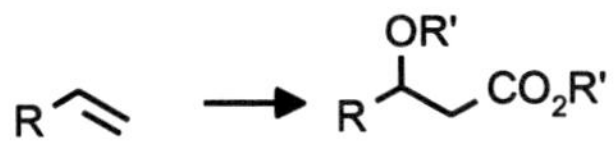	Stille J.K., *J. Org. Chem.*, 1979; 3474
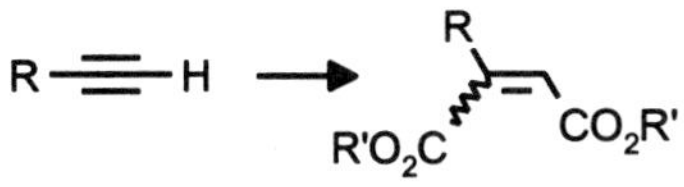	Chiusoli G.P., *J. Chem. Soc., Perkin Trans. 1*, 1994, 83
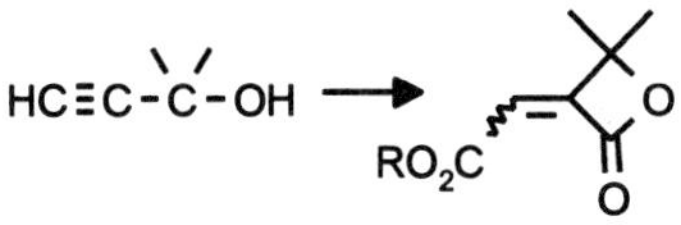	Chiusoli G.P., *J. Chem. Soc., Chem. Commun.*, 1994; 1429
	Alper H., *J. Chem. Soc., Chem. Commun.*, 1983; 1270
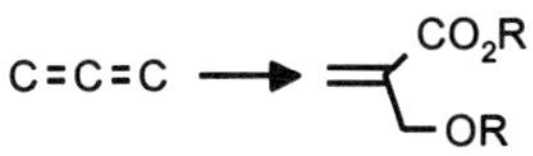	Alper H., *J. Chem. Soc., Chem. Commun.*, 1984; 905

RXN72 Addition of Pronucleophiles on 1,3-Dienes or Allenes

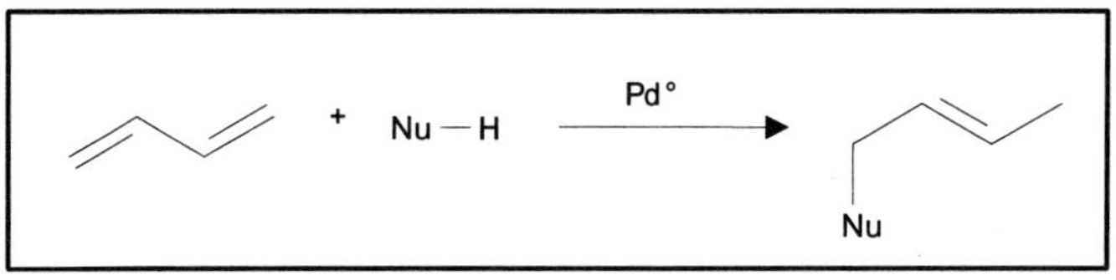

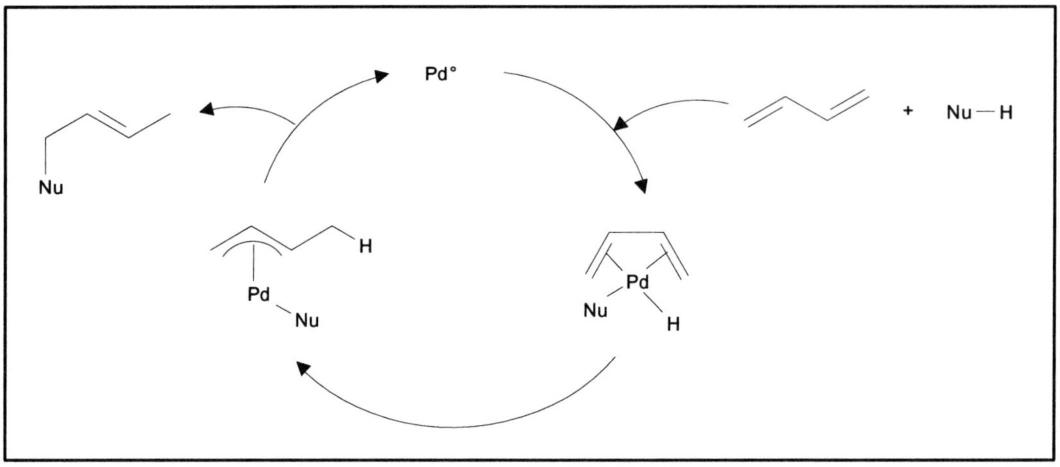

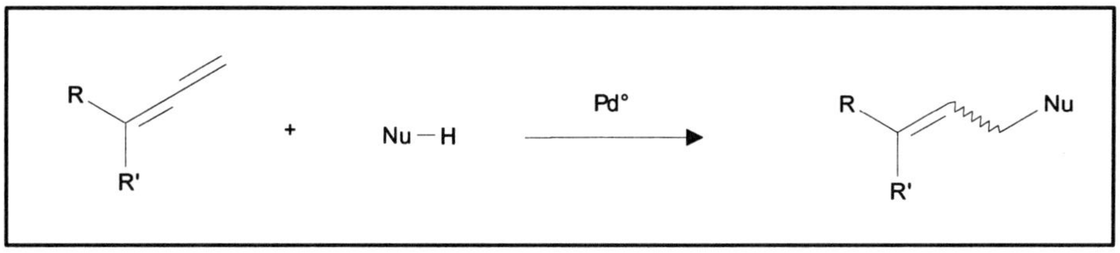

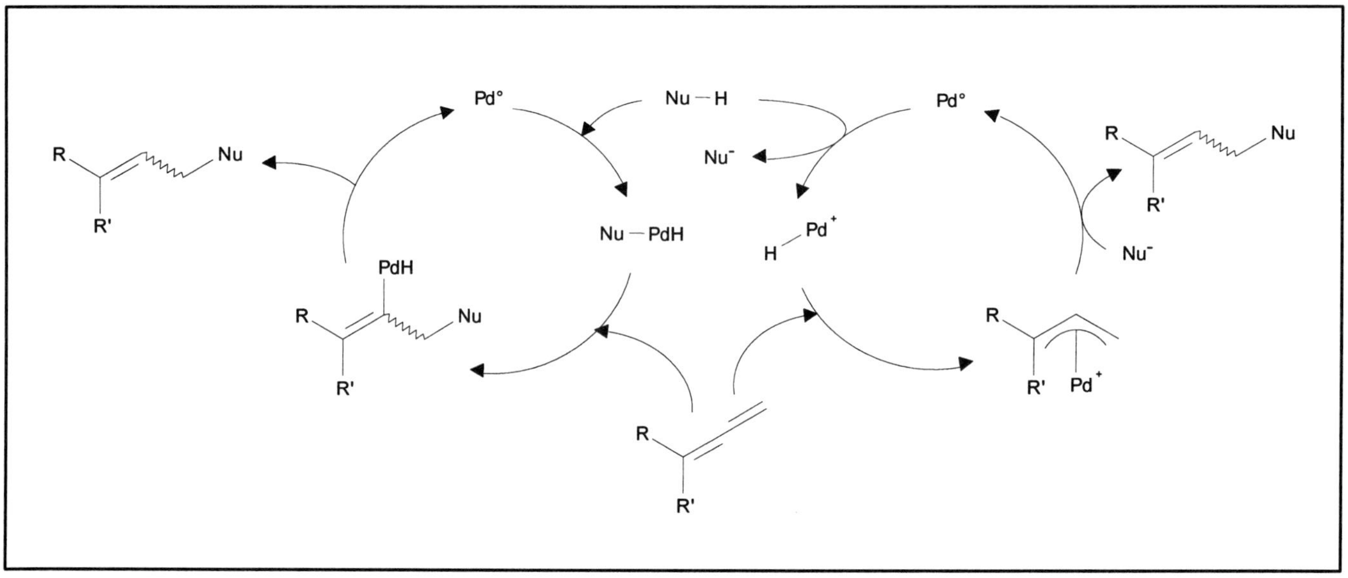

Diene	References
C=C-C=C	Trost B.M., *Tetrahedron Lett.*, 1992; 1831
C=C-C=C	Jolly P.W., *Synthesis*, 1990; 771
C=C-C=C	Trost B.M., *Tetrahedron Lett.*, 1994; 1361
	Trost B.M., *Tetrahedron Lett.*, 1992; 1831
C=C=C	Yamamoto Y., *J. Am. Chem. Soc.*, 1994; 6019 Yamamoto Y., *Tetrahedron Lett.*, 1995; 2811 Cazes B., *Tetrahedron Lett.*, 1995; 3853 Trost B.M., *J. Am. Chem. Soc.*, 1995; 5156 Cazes B., *Tetrahedron Lett.*, 1995; 3857 Yamamoto Y., *Synlett.*, 1995; 969

Pronucleophile	References
$R-C(=O)-CH_2-C(=O)-R$	Jolly P.W., *Synthesis*, 1990; 771 Jolly P.W., *J. Organomet. Chem.*, 1995; 486, 163 (intramolecular)
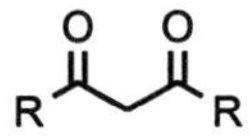	Jolly P.W., *Synthesis*, 1990; 771
CO_2R-CH_2-COR'	Trost B.M., *Tetrahedron Lett.*, 1992; 1831 Jolly P.W., *Synthesis*, 1990; 771 Cazes B., *Tetrahedron Lett.*, 1995; 3853
$PhSO_2-CH_2-CO_2R$	Trost B.M., *Tetrahedron Lett.*, 1992; 1831
$PhSO_2-CH_2-COR$	Trost B.M., *Tetrahedron Lett.*, 1992; 1831 Trost B.M., *Tetrahedron Lett.*, 1994; 1361
CN-CH(Z)(R)	Yamamoto Y., *J. Am. Chem. Soc.*, 1994; 6019 Yamamoto Y., *Synlett.*, 1995; 969
$PhSO_2-CH_2-SO_2Ph$	Trost B.M., *Tetrahedron Lett.*, 1992; 1831 Trost B.M., *J. Am. Chem. Soc.*, 1995; 5156
ROH	Jolly P.W., *Synthesis*, 1990; 771
$PhSO_2H$	Bäckvall J.E., *Acta Chem. Scand.*, 1986; 184

Pronucleophile	References
$PhSO_2Na$	Inoue Y., *Bull. Chem. Soc. Jpn.*, 1986; 3705
R-NH-R'	Cazes B., *Tetrahedron Lett.*, 1995; 3857

RXN73 Tandem Cycloisomerization-Cycloaddition of Dienynes with 1,3-Dienes via Metallodienes

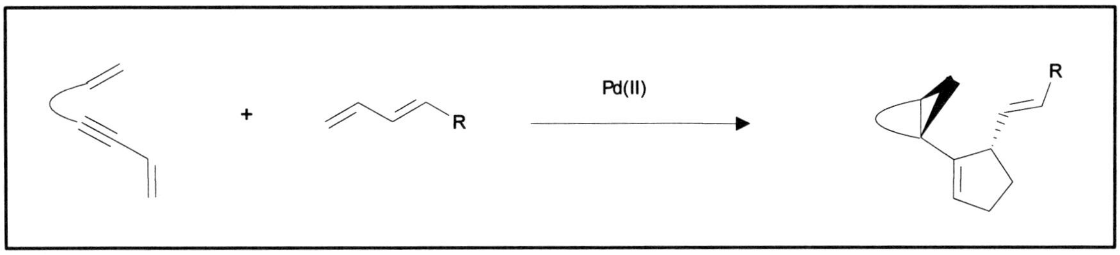

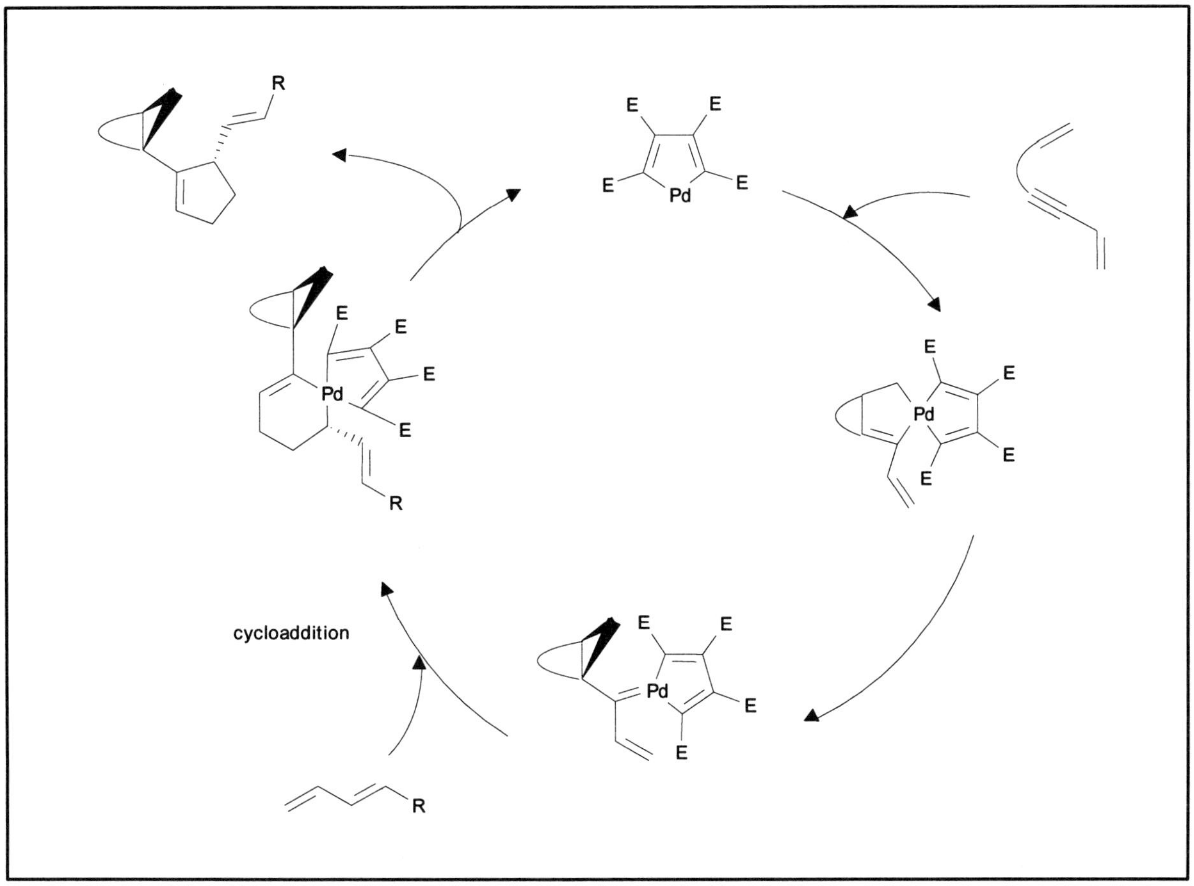

RXN73

A + B ---> C	References
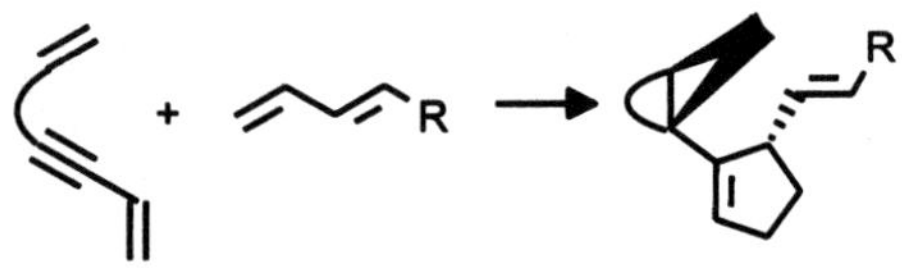	Trost B.M., *J. Am. Chem. Soc.*, 1994; 2183

RXN74 Acylcyanation of Terminal Alkynes

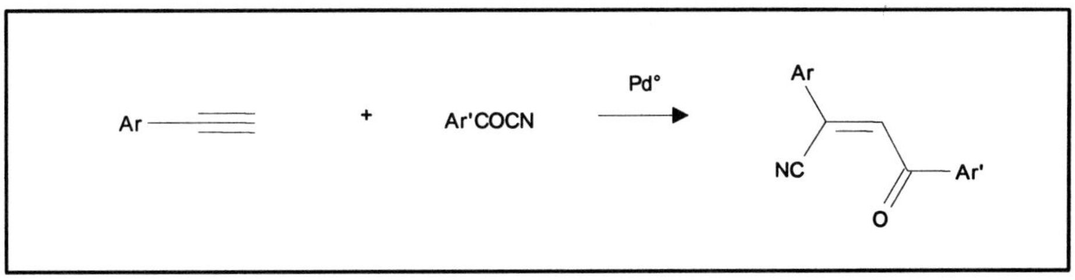

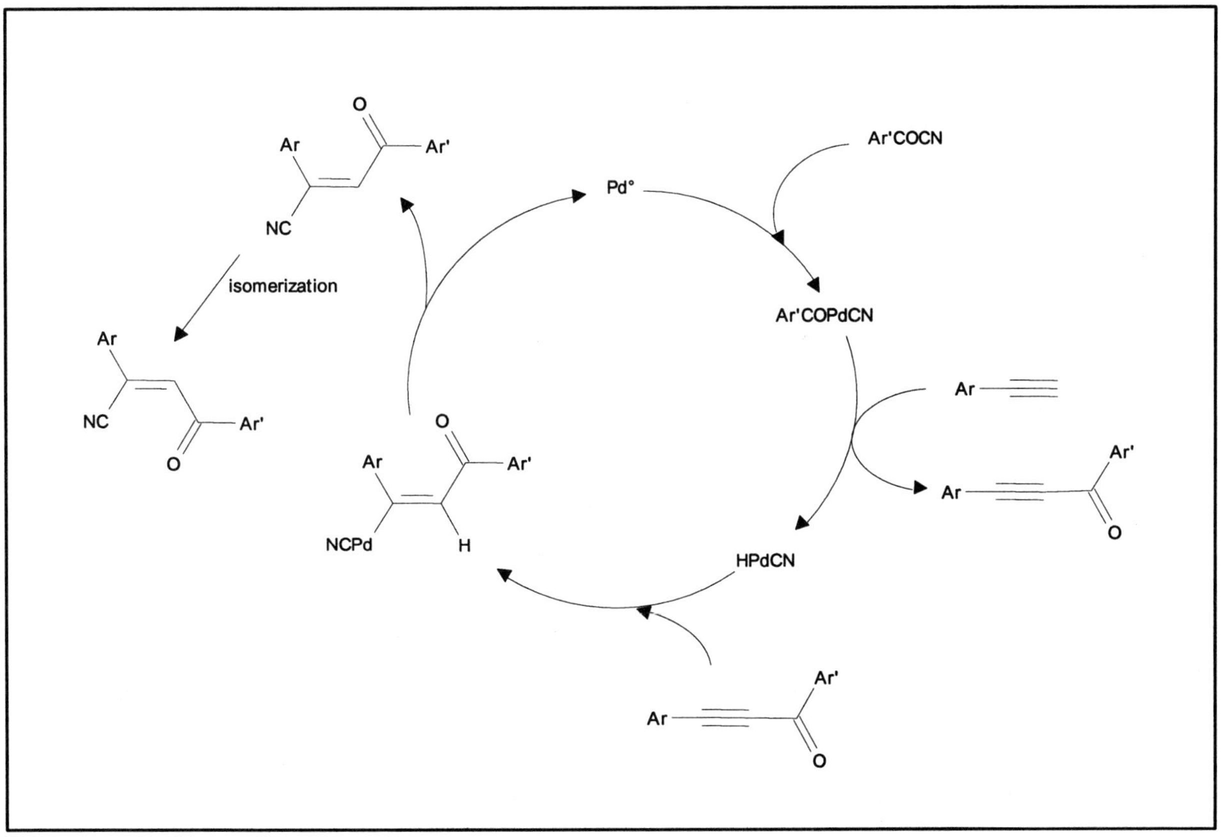

RXN74

A + B ---> C	References
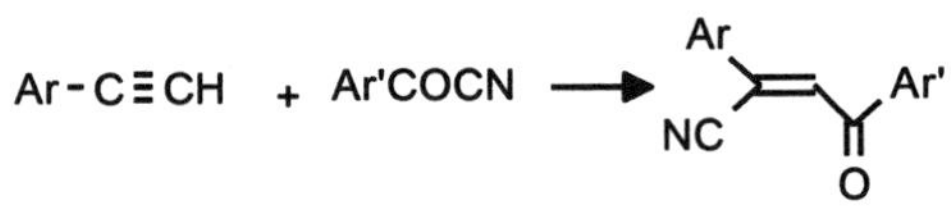	Takaya H., *J. Org. Chem.*, 1994; 2679

RXN75 1,4-Carbochlorination of 1,3-Dienes

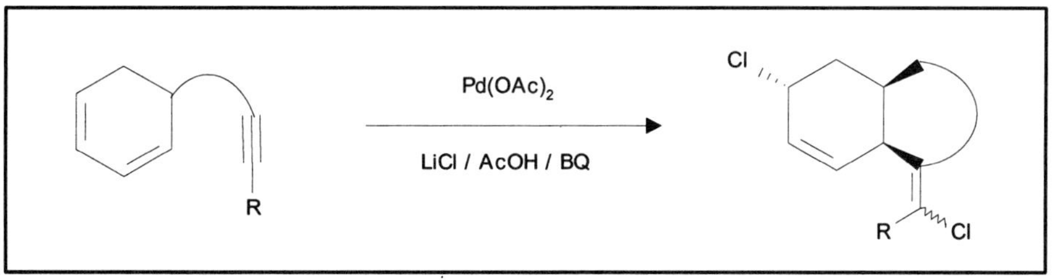

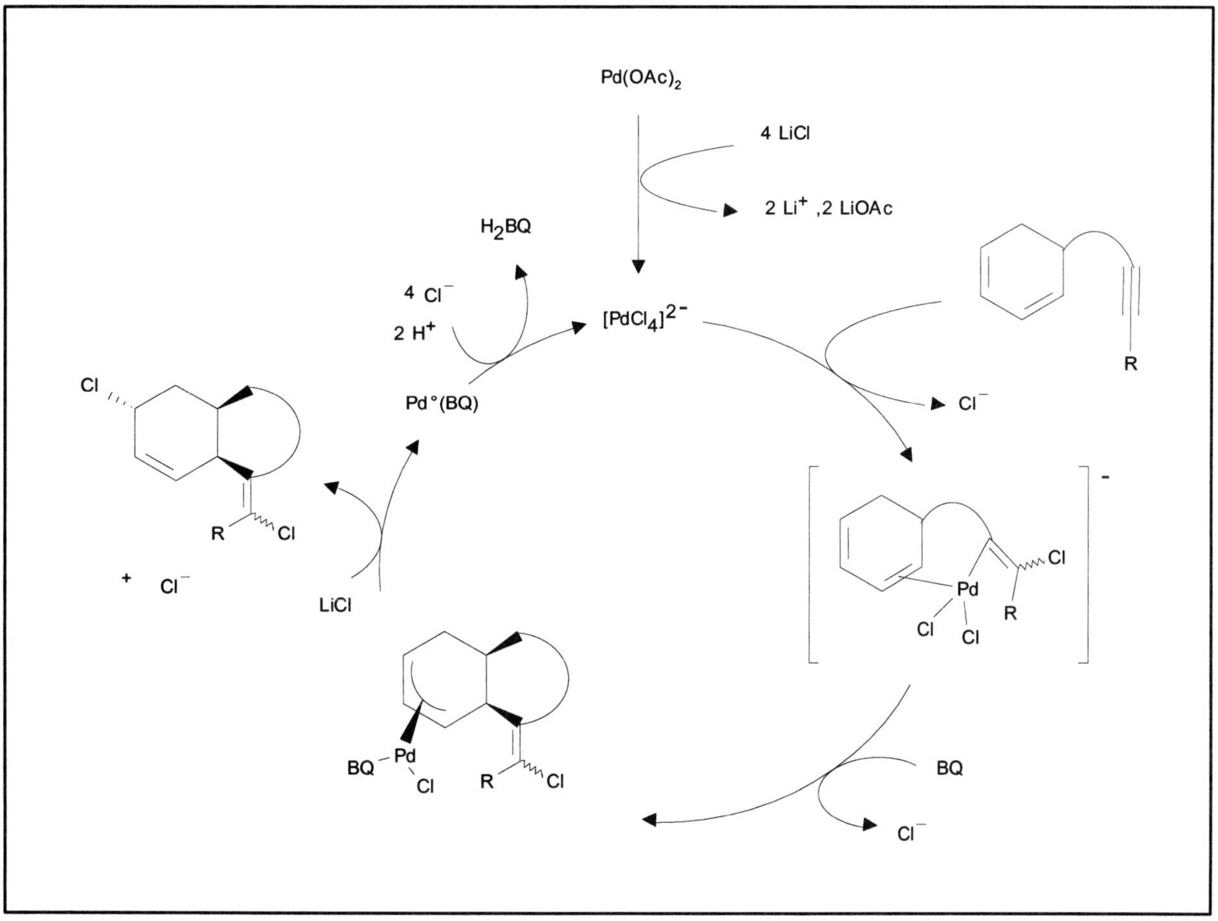

A ---> B	References
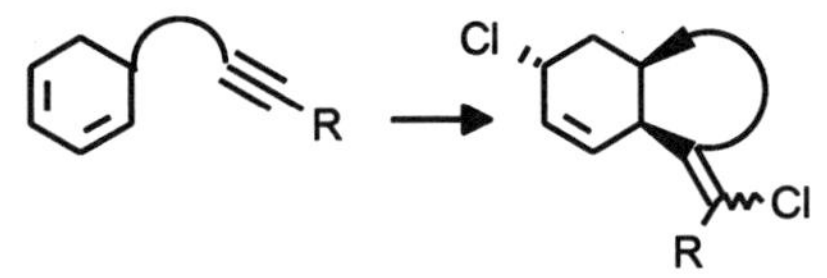	Bäckvall J.F., *Tetrahedron Lett.*, 1994; 5713
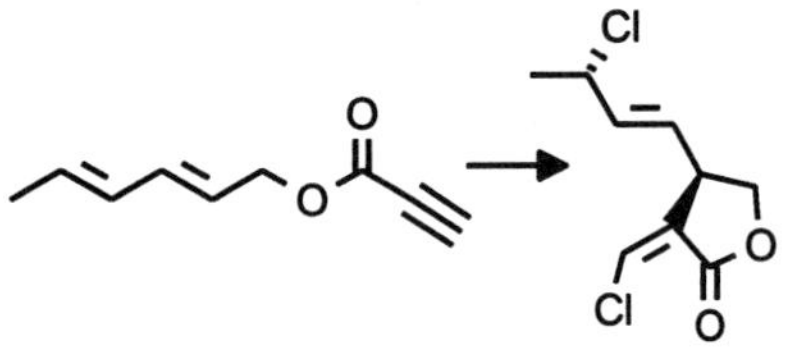	Bäckvall J.F., *Tetrahedron Lett.*, 1994; 5713

RXN76 Tandem Chlorination-Cyclization and Tandem Chlorination-Carbonylation-Cyclization of 1,6-Enynes

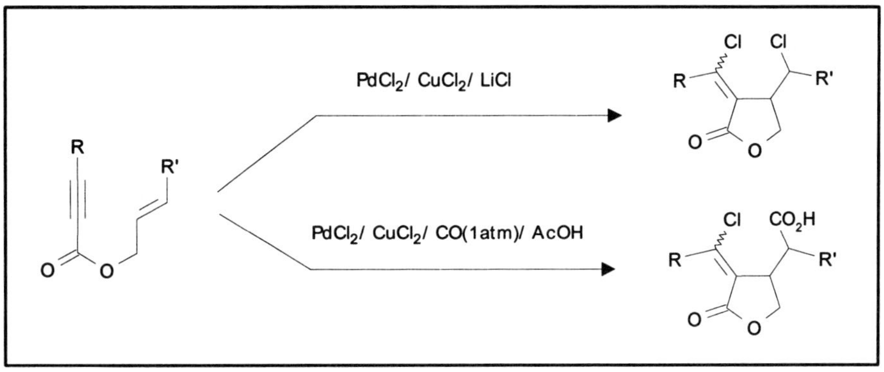

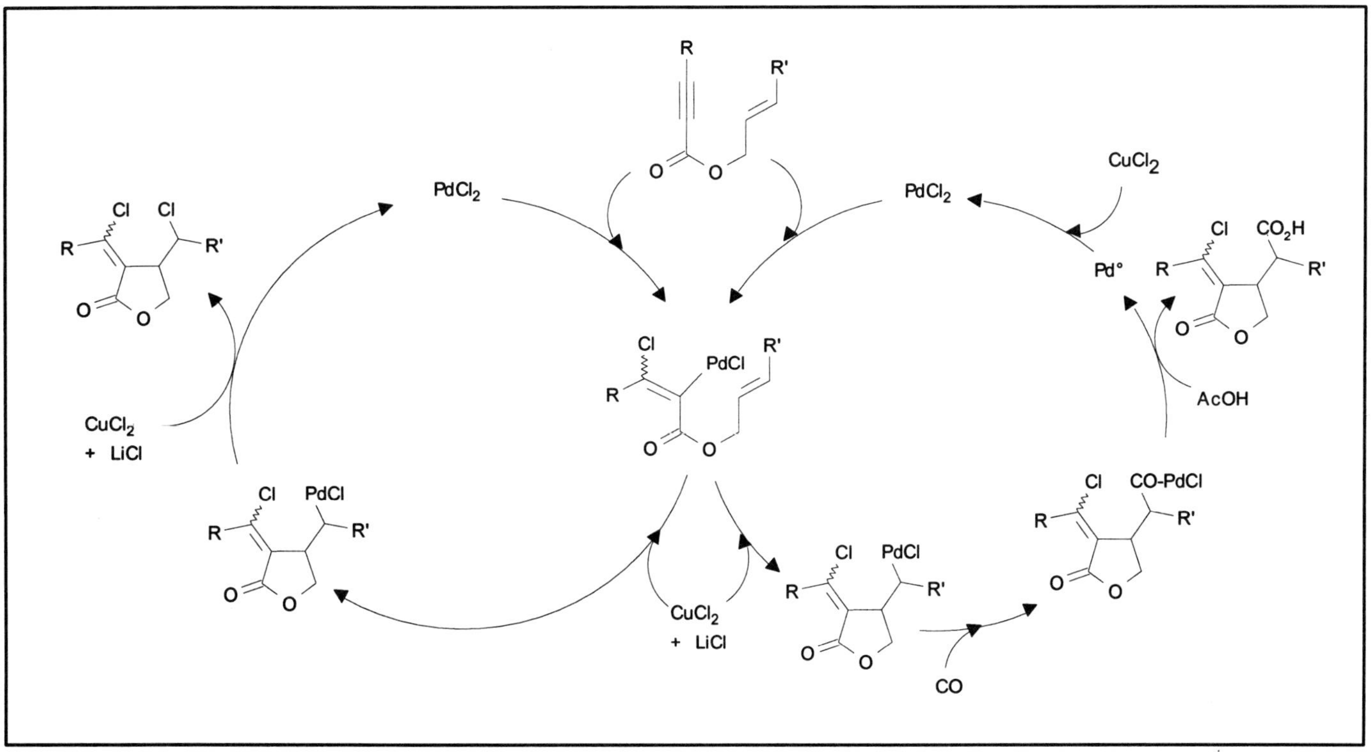

RXN76 Tandem Chlorination-Cyclization and Tandem Chlorination-Carbonylation-Cyclization of 1,6-Enynes

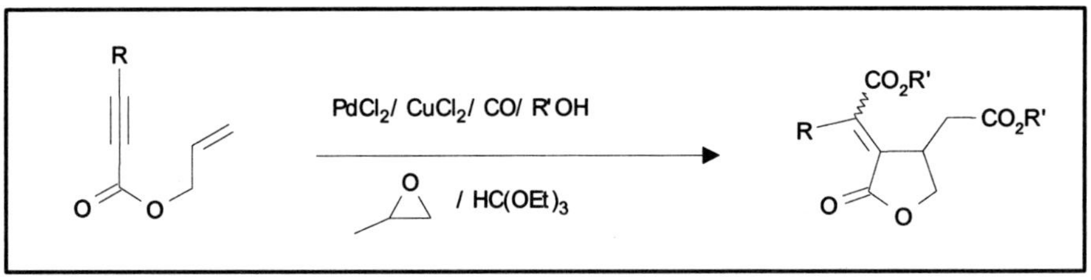

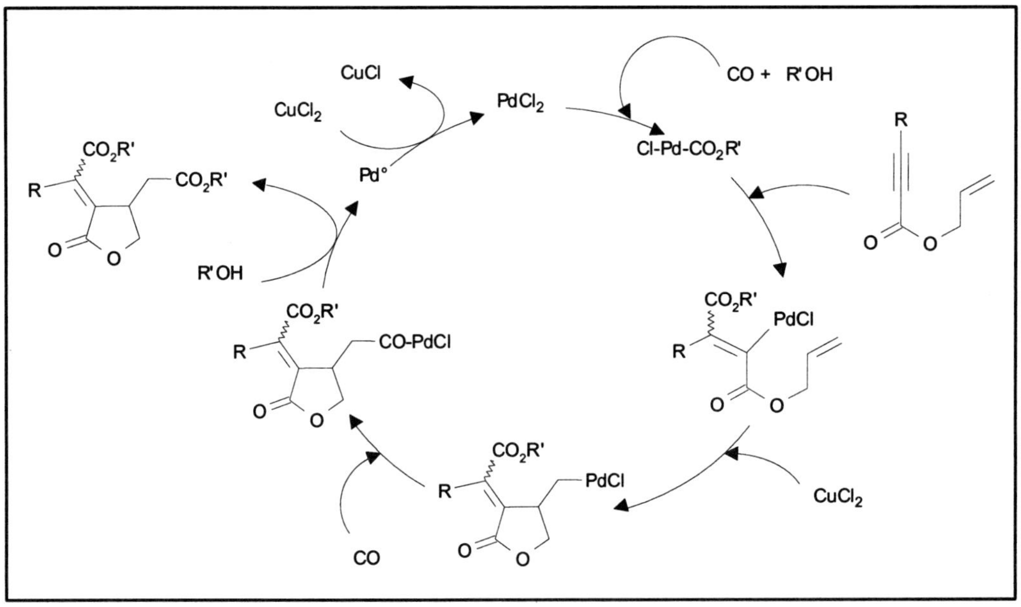

J.-L. Malleron et al

A ---> B	References
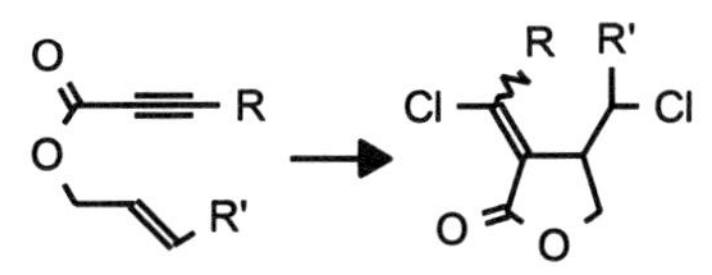	Lu X., *J. Org. Chem.*, 1993; 1245 Lu X., *Synlett.*, 1993; 68 Lu X., *J. Chem. Soc., Chem. Commun.*, 1995; 271 Lu X., *J. Org. Chem.*, 1995; 1160
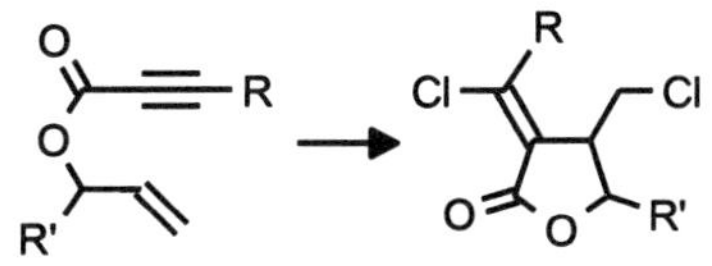	Lu X., *Tetrahedron: Asymmetry*, 1995; 345
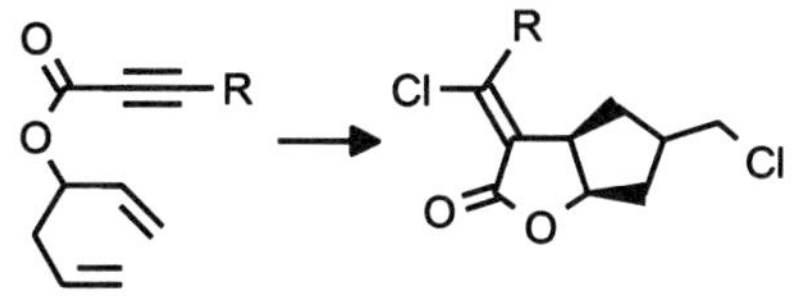	Lu X., *Synlett.*, 1993; 745
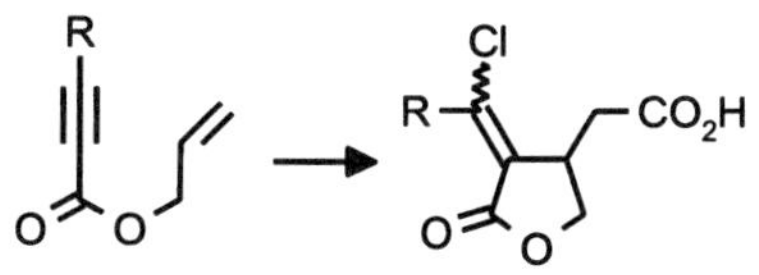	Lu X., *Tetrahedron*, 1994; 9067
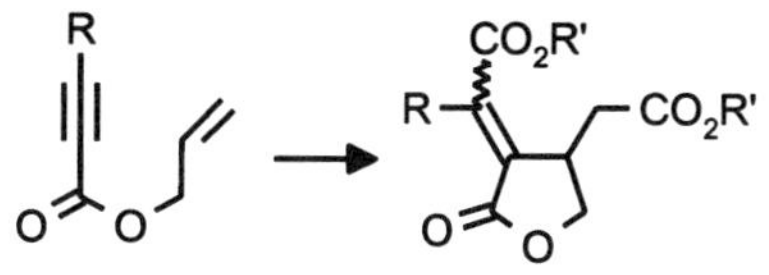	Lu X., *Tetrahedron*, 1994; 9067

RXN76

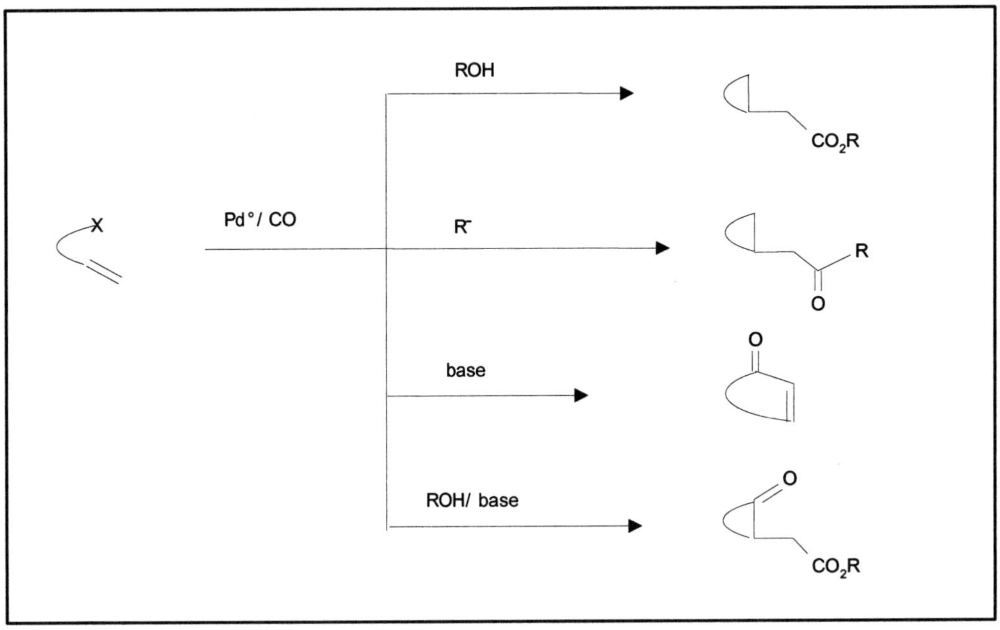

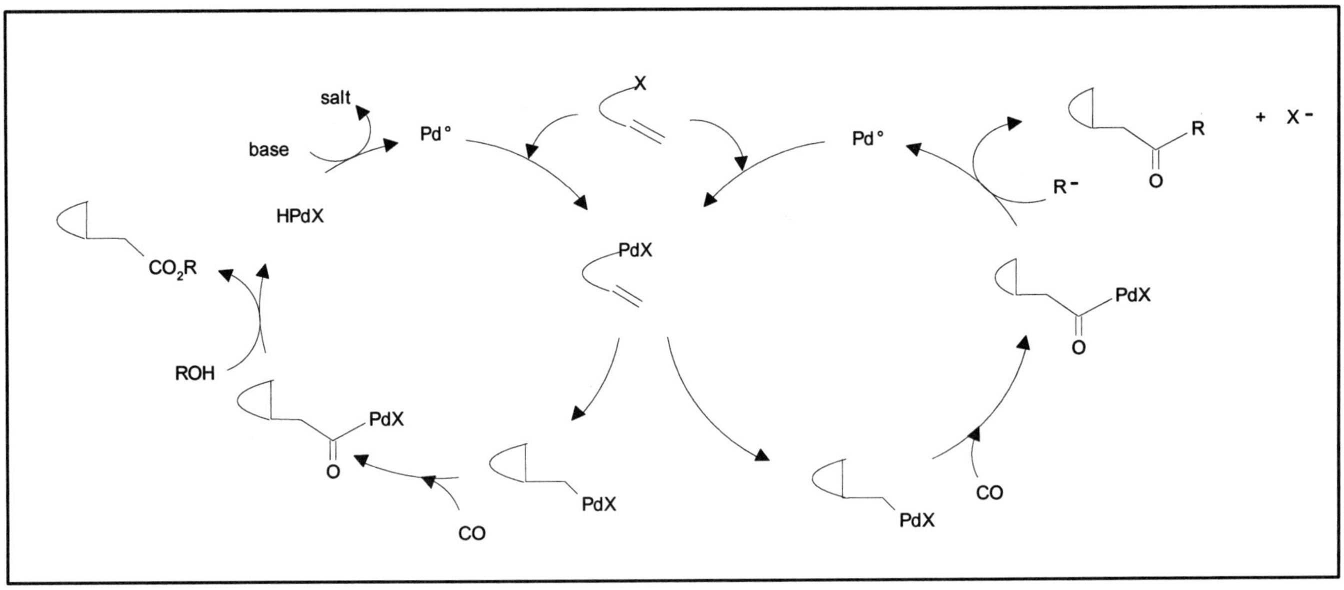

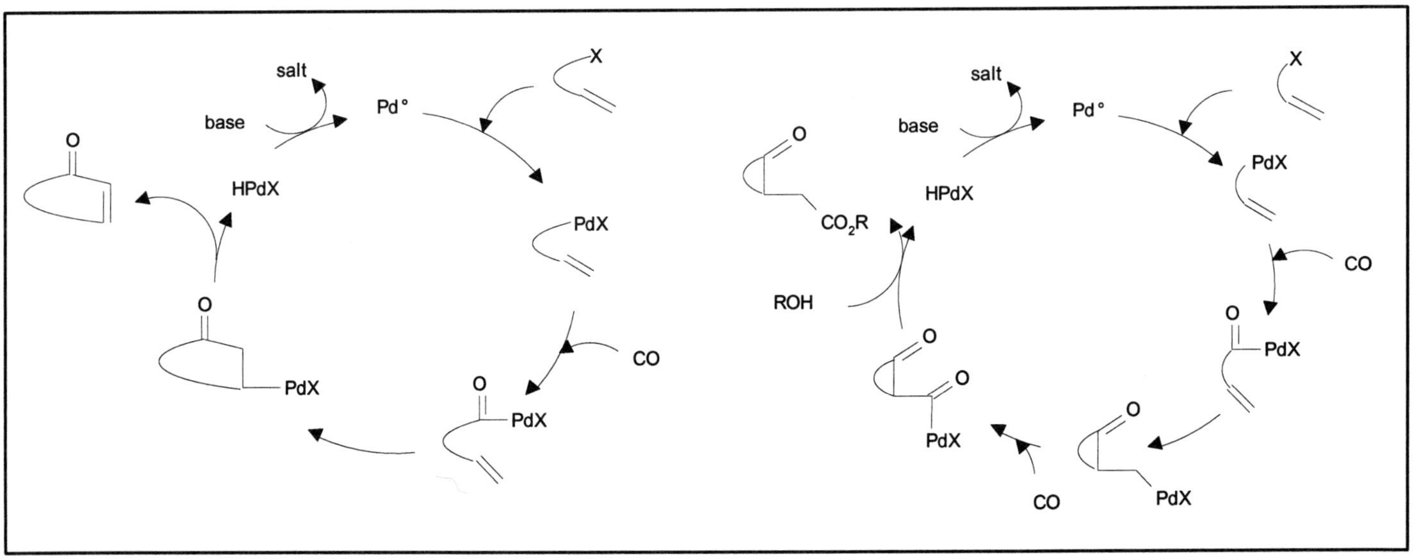

 J.-L. Malleron et al

A ---> B	References
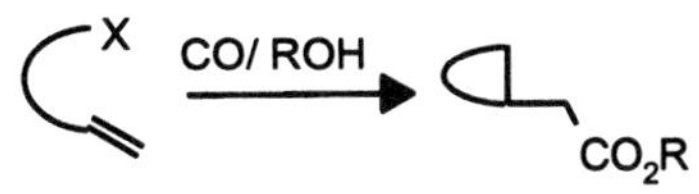	Grigg R., *Tetrahedron Lett.*, 1992; 7789 Grigg R., *Tetrahedron Lett.*, 1994; 2753 Grigg R., *J. Heterocycl. Chem.*, 1994; 631
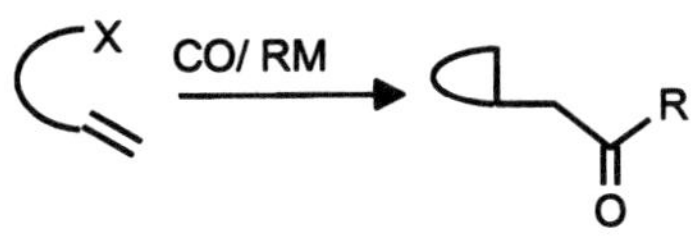	Grigg R., *Tetrahedron Lett.*, 1994; 4429 Grigg R., *Tetrahedron Lett.*, 1994; 7661 (double carbonylation) Neghishi E-I., *J. Org. Chem.*, 1991; 6506 (intramolecular) Neghishi E-I., *Tetrahedron*, 1994; 425 (intramolecular) Grigg R., *Tetrahedron Lett.*, 1993; 7471 (intramolecular) Grigg R., *J. Chem. Soc., Chem. Commun.*, 1995; 1135 (R = H) Ishikura M., *J. Chem. Soc., Chem. Commun.*, 1995; 409
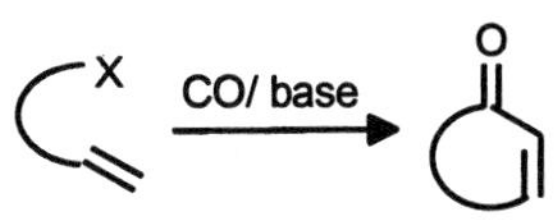	Neghishi E-I., *Tetrahedron Lett.*, 1986; 4869 Torii S., *Tetrahedron Lett.*, 1990; 7175
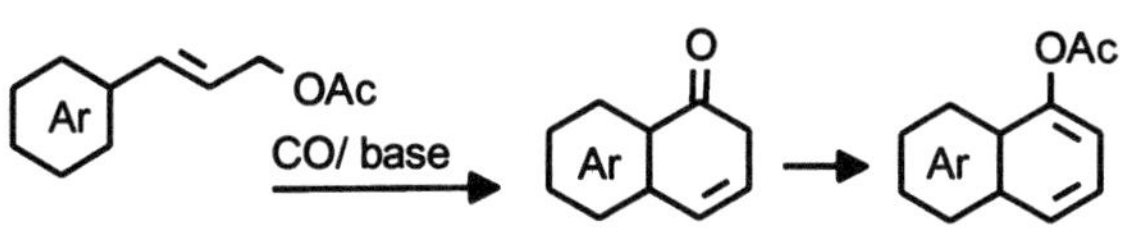	Hidai M., *Tetrahedron Lett.*, 1989; 95 Hidai M., *J. Org. Chem.*, 1988; 3832 Hidai M., *J. Org. Chem.*, 1991; 1922
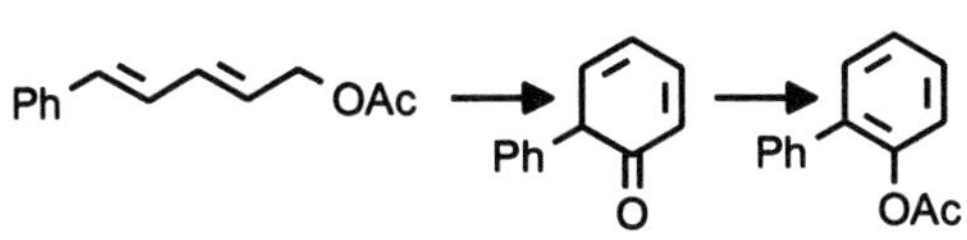	Hidai M., *J. Org. Chem.*, 1993; 6818
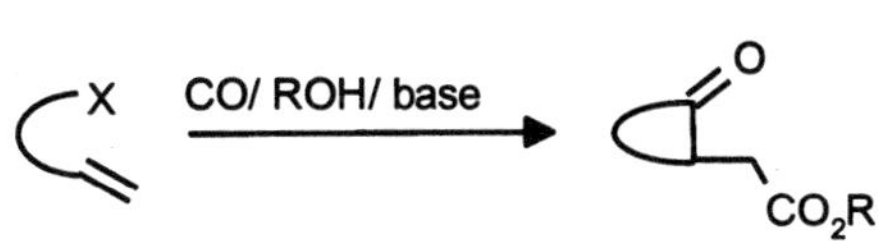	Neghishi E-I., *J. Am. Chem. Soc.*, 1985; 8289 Neghishi E-I., *Tetrahedron Lett.*, 1988; 6745

RXN78 Tandem Carbonylation-Cyclization of (Z)- β-Iodoenones and Related Reaction from Aryl Iodides and Internal Alkynes

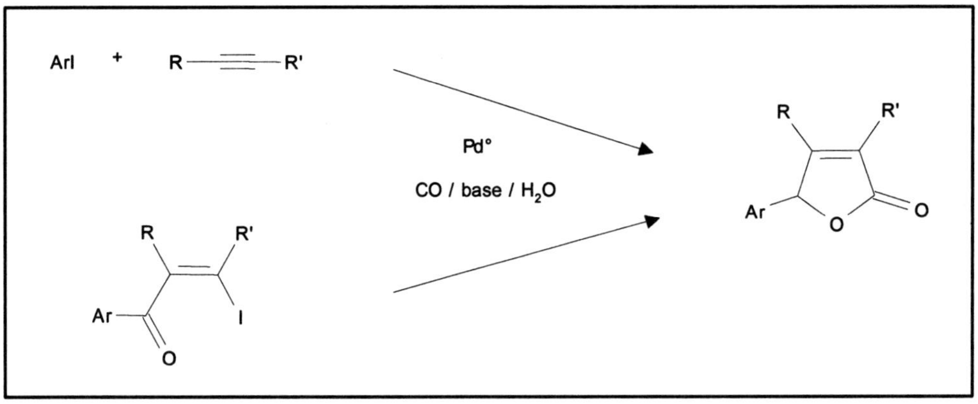

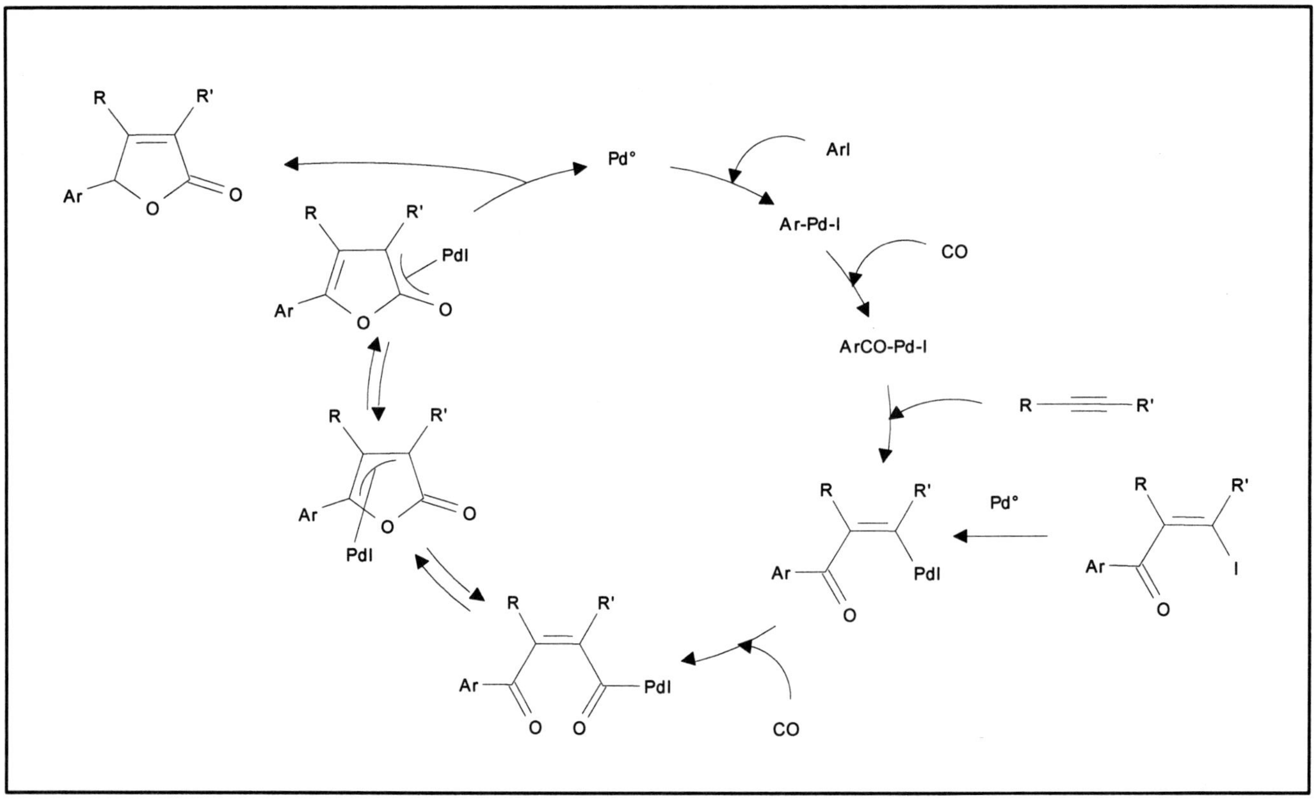

A ---> B	References
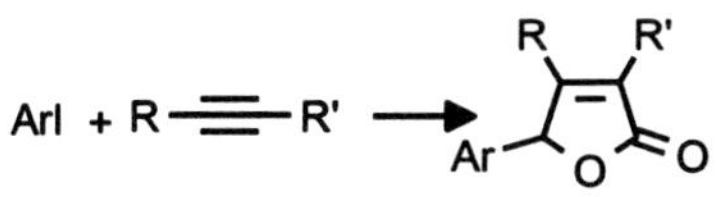	Negishi E.I., *J. Am. Chem. Soc.*, 1995; 3422
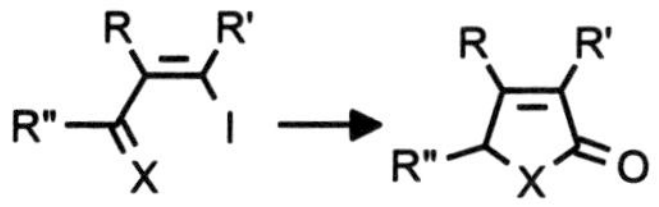	Negishi E.I., *J. Am. Chem. Soc.*, 1995; 3422
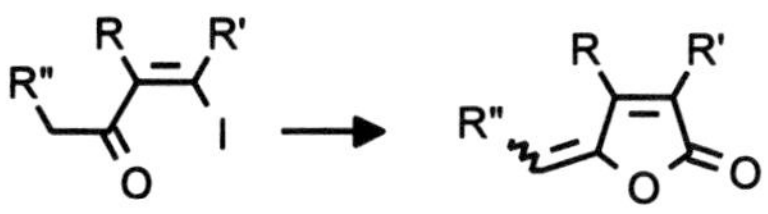	Negishi E.I., *J. Am. Chem. Soc.*, 1995; 3422

RXN79 Cross-Coupling of Organometallics with Terminal Alkynes and Iodotrimethylsilane

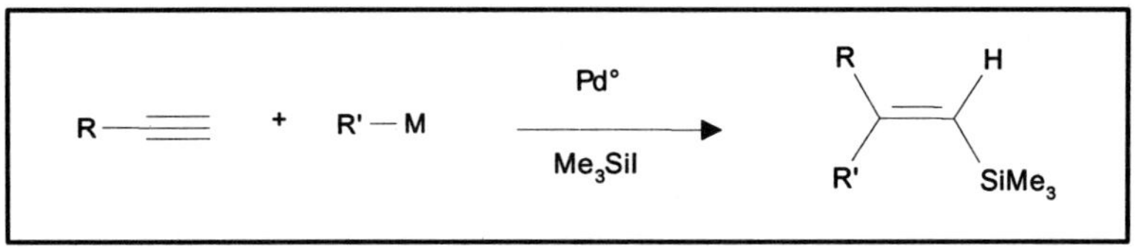

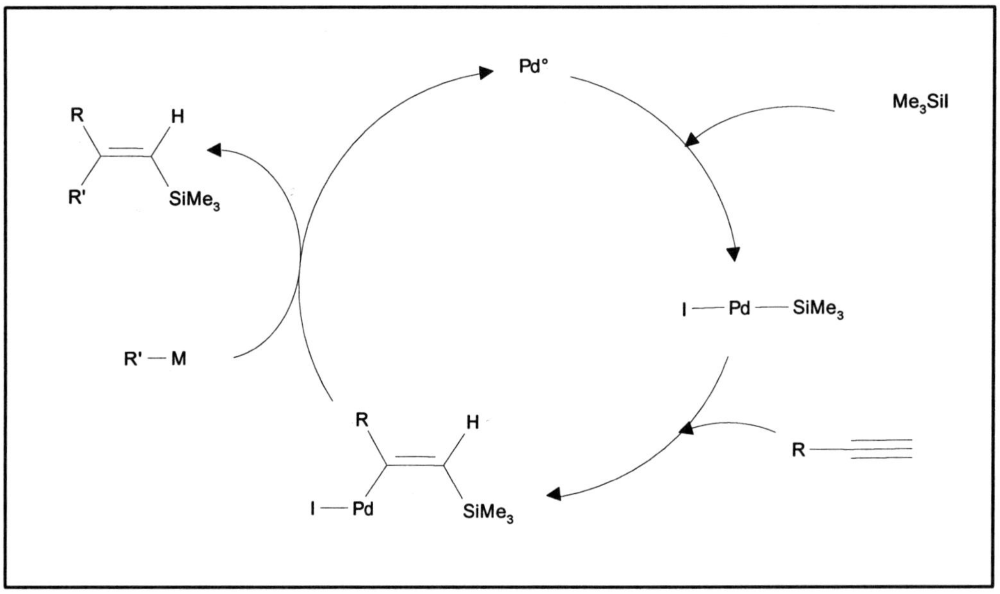

R'-M	References
R—≡—Sn	Murai S., *J. Am. Chem. Soc.*, 1991; 7778
CH_2=CH-CH_2-Sn	Murai S., *J. Am. Chem. Soc.*, 1991; 7778
CH_2=CH-Sn	Murai S., *J. Am. Chem. Soc.*, 1991; 7778
alkyl-Zn	Murai S., *J. Org. Chem.*, 1995; 1834

RXN80 Preparation of α,β-Unsaturated Carbonyl Derivatives from Enol Silanes

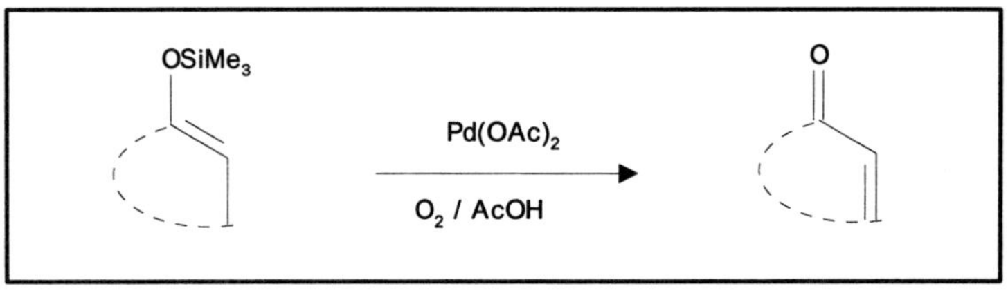

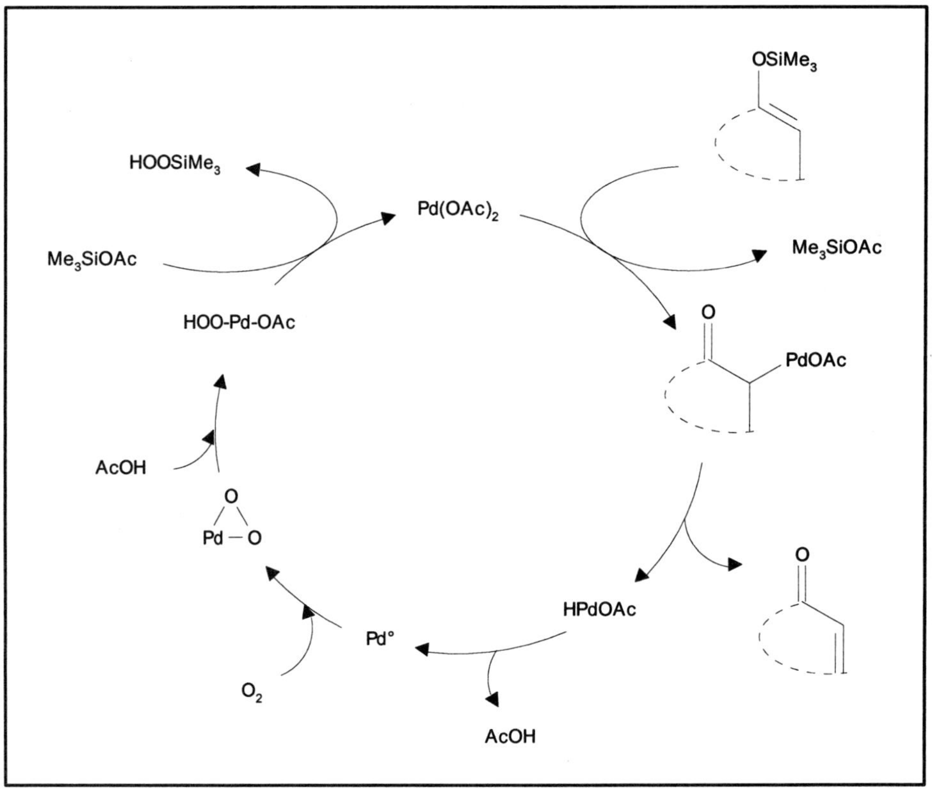

A ---> B	References
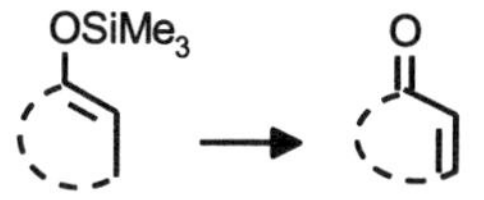	Larock R.C., *Tetrahedron Lett.*, 1995; 2423 Cossy J., *Tetrahedron Lett.*, 1995; 7877
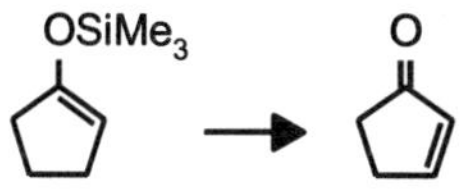	Baba T., *J. Chem. Soc., Perkin Trans. 2*, 1990; 1113 (dehydrosilylation)

RXN80

RXN81 1,3-Butadien-2-ylaminocyclization of Allenyl Tosylcarbamates

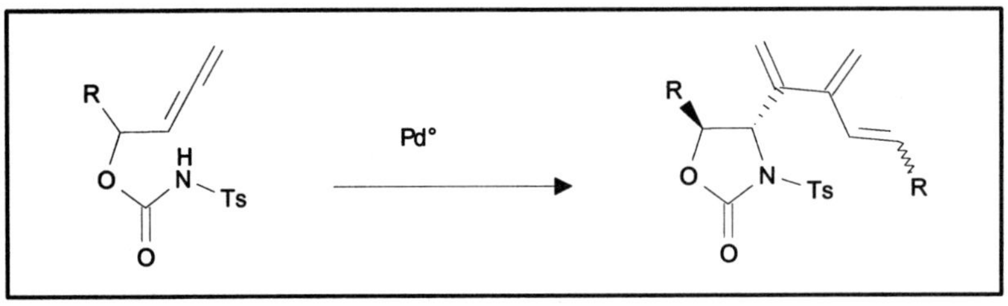

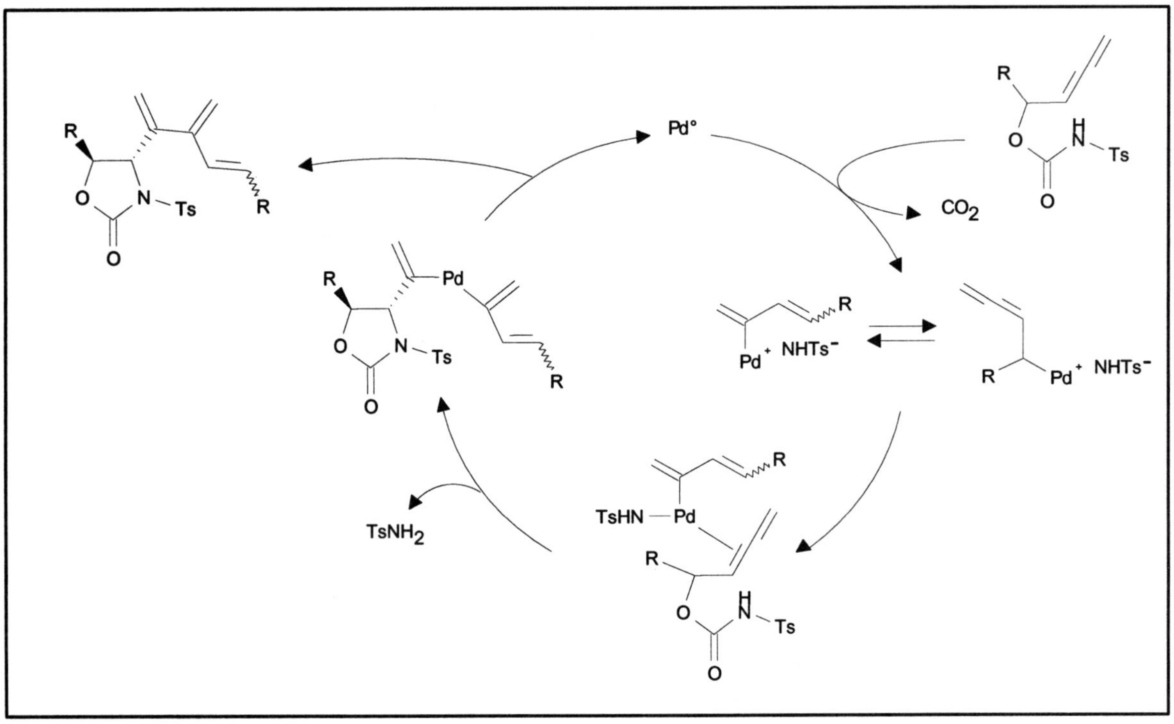

 J.-L. Malleron et al

A ---> B	References
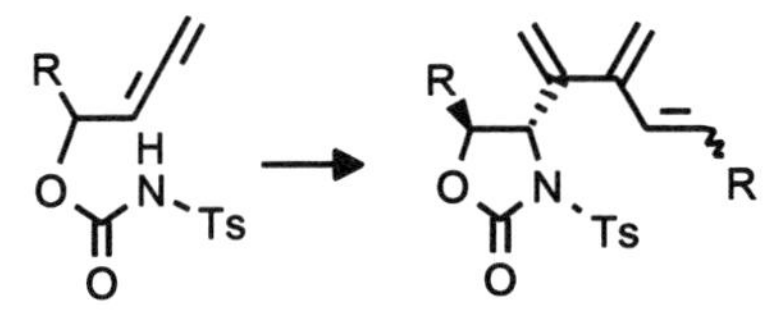	Tamaru Y., *J. Org. Chem.*, 1995; 3764

RXN81

RXN82 Addition of Pronucleophiles on Aldehydes

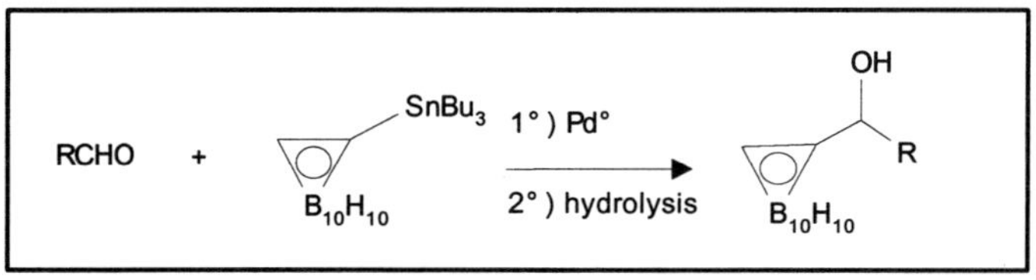

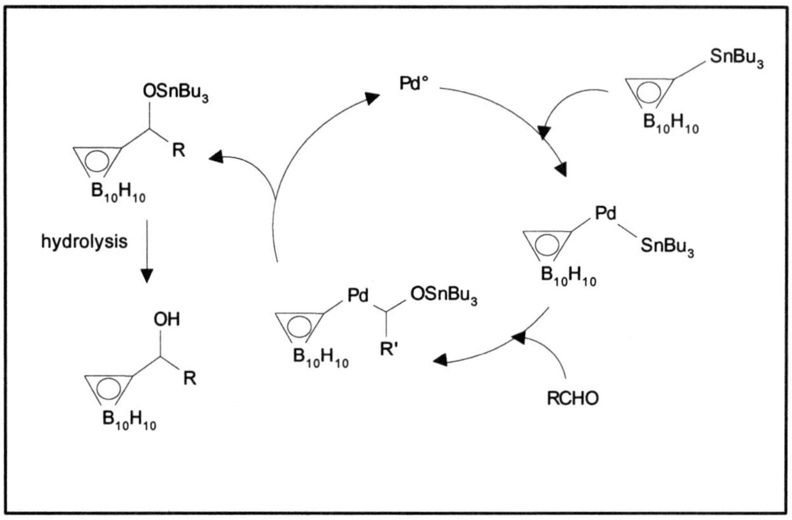

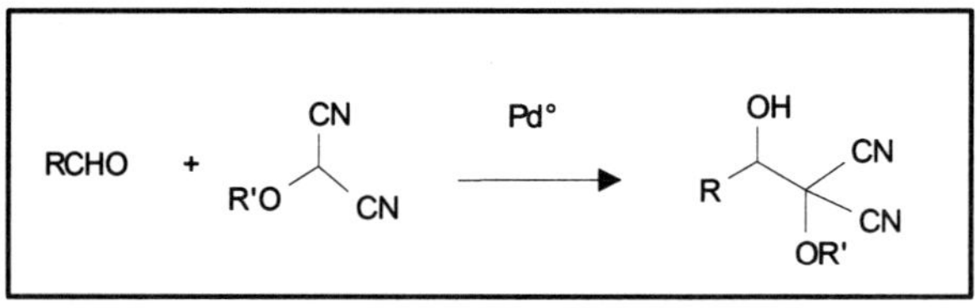

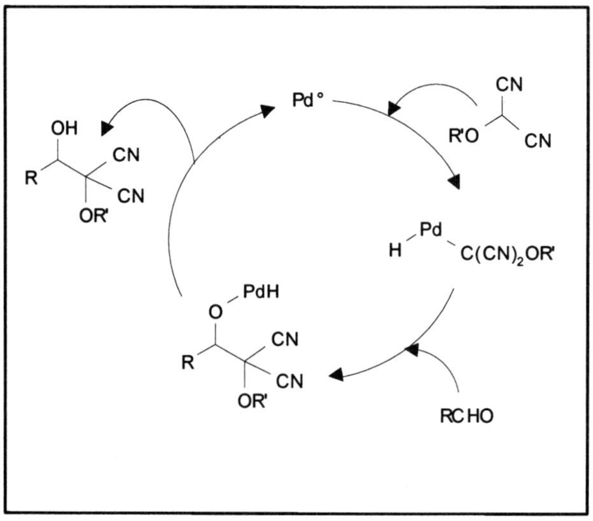

RXN82 Addition of Pronucleophiles on Aldehydes

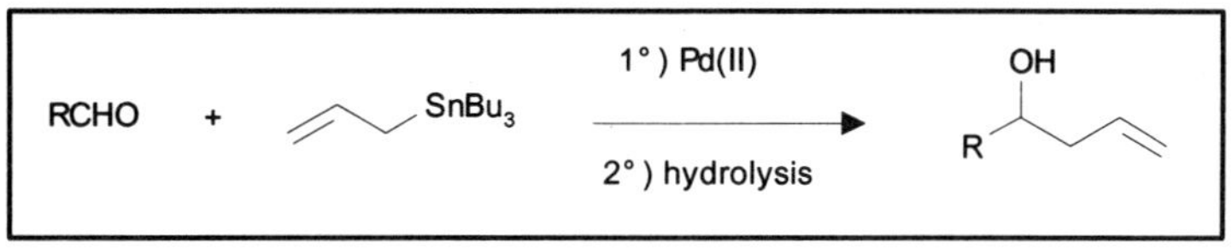

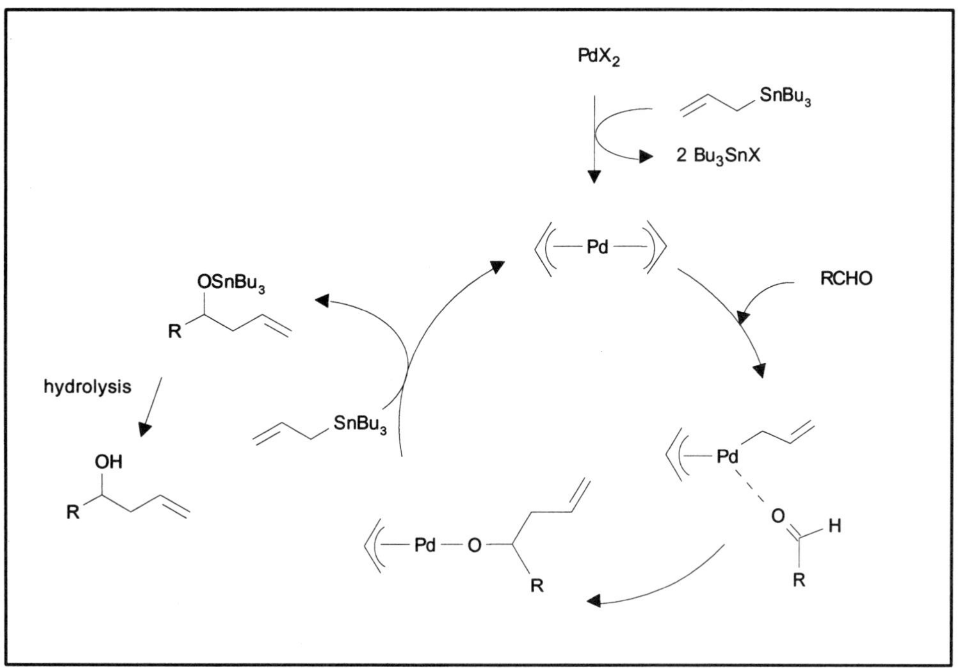

A ---> B	References
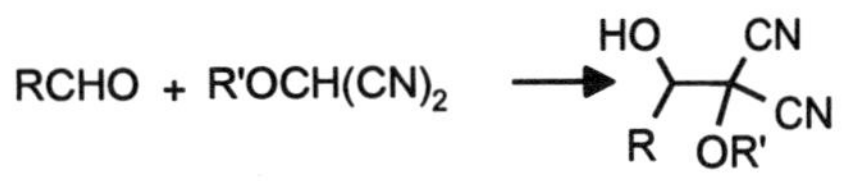	Yamamoto Y., *J. Chem. Soc., Chem. Commun.*, 1994; 1665
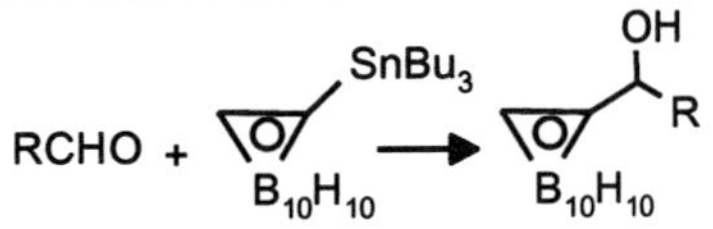	Yamamoto Y., *J. Chem. Soc., Chem. Commun.*, 1994; 2581
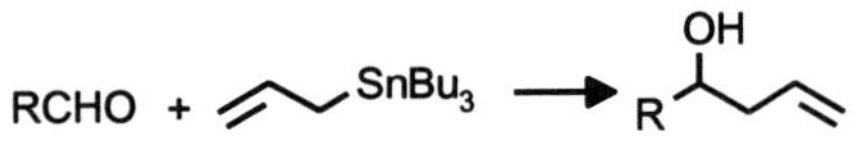	Yamamoto Y., *J. Chem. Soc., Chem. Commun.*, 1995; 1273

RXN83 Allylation of Pronucleophiles with Allylstannanes

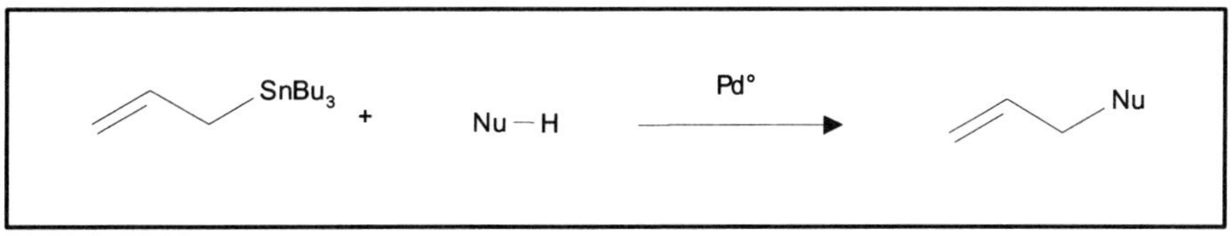

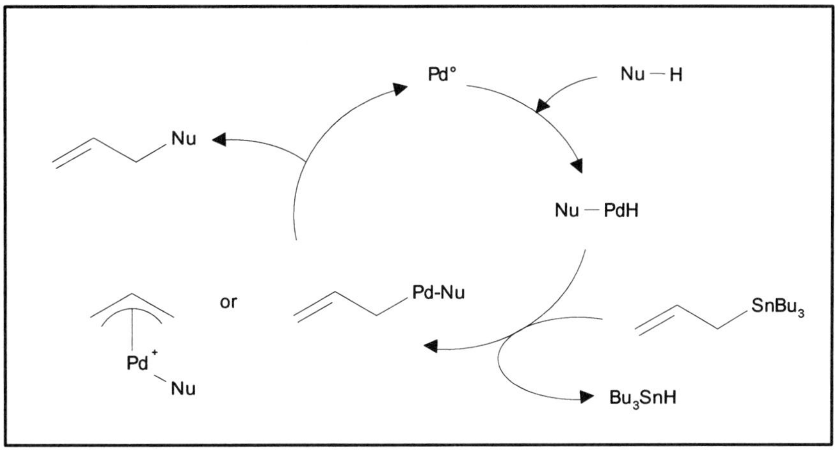

Pronucleophile	References
HC(Z)(Z')	Yamamoto Y., *J. Chem. Soc., Chem. Commun.*, 1995; 2013

RXN84 Cross-Coupling Carbonylation of Aryl Iodides with Five-Membered Cyclic Olefins

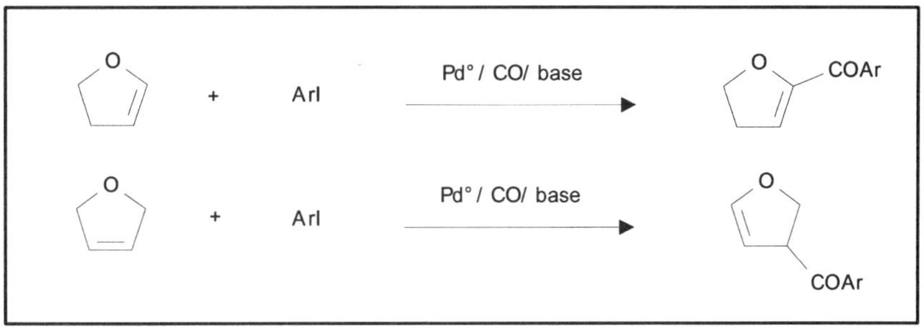

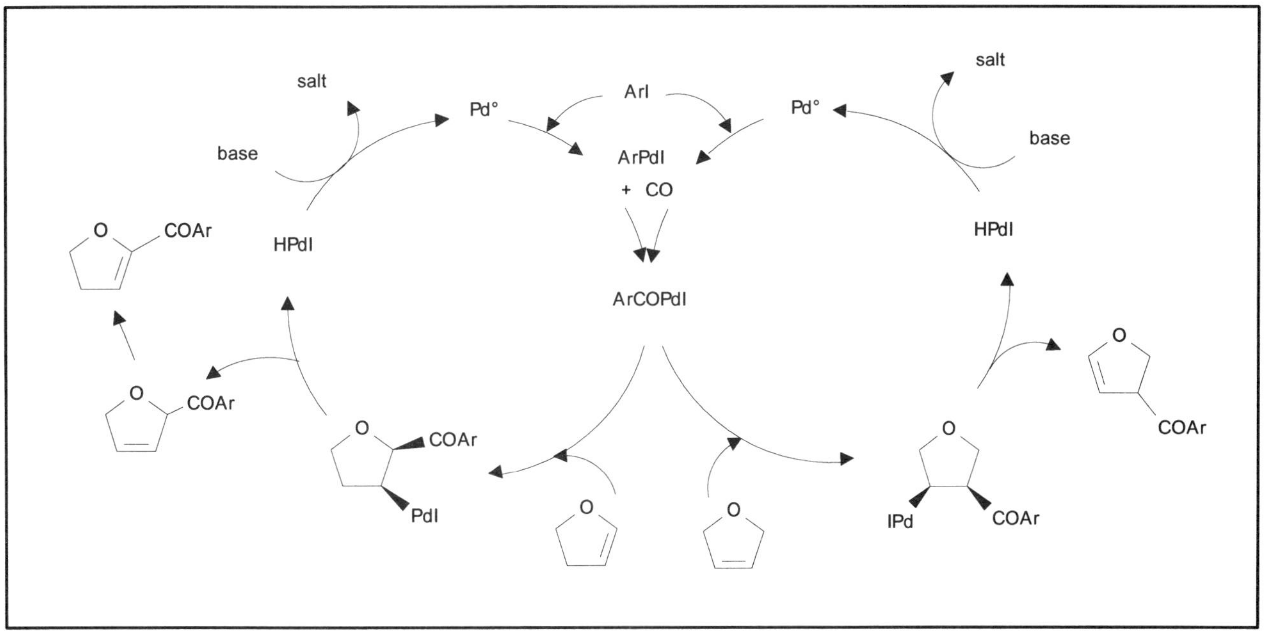

A ---> B	References
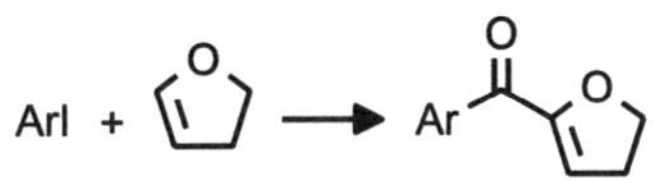	Miura M., *J. Org. Chem.*, 1995; 7267
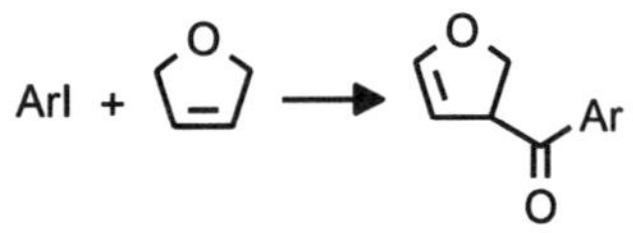	Miura M., *J. Org. Chem.*, 1995; 7267

If you care about Structure Activity Relationships, you need the new ISIS SAR Table!

Working smart means having a complete solution at your fingertips. Which is why you owe it to yourself to check out what ISIS SAR Table from MDL Information Systems, Inc. has to offer. To start with, ISIS SAR Table comes bundled with ISIS/Base, a full-functioned chemical database manager, *for about the same price as the competition charges for a SAR table application alone!* Take a look at what ISIS SAR Table can do for you.

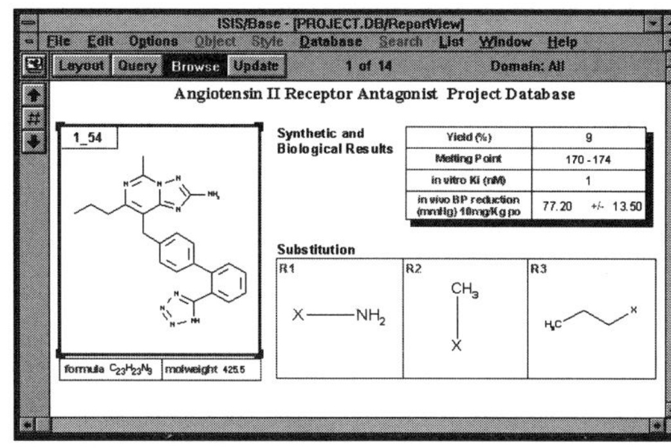

A biologist queries a project database using ISIS/Base to examine the latest test results.

1 Get the right tool for the right job.

You need more than a simple reporting tool. You need a powerful, comprehensive *investigational* tool that can solve your toughest problems. That's what we deliver. You see, ISIS is perfect for storing and searching databases of chemical structures and related data located throughout your organization. Plus, you get all the advantages of analysis with Microsoft Excel, including displaying and manipulating the SAR table.

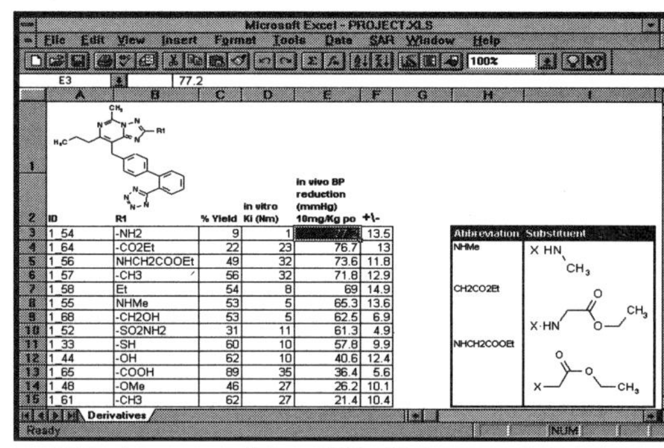

With a few mouse clicks, a SAR table is created in Excel that displays the in vivo results as a function of the change in substitution at R1.

2 Explore complex relationships.

We provide a rich set of R-group tools to help you uncover more complex relationships. In fact, the whole process of building and using the SAR table is chemically intelligent. And because ISIS SAR Table is linked directly to ISIS/Base, you'll spend less time collecting data and more time generating new ideas.

3 Make the power of ISIS SAR Table available to your whole team.

Finally, you can spend less time formatting data and more time making critical decisions. Whether you're a chemist or a biologist, you'll find that ISIS SAR Table can help you quickly communicate your findings to others. Even the most casual user can unlock the full power of this tool by accessing our extensive online help and practical application examples.

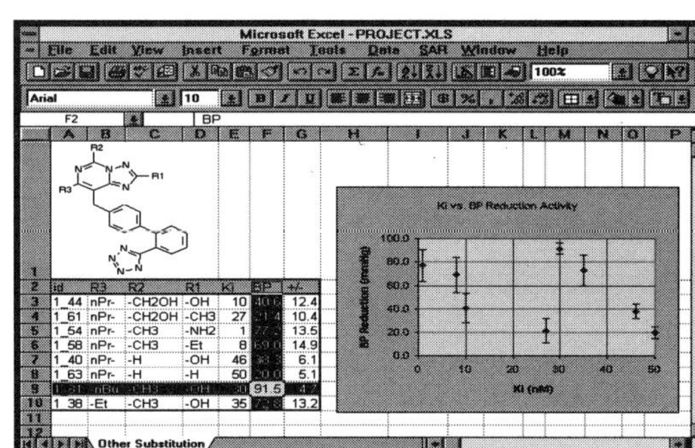

Meanwhile, a chemist decides to view the data differently by displaying all three R-groups. This new perspective helps her decide which direction her research should take.

4 ISIS SAR Table is *extremely* easy to use.

With just a few mouse clicks, the results of structure and data searches of your project or corporate databases are transferred to the SAR table. In seconds, structures and data appear in an easy-to-read Excel spreadsheet where you can take full advantage of Excel's outstanding analysis capabilities. No other product gives you the ease and power of ISIS SAR Table!

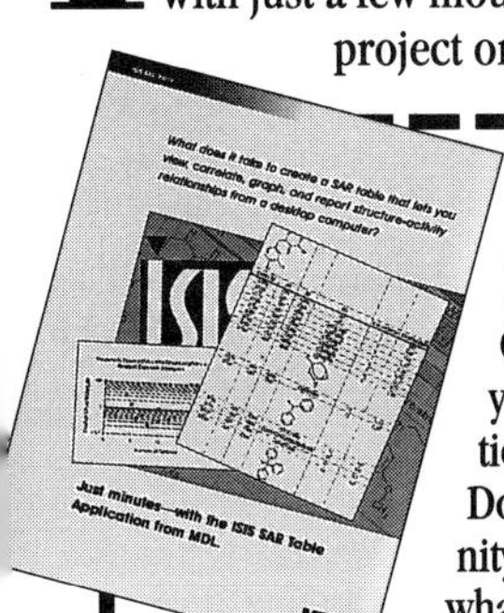

FREE

For Researchers

Call now and we'll send you important information with no obligation. Don't miss this opportunity to find out more about what ISIS SAR Table can do for *you!*

Call 1-800-635-0064 ext. 1326

Information Request Number: ISIS SAR Table

5 Work with a company you can trust.

MDL is proud of the fact that for almost two decades we've been helping tens of thousands of scientists at over 300 companies and institutions get the most from their intellectual assets. With software, services, education, support, and more.

MDL

Information Systems, Inc.

14600 Catalina Street, San Leandro, CA 94577 (510) 895-1313 Fax: (510) 483-4738
Internet: isis@mdli.com World Wide Web: http//www.mdli.com